Inorganic Chemistry Concepts
Volume 6

David L. Kepert

Inorganic
Stereochemistry

With 206 Figures and 45 Tables

Springer-Verlag
Berlin Heidelberg New York 1982

Prof. David L. Kepert

The University of Western Australia
Dept. of Physical and Inorganic Chemistry
Nedlands, Western Australia 6009

ISBN-13:978-3-642-68048-9   e-ISBN-13:978-3-642-68046-5
DOI: 10.1007/978-3-642-68046-5

Library of Congress Cataloging in Publication Data.
Kepert, David. L., 1936– Inorganic Stereochemistry.
(Inorganic chemistry concepts; v. 6) Bibliography: p.
Includes index. 1. Stereochemistry. 2. Chemistry, Inorganic.
I. Title. II. Series.
QD481.K39   541.2′23   81-5266
ISBN-13:978-3-642-68048-9 (New York) AACR2

© by Springer-Verlag Berlin, Heidelberg 1982
Softcover reprint of the hardcover 1st edition 1982

2152/3020–543210

Ubi materia, ibi geometria.

Where there is matter, there is geometry.

Johannes Kepler (1571—1630)

Preface

Molecular stereochemistry is a fundamental aspect of all areas of chemistry. It is especially important in inorganic chemistry where the coordination numbers are variable and occasionally quite high. The present book evolved naturally from a series of articles written by Professor Kepert for *Progress in Inorganic Chemistry,* elucidating aspects of the stereochemistry of inorganic compounds of coordination numbers 4—12. In the present volume, Professor Kepert has added new sections and synthesized these individual chapters into a unified treatment, updating his references when necessary to the most recent contributions in the literature, and interweaving the various themes as deemed appropriate. The result is a major contribution, describing the stereochemistry of coordination compounds having both unidentate and multidentate ligands. The viability of the repulsion approach to stereochemistry is tested to the limit in this treatise and shown to be an extremely good way of rationalizing a diverse body of data.

New York City, August 1981   Stephen J. Lippard

Contents

CHAPTER 1

Introduction

A. Historical Background

Stereochemistry occupies a central position in chemistry, and it is difficult to imagine how science would have developed in the absence of this continuous reference to molecular structure.

The idea that the distribution of atoms about a central atom is three-dimensional in nature stems from the work of van't Hoff and Le Bel, who, in 1874, independently proposed that four different groups attached to a central carbon atom were arranged at the vertices of a tetrahedron [855]. The existence of optical isomerism in such compounds, but the absence of geometrical isomerism, was then readily explained.

On the inorganic side of chemistry, it was some 20 years later in 1893 that Werner proposed that coordination complexes consisted of a central metal atom surrounded by six groups arranged at the vertices of an octahedron [633, 1084].

Since this time the usual way that the spatial arrangements of atoms have been depicted is as classical geometrical polyhedra. Although the historical development of stereochemistry has been dominated by the tetrahedron and the octahedron, it is most important to realise that polyhedra with symmetries as high as these two examples are exceedingly rare. In particular, compounds containing five, seven, eight, or nine groups attached to a central atom must have much more complicated geometries. Even the tetrahedron and octahedron will be regular only if the four or six groups surrounding the central atom are all identical. For example, the FCF angle in trifluoromethyl groups in $F_3C \cdot R$ (where R is an aliphatic group) is not the tetrahedral value of $109.5°$, but is about $106.3°$. Similarly deviations in bond angles of several degrees are frequently observed in octahedral compounds containing more than one type of ligand.

Gillespie and Nyholm in 1957 showed that such distortions may be very satisfactorily explained in a qualitative manner as arising from the repulsion between the pairs of electrons in the bonds about the central atom [473, 475]. The more electronegative groups attract the electron pairs in their bonds further away from the central atom, lessening their interaction with the other bonding electrons pairs, and decreasing the bond angles. This concept in particularly important for many compounds in which Sidgwick and Powell earlier showed could be considered to contain a non-bonding pair of electrons [988]. In this case the lack of any group at the end of the electron pair results in it being much closer to the central atom, resulting in substantially greater repulsion from this nonbonding electron pair than from the bonding electron pairs.

In 1963 Hoard and Silverton used a more quantitative approach when considering the arrangement of the eight cyanide groups about the metal atom in $K_4[Mo(CN)_8]2H_2O$ [565]. They considered that the stereochemistry was determined by the repulsion between all carbon atoms, the repulsion between any two carbon atoms being inversely proportional to the seventh power of the distance between them. They also calculated the stereochemical constraints arising from the introduction of chelate rings, as in the eight-coordinate acetylacetonate complex [Zr(acac)$_4$], by assuming that each chelate ring was rigid, and that the distance between its donor atoms was fixed.

A combination of these simple ideas is the best way, and in most cases the only way, that the stereochemistry of metal atoms with seven or more ligands can be calculated. It is also the most satisfactory way of calculating the steric changes brought about when two or more of the ligand groups are connected to form ring systems, such systems being extremely important throughout chemistry.

The development of these themes into a more comprehensive scheme of stereochemistry is described in this work. It follows a series of related works on aspects of the stereochemistry of four- and five-coordination [426], six-coordination [637], seven-coordination [639], eight-coordination [638], and nine-, ten-, and twelve-coordination [427].

It must be remembered that substantial assumptions are incorporated in the theoretical model, and that this model bypasses much of the theoretical basis of chemical bonding that has been progressively developed over the last 50 years. The model has therefore often been prematurely rejected, in spite of its undoubted utility.

The simple theoretical basis for this work has been available for many years, but the recent advances have been made possible by the acquisition of a large amount of structural data by the routine X-ray analysis of crystals using automatic diffractometers and associated computer technology. This work on the solid state has been augmented by the parallel increase in the understanding of the stereochemical rearrangements that occur in solution, which have been studied by n.m.r. techniques over a range of temperatures (for example, see Ref. [812])

B. Theoretical Background

1. The Repulsion Law

The basis of this work is the assumption that the stereochemical arrangement of ligands about a central atom is determined by the mutual repulsion between all bonds. The central atom is considered to play no role in determining stereochemistry, either through some preferred bonding directions, or even by shielding some of the surrounding bonds from each other. For the sake of convenience, the central atom will be referred to as the metal atom, and the surrounding atoms as the ligands.

The calculations are based on the minimization of the total repulsion energy, U, obtained by summing the repulsion over every pair of metal-ligand bonds. It can be considered that this repulsion originates solely from the metal-ligand bonds, or from both the metal-ligand bonds and the donor atoms, the relative importance not necessarily being the same for all compounds.

It is assumed that this very complicated repulsion can be simply represented by a model in which each metal-ligand bond is considered to effectively act as a point. The repulsion u_{ij} between two effective centers of repulsion, i and j, is inversely proportional to some power n of the distance d_{ij} between them:

$$u_{ij} = \frac{a_n}{d_{ij}^n}$$

where a_n is the proportionality constant.

If all bonds are equal, that is, if the effective centers of all bonds lie on the surface of a sphere of radius r, then the results can be expressed in the following form:

$$U = \sum_{ij} u_{ij} = \sum_{ij} a_n d_{ij}^{-n} = a_n X r^{-n}$$

where X is the numerical repulsion energy coefficient, which is a function of n and the geometry of the coordination polyhedron. The most stable stereochemistry is simply calculated by the minimization of X as a function of geometry.

The most appropriate value of n to use in the repulsion law cannot be known exactly. The lower limit is $n = 1$, which unrealistically assumes coulombic repulsion among bonds considered as points. The geometry obtained as n approaches infinity is the same as that obtained for the hard sphere model, where incompressible spheres are at each vertex of the coordination polyhedron, the spheres touching each other at the midpoint of the short edges. Comparison between calculated and experimental structures shows that the best agreement is for $n \sim 6$, this value being reasonable for repulsion between electron clouds. The usual custom is to report calculations using $n = 1$, $n = 6$ and $n = 12$, and for the hard sphere model. It is fortunate that the calculated stereochemistry is usually fairly independent of the exact value chosen for n.

For molecules containing only one type of metal-ligand bond, that is, complexes of the type [M(unidentate ligand)$_x$] and [M(symmetric bidentate ligand)$_x$], this energy relation is sufficient to enable very accurate stereochemical predictions to be made.

2. Effective Bond Length Ratios

For complexes containing different donor atoms it is necessary to introduce an empirical parameter to account for the small distortions observed compared with compounds containing only one type of metal-ligand bond. The parameter used is the distance between the central atom and the effective center of the bond. That is, the relative repulsion of the metal-ligand bond M-i and the metal-ligand bond M-j is determined by the "effective bond length ratio", R(i/j).

These empirical parameters can be successfully transferred from one stereochemistry to another, and even between different coordination numbers.

Some values for R(i/j) obtained for n = 6 will be summarised here, and more detail will be found throughout this work.

For groups attached to nonmetallic elements, it is found that the order of these empirical R(i/j) values is not simply related to the relative bond lengths. The largest R(i/j) value is observed for the very electronegative fluorine atom, but the order does not simply follow an electronegativity order, the position of hydrogen being anomalous. If the alkyl group (j = R) is taken as unity, the following values are obtained:

	R(i/R) (n = 6)
i = Me	1.00
i = Cl	1.05
i = H	1.09
i = F	1.16

For ligands attached to transition metals in moderately high oxidation states, $M^{\geq II}$, it is found that the effective bond lengths of charged ligands are about 20% shorter than uncharged ligands:

$$R(\text{ligand}/X^-) \sim 1.2 \quad \text{or} \quad R(X^-/\text{ligand}) \sim 0.8 \quad (n = 6).$$

Any physical meaning that is attached to these R(i/j) values is more uncertain for groups which can be considered as potentially bonding through multiple bonds, for example, oxygen can be considered to be bonded as $X = O$, $X \to O$, or $X^{2+} \leftarrow O^{2-}$. It is most convenient to treat these bonds in the same way as before, to yield, for example $R(O^{2-}/R) \sim 0.9$ (n = 6) for central nonmetallic atoms, or $R(O^{2-}/\text{ligand}) \sim 0.7$ (n = 6) for transition metal complexes.

A particularly interesting class of compounds is where there is a non-bonding pair of electrons attached to the central atom. In this case the R(lone pair/j) values found are not constant, the lone pair of electrons being constricted closer to the central atom for atoms further down the Periodic Table. For example:

	R(lone pair/Cl⁻) (n = 6)
$N^{III}Cl_3$	0.90
$P^{III}Cl_3$	0.59
$As^{III}Cl_3$	0.53
$Sb^{III}Cl_3$	0.47

The lone pair of electrons is forced closer to the central atom as the coordination number increases, although the R(lone pair/j) value may increase if the steric crowding is reduced by replacing the unidentate ligands with bidentate ligands which hold the donor atoms closer together. For example:

	R(lone pair/j) (n = 6)
$[Sb^{III}Cl_6]^{3-}$	0.0
$[Sb^{III}(S_2COEt)_3]$	0.1_5

3. Repulsion Energy Coefficients

The minimum value for the repulsion energy coefficient X obtained for the most stable stereochemistry of each coordination number is shown:

Coordination Number	X (n = 6)	Coordination Number	X (n = 6)
2	0.016	8	5.185
3	0.111	9	8.105
4	0.316	10	12.337
5	0.877	11	18.571
6	1.547	12	23.531
7	3.230		

A plot of X against coordination number shows the regular tetrahedron, octahedron, and icosahedron are $10-20\%$ more stable than would be predicted from the smooth curve passing through the remaining points corresponding to the semi-regular and nonuniform polyhedra.

Precise energy units cannot be associated with these repulsion energy coefficients, but it is generally found that distortions corresponding to an increase in X by $0.01-0.05$ units (for n = 6) are of chemical significance. For example an increase of 0.01 units corresponds to a distortion of a regular tetrahedron by about $4°$. In seven-coordinate structures a change in X by as little as 0.001 units is sufficient to allow interconversion between different stereochemistries, with very rapid intramolecular rearrangements occurring in solution. As the coordination number increases the value of r increases slightly, making small differences in X slightly less significant. It is for this reason that the calculated potential energy surfaces for coordination numbers < 6 are drawn as 0.01 energy contours, whereas those for coordination numbers ≥ 6 are drawn as 0.02 energy contours.

The total repulsion energy U can be divided into the repulsions V_i experienced by each ligand:

$$U = \sum_i V_i.$$

Similarly the repulsion energy coefficient X can be separated into the individual atom repulsion energy coefficients Y_i:

$$X = \sum_i Y_i.$$

It is predicted that those atoms with the largest individual atom repulsion energy coefficients, that is those which are the most sterically crowded, will have the longest bond lengths. Differences in repulsion energy can lead to differences in bond length of about 5%.

4. Multidentate Ligands

It is assumed that each bonded bidentate ligand is sufficiently rigid that interaction between its metal-ligand bonds can be considered to be constant, and this interaction is therefore omitted when summing the electron-pair repulsions. The chelate geometry is given by the normalized "bite", b, which is defined as the distance between the effective bond centers divided by r. For both symmetrical and unsymmetrical bidentate ligands the normalized bite can be defined in terms of the bond angle iMj subtended by the ligand, that is, $b = 2 \sin (iMj/2)$. The normalized bite is a measure of ring geometry and is particularly dependent on the number of atoms in the chelate ring and to a lesser extent on the size of those atoms.

The geometry of a symmetric tridentate ligand with donor atoms A, B, and C is given by two variables, the normalized bite, b, of each chelate ring, and the tridentate angle ABC. Two limiting assumptions can be made about the rigidity of the tridentate ligand with respect to the tridentate angle ABC and the interaction between A and C:

a) The tridentate ligand is completely flexible. The two arms of the ligand are freely hinged at B, and the full repulsion between A and C is included in the total repulsion energy.

b) The tridentate ligand is completely rigid. The tridentate angle ABC is fixed, and hence u_{AC} is constant and may be omitted from the total repulsion energy.

Only single-chain tridentate ligands are considered, and ligands with more complex branching, such as 'tripods', are excluded.

5. Trigonometry

The method of calculation involves the above very simple energy expressions coupled with elementary trigonometry. Each center of repulsion is located by its distance from the central atom, r_i, and by its spherical coordinates, ϕ_i and θ_i (Fig. 1.1). The distance between two such centers i and j is given by:

$$(d_{ij})^2 = (r_i - r_j)^2 + r_i r_j [2 - 2 \cos \phi_i \cos \phi_j - 2 \sin \phi_i \sin \phi_j \cos (\theta_i - \theta_j)].$$

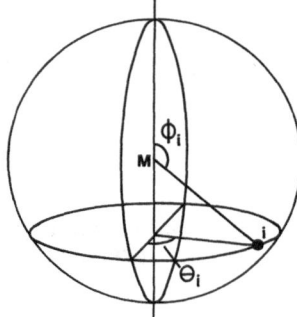

Fig. 1.1. Definition of spherical coordinates

CHAPTER 2

Polyhedra

A. Classic Geometric Polyhedra

The stereochemical arrangement of atoms around a central atom, or of a cluster of atoms grouped about a common center, is usually imagined as a planar polygon or as a polyhedron with an atom at each vertex.

There are an infinite number of regular polygons with identical corners, but these are of chemical importance only in three-coordinate compounds, in square planar transition metal complexes $[M(ligand)_4]^{x+}$, and in planar aromatic molecules and ions such as $C_5H_5^-$, C_6H_6, and $C_7H_7^+$.

The four-coordinate tetrahedron and the six-coordinate octahedron are the two most common polyhedra encountered in chemistry. However, very few polyhedra are as regular as these two, and for other coordination numbers a much greater variety of more complex polyhedra are observed. At first sight it is not clear how many polyhedra are geometrically possible for any particular coordination number, or how they are related to each other. As an example of the diversity available, consider the following stereochemistries which all belong to different classes of polyhedra: the square pyramid which is common for five-coordination, the pentagonal bipyramid for seven-coordination, the triangular dodecahedron for eight-coordination, the tricapped trigonal prism for nine-coordination, and the bicapped square antiprism for ten-coordination.

In this section the three groups of classic geometric polyhedra are described. The differences which arise in real molecules are described in Sect. 2.B. More attention will be focused on polyhedra with twelve or fewer vertices, as the larger examples are of less chemical interest. The classical geometric treatment is based on the construction of polyhedra using faces which are regular polygons, namely the equilateral triangle, square, regular pentagon, regular hexagon, and so on.

The first group are the completely regular polyhedra, in which all faces are identical, all edges are identical, and all vertices are identical. Secondly there are the semiregular polyhedra that contain more than one type of face.

The regular and semiregular polyhedra are of particular chemical interest since they have all vertices identical. It is also important to note that they all have an even number of vertices, and are not relevant to coordination numbers of five, seven, nine, and eleven.

The third group considered are the non-uniform polyhedra, which contain more than one type vertex, and are particularly important for the odd coordination numbers.

1. Regular Polyhedra

There are only five regular polyhedra, the well-known Platonic solids described by Euclid in about 300 B. C. (Fig. 2.1):

a) The tetrahedron, with four vertices.
b) The octahedron, with six vertices.
c) The cube, with eight vertices.
d) The icosahedron, with twelve vertices.
e) The pentagonal dodecahedron, with twenty vertices.

Those composed of equilateral triangular faces are the tetrahedron with three faces meeting at each vertex, the octahedron with four faces meeting at each vertex, and the icosahedron with five faces meeting at each vertex. Each of these polyhedra dominates large areas of stereochemistry. A polyhedron cannot have less than three faces meeting at each point, and since the internal angle of an equilateral triangle is 60°, it is impossible to construct a regular polyhedron with six or more equilateral triangles meeting at a point, as this would simply create a plane ($6 \times 60 = 360°$) (at least for a convex polyhedron). Thus, only three polyhedra can be formed from equilateral triangular faces. Three squares meeting at each vertex forms a cube. Three regular pentagons meeting at each vertex form the pentagonal dodecahedron but since it has 20 vertices it is of no importance in the stereochemistry of coordination polyhedra. It is again clear that it is not possible to construct polyhedra with four squares or four regular pentagons meeting at each vertex ($4 \times 90 = 360°$ and $4 \times 108 = 432°$, respectively) or to have faces that are regular hexagons ($3 \times 120 = 360°$) or larger polygons.

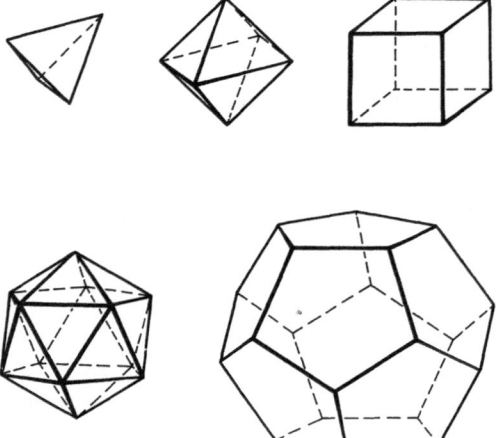

Fig. 2.1. The five regular polyhedra: tetrahedron, octahedron, cube, icosahedron and pentagonal dodecahedron

It may be of interest to remember that the first four Platonic polyhedra, the tetrahedron, octahedron, icosahedron, and cube, represented the shapes of the atoms of the four original elements, Fire, Air, Water and Earth, respectively. The fifth polyhedron, the pentagonal dodecahedron, was taken as the shape of the universe, and the most essential part of all matter, the quintessence.

The tetrahedron, octahedron, and icosahedron are common stereochemistries. However, it does not follow that merely being regular is sufficient to create a favourable structure. The cube is rare, except in three-dimensional structures where it is the only eight-coordinate polyhedron that will completely fill space by itself.

2. Semiregular Polyhedra

A second important class of polyhedra are the semiregular polyhedra. All vertices are identical, all edges are of the same length, and all faces are regular polygons, but now not all faces are identical.

These semiregular polyhedra are classified into three groups:
a) The prisms.
b) The antiprisms.
c) Thirteen Archimedean polyhedra.

a) The Prisms

The prisms are shown in Fig. 2.2. The triangular prism contains two equilateral triangular faces, and three square faces. The square prism is the cube, a regular polyhedron described above. The pentagonal prism has two regular pentagonal and five square faces, the hexagonal prism has two regular hexagonal and six square faces, and so on. There are an infinite number of examples.

b) The Antiprisms

The second group consists of the antiprisms (Fig. 2.2), which are more common than the prisms for discrete molecules and ions. The antiprisms are formed by taking the two unique, non-square faces of a prism and rotating them relative to each other until they are in a staggered rather than an eclipsed configuration, and replacing each square face with two equilateral triangular faces. Thus the trigonal

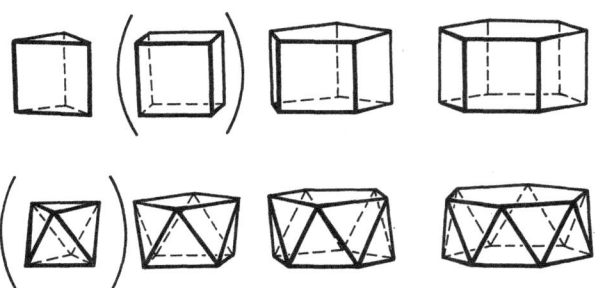

Fig. 2.2. Prisms and antiprisms

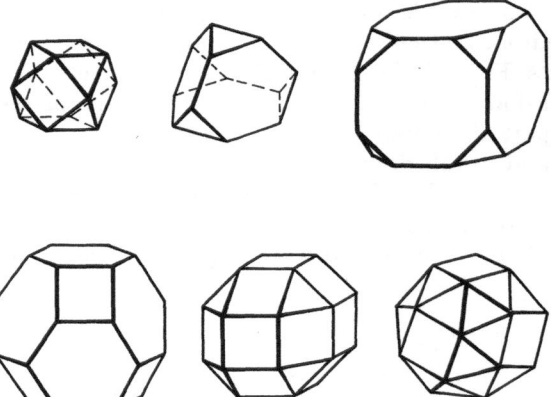

Fig. 2.3. Archimedean polyhedra: cuboctahedron, truncated tetrahedron, truncated cube, truncated octahedron, rhombicuboctahedron and snub cube

prism forms the octahedron, described above, which is important in six-coordination, the cube forms the square antiprism which is important in eight-coordination, the pentagonal prism forms the pentagonal antiprim, which is the structure observed for ferrocene, and so on. There are an infinite number of antiprisms corresponding to the infinite number of prisms.

c) Thirteen Archimedean Polyhedra

The third group of semiregular polyhedra consists of thirteen unique polyhedra. The cuboctahedron and truncated tetrahedron have twelve vertices, the truncated cube, truncated octahedron, rhombicuboctahedron and snub cube have twenty-four vertices, but the remaining seven have an ever higher number of vertices and are of no chemical importance. The six small polyhedra are shown in Fig. 2.3.

3. Non-uniform Polyhedra

There are 92 convex, non-uniform polyhedra possible [622]. All have regular polygons as faces, but all have at least two different kinds of vertex.

The non-uniform polyhedra are particularly important to the chemistry of the odd coordination numbers, five, seven, nine, and eleven.

The presence of geometrically different vertices leads directly to three important chemical consequences:

(i) In many non-uniform polyhedra, all vertices do not lie on the surface of a sphere. In compounds of the type [M(ligand)$_x$], where each ligand is situated at a vertex, the polyhedron will be substantially distorted to equalize, or nearly equalize, the metal-ligand bond lengths. These distortions are considered in more detail in Sect. 2.B.

(ii) In molecules of the type [M(ligand)$_x$], the ligands in different sites should have different properties. In many instances, only an averaged property is observed, due to the rapid intramolecular rearrangements which exchange ligands between different sites. Such behaviour is dependent on temperature.

(iii) In molecules of the type [M(ligand A)$_x$(ligand B)$_y$], the chemical differences between ligand A and ligand B will lead to a preference by each ligand to a particular vertex, leading to the stabilisation of particular structural isomers.

Of the 92 convex non-uniform polyhedra, only those 29 with twelve or fewer vertices are described below. They are divided into four groups:

a) Pyramids, bipyramids, capped prisms, and capped antiprisms.
b) Cupolas and bicupolas.
c) Diminished icosahedra.
d) Five unique polyhedra.

a) Pyramids, Bipyramids, Capped Prisms, and Capped Antiprisms

There are three pyramids, namely the tetrahedron which has been considered above, the square pyramid, and the pentagonal pyramid. The hexagonal pyramid and higher members of the series cannot be constructed using equilateral triangles.

The pyramids can be considered alone, or as two units sharing a common base to form bipyramids. The square bipyramid is the octahedron which has also been considered above. The pyramids can also share a common face with prisms, adding onto one or both of the nonsquare end faces to form end-capped prisms, or onto one or more of the square side faces to form square-capped prisms. Adding a pyramid onto one or both of the nontriangular end faces of an antiprism leads to the capped antiprisms. The bicapped pentagonal antiprism is the icosahedron referred to above. Capping the triangular face of an antiprism, or the triangular face of an octahedron, leads, for example, to the "monocapped octahedron". This polyhedron is not convex, but will be referred to on Sect. 2.B on chemical coordination polyhedra.

These polyhedra are listed in Table 2.1 and shown in Fig. 2.4.

b) Cupolas and Bicupolas

There are two cupolas with twelve or fewer vertices, the triangular cupola and the square cupola (Fig. 2.5). In the same way as the pyramids, the cupolas can share basal faces with each other to form bicupolas. If two triangular cupolas join so that the opposing triangular faces are staggered with respect to each other, a cuboctahedron is formed, but if they join so that the opposing faces are eclipsed with respect to each other, an anticuboctahedron is formed. The anticuboctahedron is a fragment of hexagonal close packing, whereas the cuboctahedron is a fragment of cubic close packing.

Polyhedra in this group with twelve or fewer vertices are listed in Table 2.2, and illustrated in Fig. 2.5. The sharing of faces between cupolas and prisms or between cupolas and antiprisms leads to larger polyhedra which are not considered here. (Two square cupolas and an octagonal prism may form the rhombicuboctahedron, one of the Archimedean semi-regular polyhedra with 24 vertices described above.)

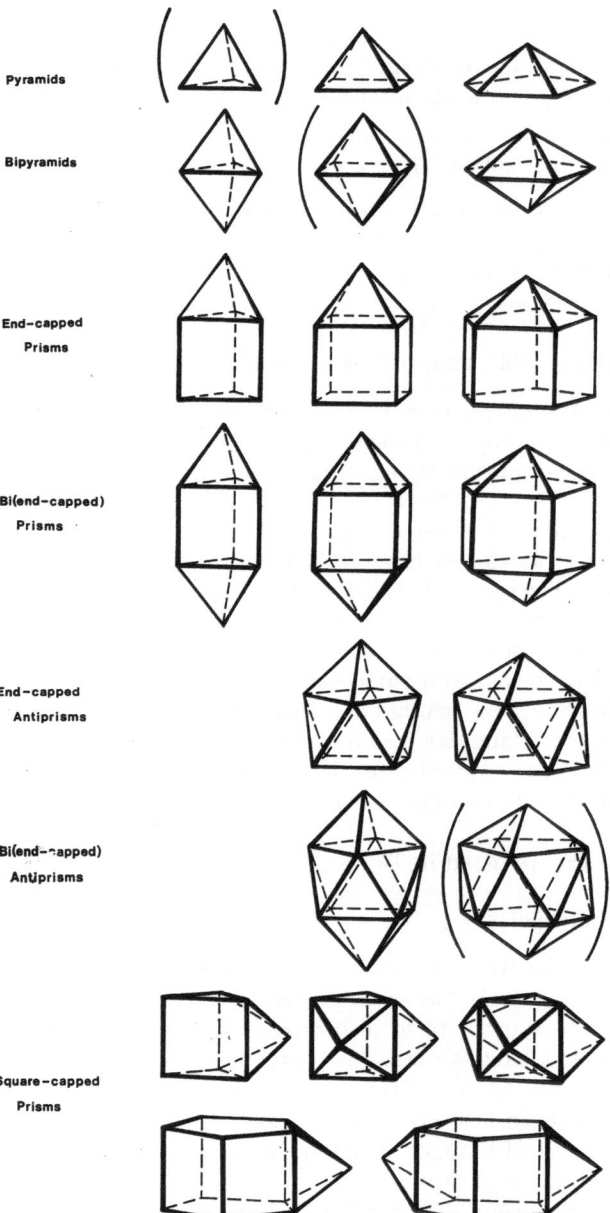

Fig. 2.4. Pyramids, bipyramids, capped prisms and capped antiprisms

Table 2.1. Pyramids, bipyramids, capped prisms, and capped antiprisms

Polyhedron	Faces			No. of vertices	Types of vertex
	Triangles	Squares	Pentagons		
Pyramids					
(Tetrahedron)	(4)			(4)	(1)
Square pyramid	4	1		5	2
Pentagonal pyramid	5		1	6	2
Bipyramids					
Trigonal bipyramid	6			5	2
(Octahedron)	(8)			(6)	(1)
Pentagonal bipyramid	10			7	2
End-capped Prisms					
End-capped trigonal prism	4	3		7	3
Capped cube	4	5		9	3
End-capped pentagonal prism	5	5	1	11	3
Bi(end-capped) trigonal prism	6	3		8	2
Bi-capped cube	8	4		10	2
Bi(end-capped) pentagonal prism	10	5		12	2
End-capped Antiprisms					
End-capped square antiprism	12	1		9	3
End-capped pentagonal antiprism	15		1	11	3
Bi(end-capped) square antiprism	16			10	2
(Icosahedron)	(20)			(12)	(1)
Square-capped Prisms					
Capped trigonal prism	6	2		7	3
Bicapped trigonal prism	10	1		8	3
Tricapped trigonal prism	14			9	2
Capped pentagonal prism	4	4	2	11	4
Bicapped pentagonal prism	8	3	2	12	4

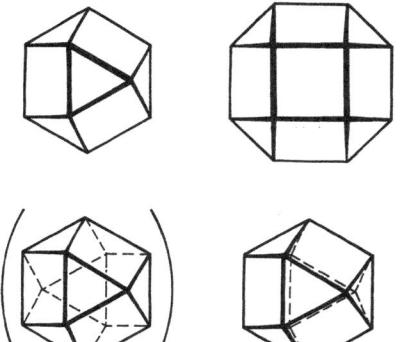

Fig. 2.5. Triangular cupola, square cupola, cuboctahedron and anticuboctahedron

Table 2.2. Cupolas and bicupolas

Polyhedron	Faces			No. of vertices	Types of vertex
	Triangles	Squares	Others		
Triangular cupola	4	3	1 Hexagon	9	2
Square cupola	4	5	1 Octagon	12	2
(Cuboctahedron)	(8)	(6)		(12)	(1)
Anticuboctahedron	8	6		12	3

c) Diminished Icosahedra

The removal of one pentagonal pyramid from an icosahedron forms the end-capped pentagonal antiprism considered above. Similarly removal of two penta-gonal pyramids *trans*- to each other simply forms a pentagonal antiprism. How-ever, removal of two or three pentagonal pyramids from other than mutually *rans*-sites leads to three additional polyhedra (Table 2.3 and Fig. 2.6).

 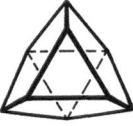 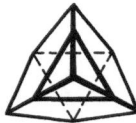

Fig. 2.6. Bidiminished icosahedron, tri-dimished icosahedron and capped tridiminished icosahedron

Table 2.3. Diminished icosahedra

Polyhedron	Faces		No. of vertices	Types of vertex
	Triangles	Pentagons		
Bidiminished icosahedron	10	2	10	4
Tridiminished icosahedron	5	3	9	3
Capped tridiminished icosahedron	7	3	10	4

d) Five Unique Polyhedra

In addition to the above groups, there are an additional five unique non-uniform polyhedra with twelve or fewer vertices. These are listed in Table 2.4 and illu-strated in Fig. 2.7. Only the double triangular prism is clearly related to any of the polyhedra described above.

Table 2.4. Five unique, non-uniform polyhedra

Polyhedron	Faces		No. of vertices	Types of vertex
	Triangles	Squares		
Double triangular prism	4	4	8	2
Triangular dodecahedron	12		8	2
Sphenocorona	12	2	10	4
Capped sphenocorona	16	1	11	7
Sphenomegacorona	16	2	12	5

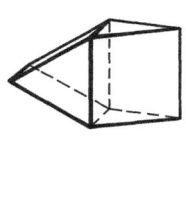  

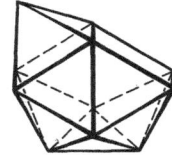 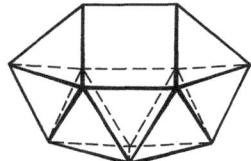

Fig. 2.7. Double triangular prism, triangular dodecahedron, sphenocorona, capped sphenocorona and sphenomegacorona

The triangular dodecahedron completes the series of *deltahedra*, whose faces consist entirely of equilateral triangles. These are the tetrahedron, trigonal bipyramid, octahedron, pentagonal bipyramid, triangular dodecahedron, tricapped trigonal prism, bicapped square antiprism and icosahedron, all of which have been observed for four-, five-, six-, seven-, eight-, nine-, ten-, and twelve-coordinate compounds respectively.

B. Chemical Coordination Polyhedra

There are some differences between the classic geometric polyhedra built up from regular polygons, and the structures observed in molecules, where each atom surrounding a central atom is taken as a vertex of the coordination polyhedron. These differences are discussed under the following five headings:

1. Distortions arising from different sized faces.
2. Distortions of non-spherical polyhedra to spherical polyhedra.
3. Some additional polyhedra.
4. Distortions resulting from chelate groups.
5. Small bond-length distortions.

1. Distortions Arising from Different Sized Faces

All the semiregular polyhedra, and most of the non-uniform polyhedra, contain two or more different types of polygon as faces. Although all of the same edge length, the size of the polygon increases as the number of corners increases. In molecules these larger faces are relatively uncrowded regions, and the molecule may relax, increasing the size of the smaller triangular faces at the expense of the larger square, pentagonal, and hexagonal faces. This distortion removes the equality of the edge lengths, with the result that not all polygons remain regular.

This effect will be illustrated with two examples, the square pyramid and the square antiprism, both composed of a mixture of triangular and square faces.

 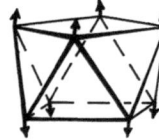

Fig. 2.8. Distortion of the square pyramid and square antiprism

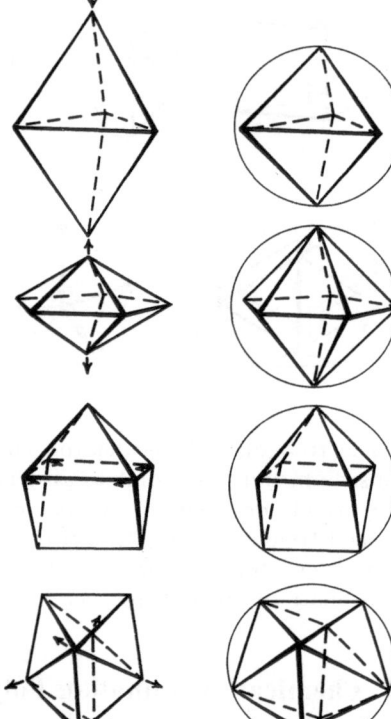

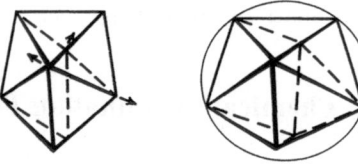

Fig. 2.9. Distortions of non-spherical polyhedra to spherical polyhedra: trigonal bipyramid, pentagonal bipyramid, capped trigonal prism and triangular dodecahedron

The undistorted square pyramid contains four equilateral triangular faces and one square face. In square pyramidal molecules [M(ligand)$_5$], the size of the square faces descreases so that the four (apical ligand)-metal-(basal ligand) angles increase from 90.0° to about 101°, with the square edges becoming about 10% shorter than the edges linking two triangles (Fig. 2.8).

In square antiprismatic molecules [M(ligand)$_8$], the square faces again contract so that the angle the metal-ligand bonds make with the eight-fold inversion axis decreases from 59.3° to about 57°, with the square edges becoming about 6% shorter than the edges linking two triangular faces (Fig. 2.8).

However not all semiregular polyhedra can distort in this manner. In the cuboctahedron, for example, *every* edge links a triangle and a square, and it is not possible to decrease simultaneously the size of one type of polygon and increase the size of the other.

2. Distortions of Non-spherical Polyhedra to Spherical Polyhedra

The non-uniform polyhedra constructed from regular polygons do not have all vertices identical. Even more important, in many cases these vertices do not lie on the surface of a sphere centered on the center of the polyhedron. This result is sharp contrast to chemical coordination polyhedra, where all metal-ligand bond lengths are equal, or at least nearly so. Thus the molecular polyhedra may differ substantially from the classic geometric polyhedra.

Four examples are shown in Fig. 2.9. The trigonal bipyramid constructed from equilateral triangles is substantially elongated along the three-fold axis compared with a trigonal bipyramidal molecule, [M(ligand)$_5$], with five equal metal-ligand bond lengths. In the latter case the faces are only isosceles triangles, with edge lengths in the ratio $2^{1/2} : 3^{1/2} = 1.000 : 0.816$.

The *reverse* distortion is found in pentagonal bipyramidal [M(ligand)$_7$] molecules, where the elongation along the five-fold axis increases the (apical ligand): (equatorial ligand) edge length ratio from $1.000 : 1.000$ to $1.203 : 1.000$.

More complex distortions occur in monocapped trigonal prismatic molecules [M(ligand)$_7$], where not only are the resulting isosceles triangles of three different types, but the square faces are converted to trapezoidal faces. The edge lengths in the polyhedron again vary by about 20% (Fig. 2.9).

A particularly important example is observed in eight-coordinate molecules with the triangular dodecahedral structure. Distortion of the polyhedron so that all ligands sit on the surface of a sphere about the metal atom (Fig. 2.9) lengthens the four polyhedral edges around the center of the molecule by about 25% This distortion creates four holes around the molecule which are coplanar with the metal atom.

3. Some Additional Polyhedra

The distortions described above requiring that all atoms lie on the surface of a sphere also allows, in a few cases, the creation of additional polyhedra.

It is usual developing the classic geometric polyhedra initially to restrict attention to *convex* polyhedra, that is, ones in which the interior dihedral angle between all pairs of faces is less than 180°. This is why there are no examples of antiprisms with tetrahedra capping one or more triangular faces described in Sect. A.3 containing the capped antiprisms. However the distortion of a polyhedron so that all vertices are equidistant from the center may convert a non-convex polyhedron into a convex polyhedron.

An important example is the capped octahedron, formed from a regular octahedron and a regular tetrahedron sharing a face. This polyhedron does not have regular polygonal faces, as the dihedral angle between the tetrahedral and octahedral faces is 180.0°, with the creation of rhomboidal faces (Fig. 2.10). However, distortion to force all seven vertices onto the surface of a sphere creates an

  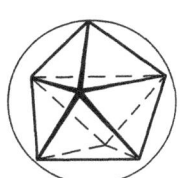

Fig. 2.10. Octahedron + tetrahedron. Octahedron + tetrahedron sharing a face. Spherical capped octahedron

irregular convex polyhedron (Fig. 2.10) in which only one of the ten triangular faces is equilateral, but which is nevertheless of considerable importance to the stereochemistry of seven-coordinate compounds.

The introduction of isosceles triangular faces also allows the formation of hexagonal (and higher) pyramids, and polyhedra where these pyramids cap hexagonal (and higher polygonal) faces of other polyhedra.

4. Distortion Resulting from Chelate Groups

An important result arising from the distortion of the classic geometric polyhedra to the molecular polyhedra is that the equality of the edge lengths is removed. For example, the square antiprism has eight edge lengths of 1.187 r, and eight edge lengths of 1.262 r (where r is the metal-ligand distance). These different polyhedral edge lengths are important when arranging chelate rings around a central atom, as the stereochemistry observed depends upon fitting the size of the chelate ring to the appropriate polyhedral edge length.

It is of much greater importance, however, to realize that in the great majority of molecules containing chelate groups, it will not be possible to find a polyhedral edge length which precisely fits the edge length required by the ligand, which is determined by the chemical design of the ring system. Such molecules will then adopt a stereochemistry which at first sight may not be readily recognized as belonging to any of the polyhedra which have been described in this section.

A major result emerging from the calculations described here is that the structures involving ring system are usually intermediate between two idealized polyhedra, and that there is a smooth and continuous change in geometry from one polyhedron into another as the geometric constraints imposed by ligand design are changed, and as the size of the metal atom is varied. These distortions can be briefly illustrated by a number of relatively simple examples:

a) [M(bidentate)₃]

For large six-membered chelate rings, where the angle about the metal atom subtended by the bidentate ligand is close to 90°, the observed stereochemistry is close to octahedral (Fig. 2.11 a). As the size of the chelate ring is reduced, there is a *continuous* change in stereochemistry until eventually a trigonal prism is formed (Fig. 2.11 c). As shown in more detail in Chap. 8, there are very many structurally characterized compounds which cover the first half of the range from a regular

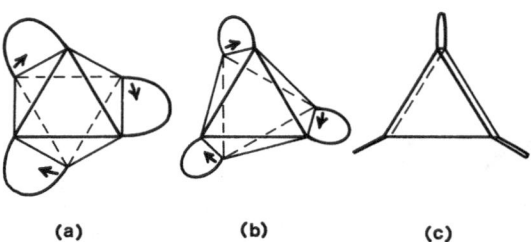

(a)   (b)   (c)

Fig. 2.11. Distortion of [M(bidentate)₃] stereochemistry from an octahedron (a) to a trigonal prism (c) as the size of the chelate ring is reduced

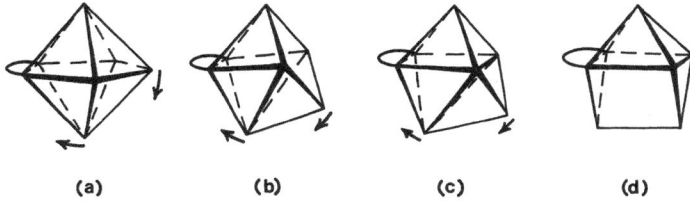

Fig. 2.12. Distortion of [M(bidentate) (unidentate)$_5$] stereochemistry from a pentagonal bipyramid (**a**), through a 4 : 3 (**b**) and capped octahedron (**c**), to a capped trigonal prism (**d**) as the size of the chelate ring is increased

octahedron to about half way between an octahedron and a trigonal prism, but there is much less information concerning compounds nearer the trigonal prism end of the range.

b) [M(bidentate)(unidentate)$_5$]

A variety of structures are possible for seven-coordinate complexes containing one bidentate ligand, one of the most common being stereochemistry B/C. For small sized chelate rings a pentagonal bipyramidal structure is observed (Fig. 2.12a) but as the size of the chelate ring increases the pentagonal plane distorts, and there is a smooth and continuous change in structure to the capped trigonal prism (Fig. 2.12d). Known compounds cover the range from one polyhedron to the other. Intermediate structures between these extremes have been claimed as "new" structures, for example, a capped octahedron (Fig. 2.12c), or a 4 : 3 "piano stool" (Fig. 2.12b).

c) [M(tridentate)$_2$(unidentate)$_2$]

As an example of a continuous change brought about by tridentate ligands, consider the eight-coordinate [M(tridentate)$_2$(unidentate)$_2$]. Figure 2.13 shows the change from a square antiprism to a hexagonal bipyramid as the size of each chelate ring is reduced. Known compounds have intermediate structures which at first sight are not obviously related to either of these extremes.

d) [M(bidentate)$_4$(unidentate)]

The nine-coordinate monocapped square antiprismatic structure containing four identical chelate groups is shown in Fig. 2.14. Reducing the size of the chelate

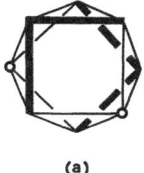

(a)   (b)

Fig. 2.13. Distortion of [M(tridentate)$_2$(unidentate)$_2$] stereochemistry from a square antiprism (**a**) to a hexagonal bipyramid (**b**) as the size of the chelate ring is reduced

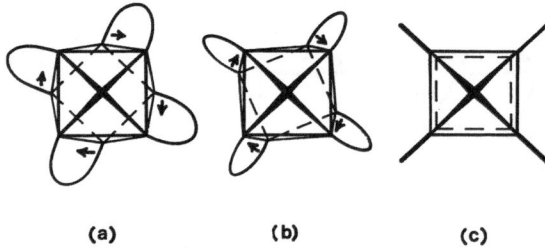

Fig. 2.14. Distortion of [M(bidentate)₄(unidentate)] stereochemistry from a capped square antiprism (**a**) to a capped cube (**c**) as the size of the chelate ring is reduced

groups twists the square face until eventually a monocapped cube is formed (Fig. 2.14), with real molecules having intermediate structures. This twisting is very similar to that described for [M(bidentate)₃] complexes above.

5. Small Bond-length Distortions

It has been seen above that many coordination polyhedra contain different types of vertex, either because they are non-uniform polyhedra, and/or because they are chemically distorted. The chemical distortion may arise from the polyhedron having different sized faces, by placing an atom at the center bringing all vertices onto the surfaces of a sphere, or by introducing chelate groups into the molecule.

Atoms at different types of vertex will interact with their neighbours to a different extent, with the result that atoms at the more sterically crowded sites will be expected to have longer metal-ligand bonds.

The changes in bond length are particularly noticeable for the relatively large polyhedral distortions created by the introduction of chelate groups, and will be illustrated with some simple examples:

a) [M(unidentate)₉]

In the tricapped trigonal prism, each capping atom has six close neighbours (the four atoms of its square face plus the other two capping atoms), whereas each prism atom has only five close neighbours (three prism atoms and two capping atoms). The capping atoms are therefore more sterically crowded, and have metal-ligand bonds about 4% longer than those to the prism atoms.

b) [M(bidentate)₃(unidentate)₃]

A similar bond length distortion is observed if each capping atom of a tricapped trigonal prism is connected to a prism atom by a bidentate ligand. In these cases the bidentate ligand is not coordinated symmetrically to the metal atom, the bond to the capping atom again being about 4% longer than the bond to the prism atom.

c) [M(bidentate)₄]

Eight-coordinate compounds containing four bidentate ligands with the triangular dodecahedral structure show these bond length changes very clearly. The structure can be considered as two trapezoids (Fig. 2.15) which are at right angles to each

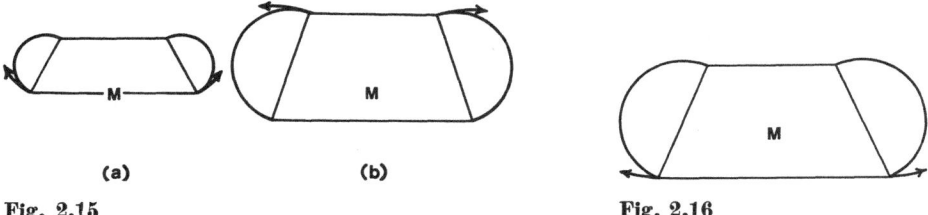

Fig. 2.15   Fig. 2.16

Fig. 2.15a, b. Trapezoidal planes in dodecahedral [M(bidentate)₄] complexes, showing bond length distortions

Fig. 2.16. Trapezoidal plane in *trans*-[M(bidentate)₂(unidentate)₂] complexes, showing bond length distortions

other. For small chelate rings it is the atoms at the ends of the long edges of the trapezoids which are subjected to the greatest repulsion, and these metal-ligand bonds are about 3% longer than those to the other vertices (Fig. 2.15a). This distortion is reversed for large chelate rings, where it is the atoms forming the short edges of the trapezoids which are subjected to the greatest repulsion, and it is these bonds which are now about 3% longer (Fig. 2.15b).

d) [M(bidentate)₂(unidentate)₂]

Unsymmetrical bonding of chelate rings is observed even in the classic case of *cis*- and *trans*-isomers of six-coordinate [M(bidentate)₂(unidentate)₂]. For example, in the *trans*-isomer, the plane formed by the metal atom and the two chelate rings changes from a square for large chelate rings to a trapezoid for small chelate rings (Fig. 2.16). The bonding of the bidentate ligands becoming grossly unsymmetrical, one bond being 10—20% longer than the other.

CHAPTER 3

Four-Coordinate Compounds

A. Tetrahedral [M(unidentate A)(unidentate B)₃]

The general stereochemistry for tetrahedral molecules containing one ligand
different to the other three is shown in Fig. 3.1. The effective bond length to the
unique ligand A lying on the threefold axis is R, the other three metal-ligand
effective bond lengths being defined as unity. For R = 1, the tetrahedron is
regular, with

$$AMB = BMC = 2 \text{ arc sin } (2/3)^{1/2} = 109.47°.$$

The repulsion energy calculations show that AMB increases, and BMC decreases,
as R is decreased (Fig. 3.2). The experimental bond angles for any molecule can be
fitted against these calculated bond angles and a value of R(i/j) can be obtained for
pair of ligands i and j.

It is important to note that when the distortions from the regular tetrahedron
are more than 1.5° (that is, the observed bond angles are less than 108° or greater
than 111°, Fig. 3.2), it is not possible to obtain a fit between experimental bond
angles and bond angles calculated using n = 1, demonstrating that this is an unrealistic form of the repulsion law. In this chapter, only results using n = 6 will be
quoted.

The results from this procedure are examined for the following illustrative
cases:

1. M, unidentate A, and unidentate B are p-block elements, connected by single
bonds.
2. M, unidentate A, and unidentate B are p-block elements, connected by both
single bonds and multiple bonds.
3. Unidentate A is considered to be a nonbonding pair of electrons.
4. Unidentate A is a transition metal.
5. M is a transition metal.

1. p-Block Elements Connected by Single Bonds

There are a very large number of compounds containing carbon and nitrogen
atoms bonded to carbon, nitrogen, and oxygen atoms where the distortions from
regular tetrahedral coordination are only about 2°. For such small distortions the

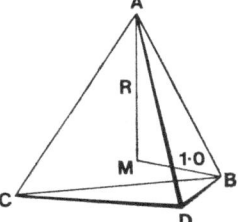

Fig. 3.1. General stereochemistry for [M(unidentate A) (unidentate B)₃]

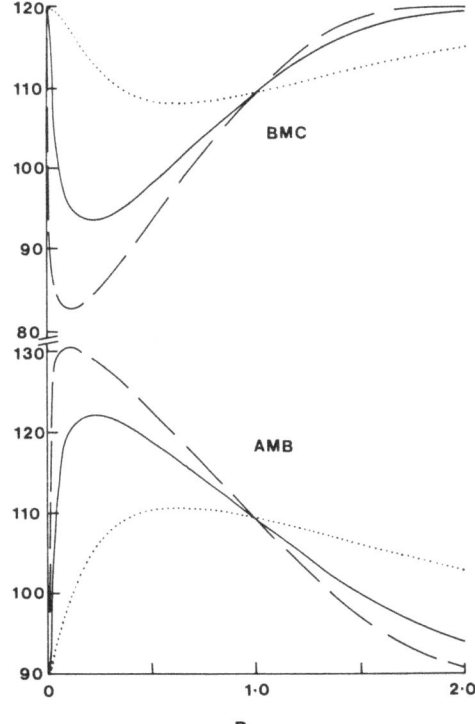

Fig. 3.2. Bond angles (degrees) for tetrahedral [M(unidentate A)(unidentate B)₃], as a function of effective bond length ratio R. *Dotted lines*, n = 1; *full lines*, n = 6; *broken lines*, n = 12

experimental uncertainties become significant, and any structural trends are not easily discerned. A selection of compounds in which larger distortions are found is shown in Table 3.1.

In trifluoromethyl groups bonded to carbon atoms, the FCC and FCF bond angles of 112.4° and 106.4° respectively correspond to R(C/F) = 0.86, or R(F/C) = 1.16.

Table 3.1. Stereochemical parameters for tetrahedral [M(unidentate A)(unidentate B)₃] complexes of the p-block elements connected by single bonds. R(A/B) values obtained for n = 6

Complex	AMB	BMC	(M—A)/(M—B)	R(A/B)	Ref.
F₃C—C (average)	112.4° (2)	106.4° (2)	1.15 (0)	0.86	[426]
F₃C—P (average)	112.2° (2)	106.6° (2)	1.42 (1)	0.86	[426]
Cl₃C—C (average)	111.4° (4)	107.5° (4)	0.86 (1)	0.91	[428]
Cl₃C—H (average)	—	110.3° (4)	—	1.04	[426]
R₃N—H (average)	—	111.4° (3)	—	1.11	[426]
K[Cl₃AlMe]	114.7°	103.6°	0.89	0.74	[59]
(Me₄As)[Cl₃GaMe]	114.3°	104.3°	0.87	0.76	[541]
(Me₄As)[Cl₃InMe]	116.3°	101.9°	0.91	0.66	[507]
[(C₆H₁₁)₃SnCl]	101.5°	116.2°	1.15	1.42	[204]

Note that the carbon-carbon effective bond length is approximately 14% *shorter* than the carbon-fluorine effective bond length, although the real carbon-carbon bond length is 15% *longer* than the real carbon-fluorine bond length. The same R values are obtained for trifluoromethyl groups attached to phosphorus atoms (Table 3.1), where the real carbon-phosphorus bond length is 42% longer than the real carbon-fluorine bond length. The large effective bond length of a carbon-fluorine bond is readily attributed to the high electronegativity of the fluorine atom, which attracts the bonding electron pair towards itself.

For trichloromethyl groups attached to carbon atoms, $R(C/Cl) = 0.91$, or $R(Cl/C) = 1.09$; that is, the electron pair in the carbon-chlorine bond is considered to be closer to the carbon atom than is the electron pair in the carbon-fluorine bond, in spite of the longer carbon-chlorine bond.

The effective bond length ratios obained for Cl_3C-H [$R(H/Cl) = 1.04$] and for R_3N-H [$R(H/C) = 1.11$] (Table 3.1), are internally consistent with those obtained above for the Cl_3C-C group [$R(C/Cl) = 0.91$, compared with $1.04/1.11 = 0.94$].

These $R(X/C)$ values, taking that for an alkyl group as unity, may be summarized:

X	R(X/C)
Me	1.00
Cl	1.09
H	1.11
F	1.16

Much greater distortions from regular tetrahedral bond angles are observed for $[AlCl_3Me]^-$, $[GaCl_3Me]^-$, $[InCl_3Me]^-$, and $[SnCl(C_6H_{11})_3]$, implying that metal-alkyl effective bond lengths are much shorter than the metal-chlorine effective bond lengths (Table 3.1). This point is discussed again later.

2. p-Block Elements Connected by Multiple Bonds

Structural details of some compounds in which one of the unidentate ligands can be formally regarded as being connected to a phosphorus atom by a double bond are given in Table 3.2. The corresponding arsenic compounds are very similar [426].

Triphenylphosphine oxide is significantly distorted from regular tetrahedral geometry, $R-P-O = 112.0°$, $R-P-R = 106.8°$, corresponding to an effective bond length ratio for the phosphorus-oxygen bond of $R(O^{2-}/C) = 0.87$. Slightly higher values are obtained when the oxygen atom is also coordinated to a metal atom [426]. Trialkylphosphine oxides are similar [426].

The analogous phosphine sulfides and phosphine selenides are similar [$R(S^{2-}/C) = R(Se^{2-}/C) = 0.84$ (Table 3.2)], again showing that the distance between the effective center of repulsion and the central atom is not a simple function of bond length.

Tribromophosphine oxide has an $R(O^{2-}/Br)$ value approximately 0.05 lower

Table 3.2. Stereochemical parameters for tetrahedral [P(unidentate A) (unidentate B)₃] complexes containing one double bond. R(A/B) values obtained for n = 6

Complex	AMB	BMC	(M—A)/(M—B)	R(A/B)	Ref.
Ph₃PO (ortho)	111.7°	107.2°	0.83	0.89	[69]
Ph₃PO (mono)	112.3°	106.5°	0.83	0.86	[947]
(o-MeC₆H₄)₃PO	113.0°	105.7°	0.80	0.83	[209]
Ph₃PS	113.1°	105.6°	1.07	0.82	[248]
(o-MeC₆H₄)₃PS	112.6°	106.2°	1.07	0.85	[209, 211]
(m-MeC₆H₄)₃PS	112.3°	106.5°	1.07	0.86	[210]
(p-MeC₆H₄)₃PS	114.2°	104.4°	1.08	0.77	[210, 211]
Me₃PS	113.2°	105.5°	1.09	0.82	[401]
(C₆H₁₁)₃PS	110.9°	108.0°	1.07	0.93	[645]
Ph₃PSe	113.1°	105.6°	1.15	0.82	[249]
(o-MeC₆H₄)₃PSe	112.4°	106.4°	1.15	0.86	[209]
(m-MeC₆H₄)₃PSe	112.2°	106.6°	1.16	0.87	[210]
Me₃PSe	113.1°	105.7°	1.18	0.82	[252]
Br₃PO	113.3°	105.4°	0.69	0.81	[1043]
Ph₃P=CR₂	111.8°	107.0°	0.95	0.89	[426]

than the trialkylphosphine oxides, as expected. Similarly, as is predicted above, the most distorted tetrahedron is the trifluorosulphonium oxide ion [698]:

$$(F_3SO)^+: \quad FSO = 116.0°, \qquad FSF = 102.2°$$

$$(S-O)/(S-F) = 0.94, \qquad R(O^{2-}/F) = 0.68$$

This structure is also in accord with an early microwave study on F_3PO, and an electron diffraction study on F_3NO [885]:

$$F_3PO: \quad FPO = 115.8° \qquad FPF = 102.5°$$

$$(P-O)/(P-F) = 0.95, \qquad R(O^{2-}/F) = 0.69$$

$$F_3NO: \quad FNO = 117.4°, \qquad FNF = 100.5°$$

$$(N-O)/(N-F) = 0.81, \qquad R(O^{2-}/F) = 0.60.$$

The structure of the triphenylphosphine carbene complexes, $Ph_3P = CR_2$ (Table 3.2), are similar to triphenylphosphine oxide and the related compounds cited above.

The effective bond length of a singly bonded group relative to that of double bonded group can also be obtained for compounds of the type (X=)₃M(unidentate) (Table 3.3). Thus in the alkyl perchlorate O_3ClOR, R(OR/O²⁻) = 1.29, the reciprocal of which is R(O²⁻/OR) = 0.78 in reasonable agreement with the value obtained for compounds containing a single M=O group given in Table 3.2. The decrease in R(OX/O) values from O_3Cl-OX (1.29) to $^-O_3S-OX$ (1.20) to $^{2-}O_3P-$ OX (1.16) reflects the decrease in formal bond order to the nonbridging oxygen atoms from 2.00 to 1.67 and 1.33, respectively (Table 3.3).

The greatest departure from regular tetrahedral geometry is again observed for the oxofluoro complexes $(O_3SF)^-$ and $(O_3PF)^{2-}$, details being given in Table 3.3. Electron diffraction studies on gaseous O_3ClF [239] and O_3BrF [51] are again in

Table 3.3. Stereochemical parameters for tetrahedral [MO$_3$(unidentate)]$^{x-}$ complexes of the p-block elements. (X=H, R, PO$_3^{2-}$, etc; R = alkyl, aryl). R(i/O) values obtained for n = 6

M—O Bond Order = $1^1/_3$	M—O Bond Order = $1^2/_3$	M—O Bond Order = 2
[O$_3$POX]$^{2-}$	[O$_3$SOX]$^-$	O$_3$ClOX
O—P—OX = 106.3° O—P—O = 112.4° (P—OX)/(P—O) = 1.06 R(OX/O) = 1.16 (Ref. [426])	O—S—OX = 105.5° O—S—O = 113.1° (S—OX)/(S—O) = 1.11 R(OX/O) = 1.20 (Ref. [426])	O—Cl—OX = 103.8° O—Cl—O = 114.5° (Cl—OX)/(Cl—O) = 1.18 R(OX/O) = 1.29 (Ref. [408])
[O$_3$PR]$^{2-}$	[O$_3$SR]$^-$	—
O—P—R = 107.8° O—P—O = 111.1° (P—R)/(P—O) = 1.17 R(R/O) = 1.08 (Ref. [582])	O—S—R = 106.4° O—S—O = 112.4° (S—R)/(S—O) = 1.22 R(R/O) = 1.16 (Ref. [426])	
[O$_3$PF]$^{2-}$	[O$_3$SF]$^-$	—
O—P—F = 104.3° O—P—O = 114.1° (P—F)/(P—O) = 1.06 R(F/O) = 1.26 (Refs. [372, 859, 862])	O—S—F = 103.4° O—S—O = 114.8° (S—F)/(S—O) = 1.08 R(F/O) = 1.31 (Ref. [1114])	

good agreement with the increase in oxygen-halogen bond order to 2.00:

$$O_3ClF: \quad OClF = 100.8° \qquad OClO = 116.6°$$
$$(Cl—O)/(Cl—F) = 1.15, \quad R(F/O^{2-}) = 1.46$$
$$O_3BrF: \quad OBrF = 103.3°, \qquad OBrO = 114.9°$$
$$(Br—O)/(Br—F) = 1.08, \quad R(F/O^{2-}) = 1.32.$$

3. Compounds Containing a Non-Bonding Pair of Electrons

The Group V compounds [M(unidentate)$_3$(lone pair)] are a stereochemically important class of compounds. In Table 3.4, the X-ray data have been augmented by recent electron diffraction and microwave data on the gasous molecules, the agreement between the two sets of results being very good for the smaller Group V atoms. However the interpretation of the X-ray results for the larger atoms becomes increasingly complicated by the presence of short intermolecular contacts. For example, in SbF$_3$ the three Sb—F bonds are 1.92 Å long, but distorted octahedral coordination about the antimony atom is completed by three Sb——F contacts of 2.61 Å. Similarly, in SbCl$_3$ there are three Sb—Cl bonds of 2.75 Å, with an additional five Sb——Cl contacts at about 3.6 Å. Antimony trichloride forms adducts with a large number of organic molecules, in which low ClSbCl angles are observed. For example in 2SbCl$_3$·C$_6$H$_3$(COMe)$_3$, ClSbCl = 93.5°, and in addition

Table 3.4. Stereochemical parameters for Group V compounds [MX$_3$(lone pair)]. R(:/X) values obtained for n = 6

Compound	Crystal			Vapour		
	XMX	R(:/X)	Ref.	XMX	R(:/X)	Ref.
NF$_3$				102.4°	0.68	[849]
NCl$_3$	106.8°	0.88	[536]	107.4°	0.91	[184, 224]
NBr$_3$						
NI$_3$						
NPh$_3$						
PF$_3$				97.4°	0.47	[562, 806]
PCl$_3$	100.1°	0.58	[535]	100.1°	0.58	[223]
PBr$_3$	100.5°	0.60	[411]	101.0°	0.62	[679]
PI$_3$	102.0°	0.67	[689]			
PPh$_3$	102.4°	0.68	[209, 210, 293]			
AsF$_3$				96.0°	0.41	[247, 668, 669]
AsCl$_3$	98.2°	0.51	[410]	98.6°	0.53	[668, 669]
AsBr$_3$	97.7°	0.50	[1056]	99.7°	0.58	[937, 954]
AsI$_3$	102.0°	0.67	[1055]	100.2°	0.60	[807]
AsPh$_3$	102.0°	0.67	[1054]			
SbF$_3$	87.3°	—	[379]	95.0°	0.36	[669]
SbCl$_3$	94.1°	0.30	[728]	97.2°	0.47	[670]
SbBr$_3$				98.2°	0.52	[671]
SbI$_3$.3S$_8$	96.6°	0.44	[118]			
SbPh$_3$	97.3°	0.48	[1000]			
BiF$_3$						
BiCl$_3$						
BiBr$_3$						
BiI$_3$						
BiPh$_3$	94.0°	0.30	[543]			

to the three Sb—Cl bonds of 2.34 Å, three additional Sb——Cl and Sb——O contacts occur at ∼ 3.3 and ∼ 2.9 Å respectively [67]. Low ClSbCl angles are also observed in SbCl$_3$·Ph$_2$ (93.7°) [729] and SbCl$_3$·^{1}/$_2$Ph$_2$NH (93.2°) [727], where there is direct interaction between the antimony lone pair of electrons and the aromatic rings.

For these simple Group V molecules (Table 3.4), it is observed that R(:/X) decreases on descending the periodic table, which is the usual expectation regarding the spatial extent of a lone pair of electrons. The R(:/X) value also decreases along the series I > Br > Cl > F, as the bonding pairs of electrons are drawn out closer to the halogen atoms. It should perhaps be mentioned at this stage that any set of experimental bond angles is consistent with two different R(:/X) values, for example, a bond angle of 98° is consistent with R(:/X) = 0.07 as well as R(:/X) = 0.50 (Fig. 3.2). The former lower values have not been used as they appear unrealistic in absolute terms and because they produce trends that are the reverse of those normally expected for the behaviour of a lone pair as a function of the nature of the central atom and the nature of the other ligands present.

The X-ray structural parameters for the Group VI cations [MX$_3$(lone pair)]$^+$ (Table 3.5) are in good agreement with those for the uncharged Group V molecules.

Table 3.5. Stereochemical parameters for Group VI compounds [MX$_3$(lone pair)]$^+$. R(:/X) values obtained for n = 6

Compound		XMX	R(:/X)	Ref.
(SF$_3$)(BF$_4$)		97.5°	0.48	[471]
(SCl$_3$)(ICl$_4$)		101.3°	0.64	[381]
(SMe$_3$)A	(A=I, HgCl$_3$, $^1/_2$SO$_4$)	102.4°	0.68	[115, 1111, 1125]
(SeF$_3$)A	(A=NbF$_6$, Nb$_2$F$_{11}$)	94.7°	0.34	[382, 383]
(SePh$_3$)Cl·nH$_2$O	(n = 1, 2)	100.6°	0.61	[704, 796]
(TeMe$_3$) (Ph$_4$B)		91.3°	—	[1123]

Table 3.6. Stereochemical parameters for [MO$_3$(lone pair)]$^{x-}$. R(:/O) values obtained for n = 6

Compound		OMO	R(:/O)	Ref.
Bond Order = 1$^1/_3$				
A$_2$(SO$_3$)xH$_2$O	(A=Li, Na, Cs, NH$_4$)	105.3°	0.81	[52, 373, 696]
NaH$_3$(SeO$_3$)$_2$		101.0°	0.62	[234]
A$_2$(TeO$_3$)xH$_2$O	(A=K, $^1/_2$Ba)	99.3°	0.56	[618, 836]
Bond Order = 1$^2/_3$				
A(ClO$_3$)	(A=Na, K)	106.6°	0.87	[2, 83, 185]
Na(BrO$_3$)		104.1°	0.76	[2]

The structures of the oxyanions [XO$_3$(lone pair)]$^{x-}$ (Table 3.6) are again consistent with the general picture outlined above. In addition to the three short X—O bonds, the larger elements again have additional longer interactions, but there is again a clear contraction of the nonbonding electron pair, for example, from R(:/O) = 0.81 for SO$_3^{2-}$ to R(:/O) = 0.56 for TeO$_3^{2-}$. The increase in R(:/O) as the X—O bond order increases is parallel to that found above for related compounds not containing a lone pair of electrons, for example:

$$SO_3^{2-}: \quad R(:/O) = 0.81; \quad SO_3(OX)^-: \quad R(OX/O) = 1.20;$$

$$ClO_3^-: \quad R(:/O) = 0.87; \quad ClO_3(OX): \quad R(OX/O) = 1.29.$$

Table 3.7. Stereochemical parameters for coordinated and uncoordinated (from Table 3.4) Group V ligands

	Ph$_3$M(lone pair)		[(Ph$_3$M)Cr(CO)$_5$]	
	PhMPh	R(:/Ph)	PhMPh	R(Cr/Ph)
M=P	102.4°	0.68	102.6°	0.69
M=As	102.0°	0.67	101.4°	0.64
M=Sb	97.3°	0.48	99.2°	0.56
M=Bi	94.0°	0.30	98.7°	0.55

4. Atoms Bonded to Transition Metal Ions

The donation of the lone pair of electrons from a Group V ligand [MX$_3$(lone pair)] to a transition metal results in a small increase in the XMX angle [426]. This effect is particularly well shown by the closely related series of compounds [(Ph$_3$M)Cr · (CO)$_5$] (Table 3.7) [218].

A more substantial stereochemical change upon coordination is observed for the sulfite ion:

$$SO_3^{2-}: \qquad OSO = 105.3°, \qquad R(:/O) = 0.81 \qquad \text{(Table 3.6)}$$

$$M \leftarrow SO_3^{2-}: \quad OSO = 109.7°, \qquad R(M/O) = 1.02 \qquad \text{(Ref. [426])}$$

5. Transition Metal Complexes

Tetrahedral complexes of a first row transition metal ion [M^{II}X$_3$L]$^-$, where M is Ni, Zn, X=Cl, Br, I, and L is an uncharged unidentate ligand, have LMX and XMX angles of about 105° and 114° respectively (Table 3.8). The corresponding effective bond length ratio of the charged ligands and uncharged ligands, R(L/X$^-$) = 1.2 or R(X$^-$/L) = 0.8, is found for a wide range of transition metal ions M^{x+} (where x ≥ 2), for all coordination numbers and stereochemistries.

The anomalous structure of (Bu$_3$PH)[Ni(PBu$_3$)Br$_3$] must be attributed to the bulk of the tri(*tert*-butyl)phosphine ligand compressing the BrNiBr angles from the expected 114° to 109°.

B. Tetrahedral [M(unidentate A)$_2$(unidentate B)$_2$]

The general stereochemistry for tetrahedral molecules containing one pair of ligands different from the other pair is shown in Fig. 2.3. A twofold axis bisects the angle between each pair of ligands, each of which also lies on a mirror plane. The effective bond length of the M—A and M—B bonds is given by R, and that for the M—C and M—D bonds is defined as unity.

At R = 1.0, AMB = CMD = 109.47° as before. As R is decreased, AMB increases and CMD decreases as intuitively expected (Fig. 3.4), but as R continues to decreases CMD passes through a minimum until eventually AMB = CMD = 180° with the creation of a planar structure.

Table 3.8. Stereochemical parameters for transition metal complexes [M^{II}(ligand)X$_3$]$^-$, (X=Cl, Br, I). R(L/X$^-$) values obtained for n = 6

Compound	LMX	XMX	(M—L)/(M—X)	R(L/X$^-$)	Ref.
(Bu$_3$PH)[Ni(PBu$_3$)Br$_3$]	110.2°	108.7°	1.04	0.96	[39]
(Ph$_4$As)[Ni(PPh$_3$)I$_3$]	104.3°	114.1°	0.90	1.26	[1039]
K[Zn(OH$_2$)Br$_3$]H$_2$O	104.5°	113.2°	0.85	1.23	[567]
K[Zn(OH$_2$)I$_3$]H$_2$O	104.8°	113.7°	0.81	1.24	[567]
(C$_6$H$_8$N)[Zn(PPh$_3$)Br$_3$]	106.2°	112.5°	1.02	1.17	[313]

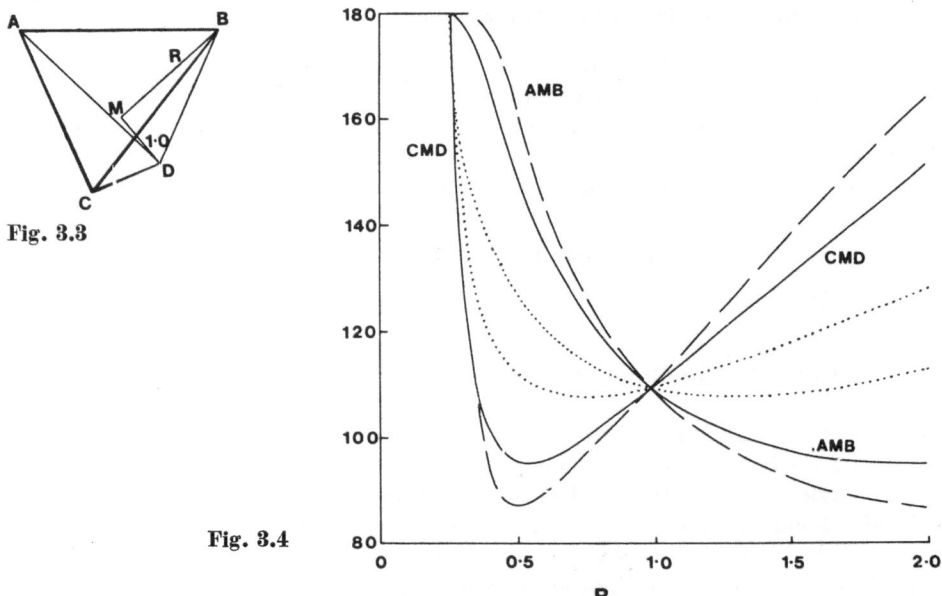

Fig. 3.3

Fig. 3.4

Fig. 3.3. General stereochemistry for [M(unidentate A)$_2$(unidentate B)$_2$]

Fig. 3.4. Bond angles (degrees) for tetrahedral [M(unidentate A)$_2$(unidentate B)$_2$], as a function of effective bond length ration R. *Dotted lines*, n = 1; *full lines*, n = 6; *broken lines*, n = 12

Values for the effective bond length ratio R(i/j) can again be obtained by fitting the experimental bond angles against the calculated bond angles. The resultes from this procedure are examined for the following cases:

1. M is a p-block element, with metal-ligand single bonds.
2. M is a p-block element, with metal-ligand double bonds as well as single bonds.
3. M is a transition metal.

1. p-Block Elements Connected by Single Bonds

Structural information on some tetrahedral compounds is summarized in Table 3.9. The effective bond length ratios R(A/C) are not very different from those obtained for [M(unidentate A)(unidentate B)$_3$], R′(A/C) (Table 3.9).

2. p-Block Elements Connected by Multiple Bonds

Structural parameters of compounds containing two double bonds and two single bonds may be summarized:

$$[SO_2Me_2]: \quad OSO = 117.9°, \ MeSMe = 103.0°$$
$$(S=O)/(S-Me) = 0.81, \ R(O^{2-}/Me) = 0.83 \ (Ref.[956])$$

Table 3.9. Stereochemical parameters for tetrahedral [M(unidentate A)$_2$(unidentate B)$_2$] complexes of the p-block elements connected by single bonds. R(A/C) values obtained for n = 6. R'(A/C) values obtained for [M(unidentate A)(unidentate B)$_3$] complexes (Table 3.1)

Compound	A, B	C, D	AMB	CMD	R(A/C)	Ref.	R'(A/C)
$C_{35}H_{74}$	H	C	—	112.0°	1.06	[980]	1.11
CH_2Cl_2	H	Cl	—	109.0°	0.99	[426]	1.04
$(Me_4As)[GaMe_2Cl_2]$	C	Cl	125.3°	99.7°	0.73	[541]	0.76
$(Me_4As)[InMe_2Br_2]$	C	Br	135.0°	98.9°	0.67	[975]	0.66
$[SnEt_2Cl_2]$	C	Cl	134.0°	96.0°	0.63	[29]	0.70
$[SnEt_2Br_2]$	C	Br	135.9°	98.5°	0.66	[29]	0.70
$[SnEt_2I_2]$	C	I	130.2°	104.0°	0.76	[29]	0.70
$[SnPh_2Cl_2]$	C	Cl	125.4°	99.8°	0.73	[501]	0.70

$$[PO_2(OX)_2]^- \quad (X=H, \text{alkyl, aryl, } PO_3^{2-}):$$
$$OPO = 117.7°, \quad XO-P-OX = 103.0°$$
$$(P=O)/(P-OX) = 0.94, \quad R(O^{2-}/OX) = 0.83 \quad (Ref.\,[426]).$$

These effective bond length ratios are again very similar to those obtained in Sect. A.2 for compounds of the type [M(unidentate A)(unidentate B)$_3$].

Large distortions from a regular tetrahedron are again found for oxyfluorides, as shown by electron diffraction studies on SO_2F_2 (OSO = 123°, FSF = 97°) and SeO_2F_2 (OSeO = 126°, FSeF = 94°) [511].

3. Transition Metal Complexes

Structural data on complexes of the first row transition metal dihalides [excluding copper(II)] and two uncharged unidentate ligands yield R(X$^-$/ligand) = 0.8$_5$ (Table 3.10). This value compares favourably with that obtained for complexes of the type [MX$_3$(ligand)]$^-$ (Sect. A.5).

C. A Comment on Square-Planar Complexes and the trans-Influence

Repulsion among four metal-ligand bonds must, in the absence of other effects, lead to a tetrahedral structure rather than a square-planar structure, the repulsion energy coefficients being:

$$X(\text{tetrahedron}) = 0.316 \ (n = 6),$$
$$X(\text{square}) = 0.531 \ (n = 6).$$

Intermediate structures can be considered as squashed tetrahedra or as puckered squares. If it is thought necessary to bring square-planar structures into the province of electron-pair repulsion theory, then it can be pointed out that it

Table 3.10. Stereochemical parameters for transition metal complexes $[M^{II}X_2(ligand)_2]$. $R(X^-/L)$ values obtained for $n = 6$

Complex	XMX	LML	$R(X^-/L)$	Ref.
$[CoCl_2(OPPh_3)_2]$	114.0°	96.4°	0.78	[759]
$[CoCl_2(OPBz_3)_2]$	113.0°	104.9°	0.91	[309]
$[CoCl_2(OAsPh_3)_2]$	109.5°	94.8°	0.79	[114]
$[CoCl_2\{OPPh_2(NHBz)\}_2]$	116.5°	106.6°	0.89	[945]
$[CoCl_2\{SC(NH_2)_2\}_2]$	107.7°	97.1°	0.86	[332]
$[CoCl_2\{SC(NHEt)_2\}_2]$	109.4°	98.6°	0.88	[135]
$[CoCl_2(N_3C_{13}H_{11})_2]$	116.6°	107.1°	0.90	[1024]
$[CoBr_2(NC_6H_9OS)_2]$	110.8°	106.9°	0.95	[203]
$[NiBr_2(PPh_3)_2]$	126.3°	110.4°	0.9	[612]
$[ZnCl_2(N_2C_3H_3)_2]$	111.5°	105.2°	0.92	[740]
$[ZnCl_2(NC_5H_4Me)_2]$	121.8°	100.6°	0.77	[741]
$[ZnCl_2(N_4CHMe)_2]$	118.4°	98.9°	0.77	[63]
$[ZnCl_2(NC_5H_5)_2]$	120.9°	106.3°	0.86	[1009]
$[ZnCl_2(NC_5H_4CHCH_2)_2]$	118.7°	101.9°	0.81	[1010]
$[ZnCl_2(NC_5H_4COMe)_2]$	123.6°	109.0°	0.86	[1010]
$[ZnCl_2(NC_5H_4CN)_2]$	125.7°	105.0°	0.80	[1010]
$[ZnCl_2(ONC_5H_3Me_2)_2]$	114.8°	94.7°	0.74	[953]
$[ZnCl_2(OC_{11}H_{12}N_2)_2]$	117.6°	97.4°	0.76	[112]
$[ZnCl_2(OCMeNMe_2)_2]$	115.7°	90.5°	—	[557]
$[ZnCl_2\{SC(NH_2)_2\}_2]$	107.3°	111.5°	1.06	[681]
$[ZnCl_2(SCMeNH_2)_2]$	109.8°	101.0°	0.89	[944]
$[ZnI_2(NC_5H_5)_2]$	120.3°	99.8°	0.77	[714]

requires only a relatively small amout of electron density above and below this square to maintain planarity.

Merely as a means of illustrating the effect some electron density lying along the z-axis normal to the plane has on the four metal-ligand bonds, imagine one-electron pair in an orbital projected along this z-axis, half situated above the metal atom and half below the metal atom (Fig. 3.5). Each half is considered to act at a point, at a distance R_e from the metal atom (defining the distance between the metal atom and each bonding pair of electrons as unity).

The dependence of AMC (or BMD) on R_e is shown in Fig. 3.6. It can be seen that as the electron pair spreads out along the z-axis, the tetrahedron becomes progressively squashed into a square, this being achieved at $R_e \geq 0.13$ for $n = 6$.

This spread of electron density is more likely to occur for the d^8 electron configuration in which a nonbonding electron pair is located in the d_{z^2} orbital, particularly for second- and third-row transition metal ions in which the d orbitals are generally expanded, and for π-bonding ligands which can further interact with this orbital. Square-planar complexes are also favored by reducing the steric crowding in the molecule as occurs for the larger second- and third-row transition metal ions and for complexes containing chelate rings of small normalized bite.

The existence of some electron density on each side of the square plane can also be used to explain the *trans*-influence in square-planar complexes, in which the bond *trans* to strong metal-ligand bond is weakened [680]. For the case where the non-bonding electron density along the z-axis is relatively expanded, $R_e \geq 0.3$,

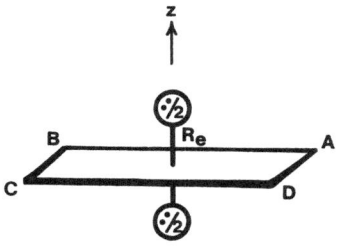

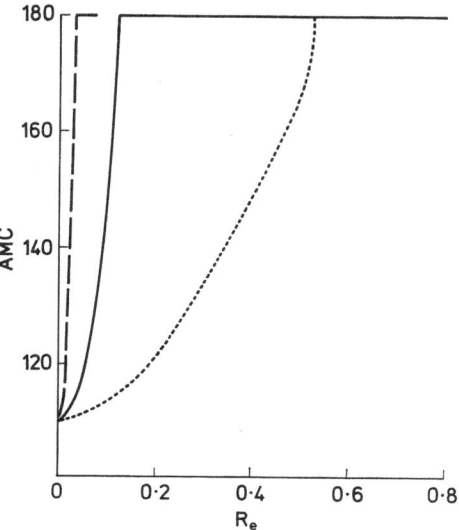

Fig. 3.5. General stereochemistry for square-planar complexes containing one nonbonding pair of electrons

Fig. 3.6. Dependence of bond angle AMC (degrees) upon distance R_e for four-coordinate complexes containing one nonbonding pair of electrons. *Dotted line*, n = 1; *full line*, n = 6; *broken line*, n = 12

a decrease in the effective bond length R(A) to ligand A shifts this region of nonbonding electron density towards atom C. Atom C therefore experiences a greater repulsion than do atoms B and D, and the M—C bond is weaker than the M—B and M—D bonds. The calculated *trans*-influence, Y_C/Y_B, is shown as a function of R_e in Fig. 3.7.

This picture of the *trans*-influence is different from the conventional π-bonding picture, where the metal π-electrons are displaced *toward* the more strongly bonding ligand A and hence cannot be used as effectively to π-bond to the *trans*-ligand C, hence weakening the M—C bond.

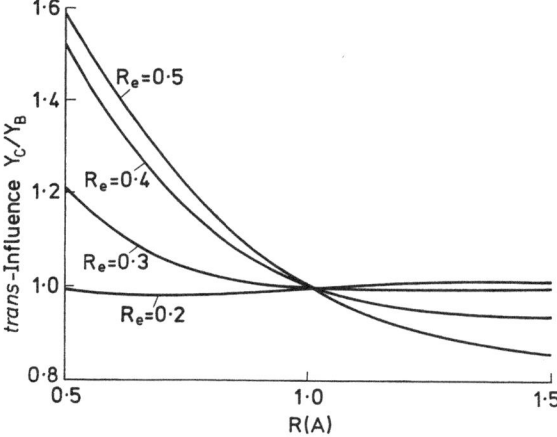

Fig. 3.7. The *trans*-influence, Y_C/Y_B, as a function of R(A) and R_e for square-planar [M(unidentate A) (unidentate B)$_3$] containing one nonbonding pair of electrons

D. Tetrahedral [M(bidentate)(unidentate)$_2$]

The general stereochemistry is that shown in Fig. 3.3, but with the addition of a bidentate ligand spanning C and D, with a normalized bite of b.

Repulsion energy calculations [519] show the angle between the two metal-unidentate ligand bonds, AMB, is fairly independent of b, but varies with R in a similar manner to that found for [M(unidentate A)$_2$(unidentate B)$_2$] (Fig. 3.4).

There are a multitude of compounds containing four-coordinate atoms incorporated into rings, but the amount of structural information is much more limited if two constraints are imposed: that the ring be symmetrical and that the two unidentate ligands be the same, but not hydrogen atoms since reasonably accurate bond angles are required. Structural details for selected classes of compounds are summarized in Table 3.11.

The effective bond length ratios for cyclic carbon compounds are in reasonable agreement with those obtained for compounds containing four unidentate ligands:

	[C(bidentate)(unidentate)$_2$]	[C(unidentate A)$_2$(unidentate B)$_2$]
R(C/C)	1.00	1.00
R(F/C)	1.12	1.16
R(Cl/C)	1.05	1.09

Table 3.11. Stereochemical parameters for [M(bidentate) (unidentate)$_2$] complexes. R(A/C) values obtained for n = 6

Complex	b	AMB	R(A/C)	Ref.
$CC\big\langle\!\begin{smallmatrix}R\\R\end{smallmatrix}$	1.3—1.6	110.8°	1.00	[426]
$CC\big\langle\!\begin{smallmatrix}F\\F\end{smallmatrix}$	1.4	105.4°	1.12	[426]
$CC\big\langle\!\begin{smallmatrix}Cl\\Cl\end{smallmatrix}$	1.4—1.7	108.0°	1.05	[426]
(Me$_2$PN)$_m$ (m = 3, 4, 5, 6, 7, 8)	~1.72	104.0°	1.17	[426]
(F$_2$PN)$_m$ (m = 3, 4, 5)	~1.74	99.5°	1.35	[426, 537]
(Cl$_2$PN)$_m$ (m = 3, 4, 5)	~1.73	102.4°	1.23	[426]
(Br$_2$PN)$_m$ (m = 3, 4, 5)	~1.72	103.1°	1.20	[426]
(PO$_3$)$_3^{3-}$	~1.55	120.1°	0.79	[426]
(PO$_3$)$_4^{4-}$	~1.56	119.2°	0.80	[426]
[Co(N$_2$C$_6$H$_{12}$O$_2$)Cl$_2$]	1.40	116.8°	0.85	[967]
[Ni(N$_2$C$_9$H$_6$)Br$_2$]	1.32	124.9°	0.72	[196]
[Ni(N$_2$C$_{26}$H$_{20}$)I$_2$]	1.33	132.0°	0.65	[196]
[Ni(N$_2$C$_{14}$H$_{12}$)I$_2$]	1.34	126.8°	0.77	[190]
[Zn(Me$_2$NCH$_2$CH$_2$NMe$_2$)Cl$_2$]	1.39	119.4°	0.80	[584]
[Zn(N$_2$C$_{14}$H$_{12}$)Cl$_2$]	1.31	120.3°	0.80	[897]
[Zn(Me$_2$NCH$_2$CH$_2$NMe$_2$)I$_2$]	1.40	114.5°	0.90	[585]
[Zn(N$_2$C$_7$H$_{18}$)I$_2$]	1.51	118.1°	0.82	[928]

A relatively large amount of structural data is available for the cyclic phospha-zenes, $(X_2PN)_m$. Structural data are available for m = 3—8, and as predicted the XPX angle does not significantly depend upon the value of m. The differences between the R(X/N) values (X=Me, F, Cl, Br) (Table 3.11) are very similar to those obtained above, again showing that these R(i/j) values can be transferred among reasonably different types of molecules. It should be noted that the R(X/N) values for the phosphazenes are approximately 0.2 higher than the R(X/C) values for the cyclic carbon compounds, which is strong evidence for multiple bonding within the phosphorus-nitrogen rings.

Conversely, the R(terminal O/ring O) value of 0.8 found for cyclic metaphos-phates (Table 3.11) is that expected for phosphorus-(terminal oxygen) multiple bonding.

The structures of complexes between the dihalides of cobalt(II), nickel(II), and zinc(II) with uncharged bidentate ligands correspond to R(X$^-$/bidentate) = 0.8, in complete accord with similar complexes of the type [M(unidentate A)(uniden-tate B)$_3$] and [M(unidentate A)$_2$(unidentate B)$_2$].

CHAPTER 4

Five-Coordinate Compounds Containing only Unidentate Ligands

A. [M(unidentate)$_5$]

There are no regular or semiregular polyhedra available for five-coordination. The starting point therefore is a consideration of two non-uniform polyhedra, the trigonal bipyramid and the square pyramid, and the relation between them. Both are examples of the more general stereochemistry shown in Fig. 4.1. A twofold axis passes through the metal atom M and the donor atom E, the other four atoms lying on a pair of vertical mirror planes. Descriptions of the stereochemical changes without enforcing full C$_{2v}$ symmetry are considered later in this section. The angles between this axis and the bonds to each of the pairs of donor atoms A, C and B, D are denoted by ϕ_A and ϕ_B, respectively. The trigonal bipyramid is defined by $\phi_A = 90.0°$ and $\phi_B = 120.0°$ (or $\phi_A = 120.0°$ and $\phi_B = 90.0°$), and the square pyramid is defined by $\phi_A = \phi_B$.

The potential energy surface shown in Fig. 4.2 shows that there is no potential energy barrier between the square pyramid at $\phi_A = \phi_B = 101.3°$ and the two trigonal bipyramids situated at the two minima. Movement along the "reaction coordinate" connecting the two trigonal bipyramids is usually described as the Berry pseudo-rotation [104].

An important feature of five-coordination is the large differences among the different ligand sites. For example, the apical site of a trigonal bipyramid has three nearest neighbours at $2^{1/2}r$, whereas the equatorial sites have only two such close neighbours. Thus the apical sites experience greater repulsion than the

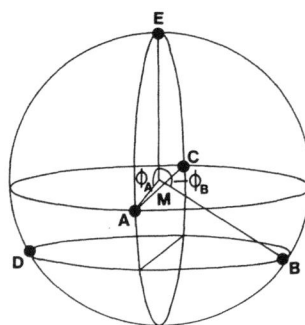

Fig. 4.1. General stereochemistry for [M(unidentate)$_5$] with C$_{2v}$ symmetry

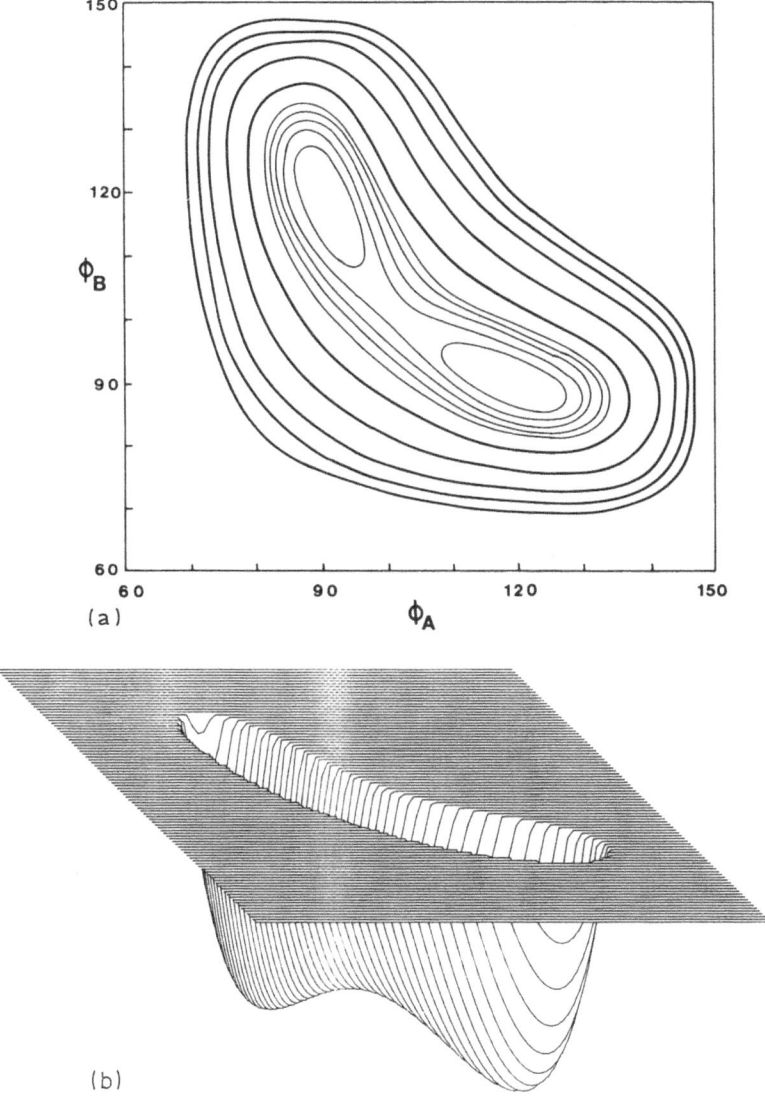

(a)

(b)

Fig. 4.2. a Projection of the potential energy surface for [M(unidentate)$_5$] onto the $\phi_A - \phi_B$ plane (in degrees). The five *faint contour lines* are for successive 0.01 increments above the minima, and the five *heavy contour lines* are for successive 0.1 increments above the minima. n = 6. **b** As in **a**, but with truncation at X = 0.05

equatorial sites, leading to the expectation of a longer metal-ligand bond:

$$\frac{Y(\text{apical})}{Y(\text{equatorial})} = 1.21 \ (n = 6).$$

Similarly, each ligand at a square basal site of a square pyramid experiences more repulsion than does the single atom at the apical site:

$$\frac{Y(\text{basal})}{Y(\text{apical})} = 1.27 \qquad (n = 6).$$

Structurally characterised molecules of the type $[M(\text{unidentate})_5]^{x\pm}$ are listed in Table 4.1, and cover the complete range from the trigonal bipyramid ($\phi_A = 90.0°$, $\phi_B = 120.0°$) to the square pyramid ($\phi_A = \phi_B \sim 100°$). Whether a compound is near the trigonal bipyramid end of the potential energy surface or near the square pyramid must be attributed to crystal-packing forces. For example, the structure of $SbPh_5$ is nearer the square-pyramidal end of the range, whereas the solvate $SbPh_5 \cdot \frac{1}{2}C_6H_{12}$ is trigonal bipyramidal. Similarly the structure of the $[Ni(CN)_5]^{3-}$ anion depends upon the choice of cation (Table 4.1).

However, a closer examination of the detailed stereochemistry reveals unexpectedly complicated behavior, particularly for those molecules nearer the square-pyramid end of the range. In Table 4.1 the compounds have been classified into two distinct types, labelled α and β, together with an intermediate type.

Type α molecules lie along the reaction coordinate connecting the trigonal bipyramid and square pyramid at $\phi_A = \phi_B \sim 100-105°$, and also have the pattern of bond lengths expected from the repulsion energy calculations, namely:

Trigonal bipyramid: $MA/ME > 1.00$, $MB/ME \sim 1.00$

Square pyramid: $MA/ME \sim MB/ME > 1.00$.

Molecules that are observed to fall into this category include nonmetallic Group V compounds MX_5 ($M = P$, As, Sb; $X = Ph$, C_6H_4Me, OPh), and metal complexes such as $(PCl_4)[VCl_5]$, $(Et_4N)_2[InCl_5]$, $[Co(MeC_5H_4NO)_5](ClO_4)_2$, $[Mg(Me_3AsO)_5](ClO_4)_2$ and $[Mg(Me_3PO)_5](ClO_4)_2$. These compounds are expected to contain relatively simple bonds between central atoms in high oxidation states and ligands that are not expected to have a high degree of π-bonding capability.

Intermediate type molecules have somewhat lower values of ϕ_A and ϕ_B than the type α molecules, and in addition have all five bond lengths approximately equal. The important group of molecules with intermediate type structures are complexes of the first-row transition metal ions with the d^8 electron configuration:

Manganese($-$I): $[Mn(CO)_5]^-$

Iron(0): $[Fe(CO)_5]$ and $[Fe(CNBu)_5]$

Cobalt(I): $[Co(CNMe)_5]^+$ and $[Co(CNPh)_5]^+$

Nickel(II): $[Ni(OAsMe_3)_5]^{2+}$ and $[Ni(PO_3C_6H_9)_5]^{2+}$.

The $[Ni(CN)_5]^{3-}$ complexes are similar, but are perhaps better classified as type β.

Type β molecules are defined as those which do not satisfy the criteria for type α molecules, which were predicted from simple repulsion energy calculations. That is, type β molecules have much lower values of ϕ_A and ϕ_B, and also have lower values of MA/ME and/or MB/ME. Molecules of this type turn out be those transition metal complexes that might have been suspected to contain a more complicated type of metal-ligand bonding. For example the metal-ligand combinations are typical of those that form square-planar complexes when the metal is

Table 4.1. Stereochemical parameters for [M(unidentate)$_5$] complexes

Complex	ϕ_A	ϕ_B	MA/ME	MB/ME	Ref.
Type α					
[SbPh$_5$]$^1/_2$C$_6$H$_{12}$	90.3°	119.2°	1.04	0.99	[142]
[AsPh$_5$]$^1/_2$C$_6$H$_{12}$	90°	120°	1.07	1.00	[152]
[PPh$_5$]	91.5°	118.7°	1.07	1.00	[1085]
(PCl$_4$)[VCl$_5$]	89.1°	116.3°	1.05	0.99	[1120]
[P(OPh)$_5$]	91.6°	116.9°	1.06	1.01	[959]
(Ph$_4$As)$_2$[Fe(N$_3$)$_5$]	93.1°	118.3°	1.04	1.00	[370]
[Sb(C$_6$H$_4$Me)$_5$]	91.1°	114.9°	1.05	1.01	[143]
[Co(MeC$_5$H$_4$NO)$_5$](ClO$_4$)$_2$	93.1°	114.5°	1.05	0.99	[275]
	93.0°	114.0°	1.05	1.00	[105]
Na$_2$[CrPh$_5$]3 Et$_2$O·C$_4$H$_8$O	99.5°	107.5°	1.07	1.00	[821]
[Nb(NMe$_2$)$_5$]	101.5°	109.1°	1.03	1.03	[551]
[SbPh$_5$]	98.3°	105.4°	1.05	1.05	[84]
[Mg(Me$_3$AsO)$_5$](ClO$_4$)$_2$	99.8°	107.6°	1.06	1.05	[834]
[Mg(Me$_3$PO)$_5$](ClO$_4$)$_2$	99.9°	106.4°	1.05	1.04	[835]
[Nb(NC$_5$H$_{10}$)$_5$]	98.4°	110.5°	1.03	1.03	[551]
	100.4°	108.4°	1.03	1.03	[551]
(Et$_4$N)$_2$[InCl$_5$]	103.0°	104.7°	1.03	1.00	[625]
[Intermediate Type					
[Fe(CO)$_5$]	88°	122°	1.03	1.02	[335, 525]
[Mn(CO)$_3$(NH$_3$)$_3$][Mn(CO)$_5$]	90.5°	122.5°	1.01	0.99	[556]
[Co(NH$_3$)$_6$][CdCl$_5$]	90.0°	120.0°	0.99	1.00	[412, 738]
[Ni(phen)$_3$][Mn(CO)$_5$]$_2$	90.9°	117.6°	1.01	1.01	[454]
	90.6°	118.3°	1.03	1.02	[454]
[Ni(PO$_3$C$_6$H$_9$)$_5$](ClO$_4$)$_2$	90.8°	118.4°	0.99	1.02	[932]
[Co(CNMe)$_5$](ClO$_4$)	89.2°	115.9°	0.98	1.00	[264]
[Ni(OAsMe$_3$)$_5$](ClO$_4$)$_2$	98.5°	104.0°	1.03	1.03	[834]
[Fe(CNBu)$_5$]	97.6°	107.3°	0.99	0.99	[80]
	96.3°	111.7°	0.99	0.98	[80]
Co(CNPh)$_5$](ClO$_4$)CHCl$_3$	101.8°	101.8°	0.98	0.98	[165]
Type β					
[Cr(NH$_3$)$_6$][CuCl$_5$]	90.0°	120.0°	0.96	1.00	[923]
[Cr(NH$_3$)$_6$][CuBr$_5$]	90.0°	120.0°	0.97	1.00	[485]
[Cr(NH$_3$)$_6$][HgCl$_5$]	90.0°	120.0°	0.95	1.00	[245]
(Me$_4$N)$_3$[Pt(GeCl$_3$)$_5$]	92.1°	109.3°	0.97	0.97	[418]
(Ph$_3$MeP)$_3$[Pt(SnCl$_3$)$_5$]	—	—	—	—	[279]
[Cr(en)$_3$][Ni(CN)$_5$]1$^1/_2$H$_2$O	93.3°	109.4°	0.92	0.96	[920]
	100.1°	100.3°	0.86	0.85	[920]
(C$_{18}$H$_{15}$N$_6$)[MnCl$_5$]Cl·H$_2$O	96.4°	103.7°	0.93	0.94	[73]
(bipyH$_2$)[MnCl$_5$]	95.1°	99.4°	0.90	0.87	[102]
[Co(CNPh)$_5$](ClO$_4$)$_2^1/_2$C$_2$H$_4$Cl$_2$	94.5°	95.6°	0.94	0.95	[627]
[Cu(OC$_3$H$_6$N$_2$)$_5$](ClO$_4$)$_2$	93.2°	93.7°	0.87	0.87	[754]
(Et$_2$NPr$_2$)$_3$[Co(CN)$_5$]	97.7°	97.7°	0.94	0.94	[175]
[Cr(C$_3$H$_{10}$N$_2$)$_3$][Ni(CN)$_5$]2 H$_2$O	99.2°		0.88		[628]
[Cr(NH$_3$)$_6$][Ni(CN)$_5$]2 H$_2$O	100.1°		0.89		[628]

four-coordinate. For those type β molecules nearer the square pyramid, the structure can be considered to be approaching square-planar four-coordination, with the addition of a fifth more weakly bonded ligand. Such (4 + 1) structures are found for $[Mn^{III}Cl_5]^{2-}$, $[Co^{II}(CN)_5]^{3-}$, and $[Ni^{II}(CN)_5]^{3-}$. In some cases the structure is further complicated by the close approach of an anion to the sixth octahedral site, as in $[Co(CNPh)_5](ClO_4)_2 \cdot {}^1/_2 C_2H_4Cl_2$, $Co-CNPh = 1.83-1.95$ Å, $Co--OClO_3 = 2.59$ Å. For those type β molecules nearer the trigonal bipyramid, the structure can be considered to be approaching linear two-coordination, with three additional more weakly bonding ligands. Such $(2 + 3)$ structures are found for $[Cu^{II}Cl_5]^{3-}$ and $[Hg^{II}Cl_5]^{3-}$.

Many five-coordinate molecules exhibit very rapid intramolecular rearrangements, leading to all five ligands being observed as equivalent over the NMR time scale, with the energy barriers to interconversion being less than approximately 20 kjoule/mole. For example the ^{19}F NMR spectrum of PF_5 shows all fluorine atoms are equivalent, over the temperature range 60 to $-197\,°C$ [811]. Proton NMR studies on $SbMe_5$ in carbon disulfide down to about $-100\,°C$ also show the presence of only one type of methyl group [816]. The ^{13}C NMR spectra of $[Fe(CO)_5]$ and $[Fe(CNBu)_5]$ show only a single resonance at temperatures as low as $-170\,°C$ and $-80\,°C$ respectively [81, 779]. Both ^{19}F NMR and ^{31}P NMR studies on $[Fe(PF_3)_5]$, $[Ru(PF_3)_5]$, and $[Os(PF_3)_5]$ down to $-160\,°C$ in $CHClF_2$ show all ligands are equivalent [779, 780]. In contrast to this behaviour, metal ions with the d^8 electron configuration bonded to five phosphite ligands are rigid trigonal bipyramids at low temperature, but show simultaneous exchange between two axial ligands and two equatorial ligands at higher temperatures, corresponding to rapid movement along the trough in the potential energy surface shown in Fig. 4.2 [409, 615, 779].

Finally, for purposes of comparison with results described in later sections, a potential energy surface is required in which the high C_{2v} symmetry enforced when calculating Fig. 4.2 is absent. The general stereochemistry is now shown in Fig. 4.3, and contains a mirror plane through MBDE as the only symmetry element. The axes are defined so that $\phi_A = \phi_B = \phi_C$, the structure being completely defined by ϕ_A, θ_A, ϕ_D, and ϕ_E.

The potential energy surface projected onto the $\phi_D - \phi_E$ plane is shown in Fig. 4.4. There are now three identical transformations starting from the central trigonal bipyramid T_0 at $\phi_D = 30.0°$, $\phi_E = 150.0°$, depending on which of the three equivalent atoms forming the trigonal plane, B, D, and E, becomes the apical atom of the square pyramid, S_1, S_2, and S_3 respectively. Each of these square

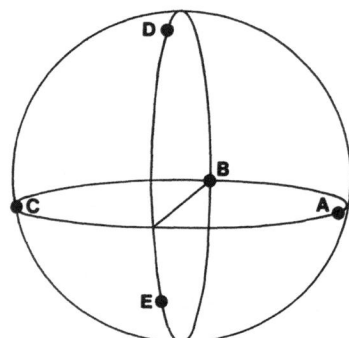

Fig. 4.3. General stereochemistry for [M(unidentate)$_5$] complexes containing one mirror plane, through BDE

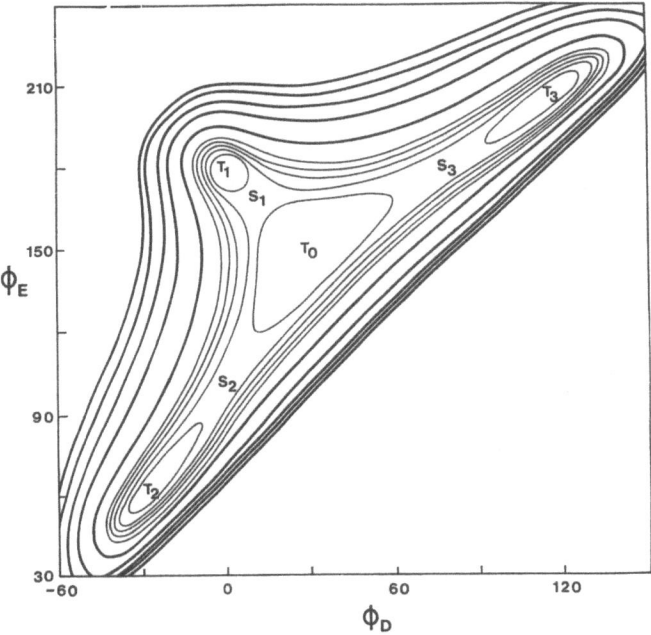

Fig. 4.4. Projection of the potential energy surface for [M(unidentate)$_5$] onto the $\phi_D - \phi_E$ plane (in degrees). The five *faint contour lines* are for successive 0.01 increments above the minima, and the five *heavy contour lines* are for successive 0.1 increments above the minima, at T$_0$, T$_1$, T$_2$ and T$_3$. n = 6. The locations of the regular stereochemistries are shown by T (trigonal bipyramid) and S (square pyramid)

pyramids exists on a saddle between the trigonal bipyramid T$_0$ and the trigonal bipyramids T$_1$, T$_2$, and T$_3$, respectively. These stereochemical changes are shown in Fig. 4.5.

B. [M(unidentate A)(unidentate B)$_4$]

The general stereochemistry for complexes with one ligand different from the other four is the same as that shown in Fig. 4.3, but now the effective bond length of the M$-$D bond is defined as R, with the other four remaining at unity.

Typical potential energy surfaces for R = 1.2, 0.8, and 0.2 are shown in Figs. 4.6$-$4.8, respectively, and should be compared with Fig. 4.4 calculated for R = 1.0.

As one center of repulsion is withdrawn from the central atom (Fig. 4.6), the trigonal bipyramids T$_1$ and T$_3$ are stabilized, and the unique ligand occupies one of the axial sites of a trigonal bipyramid.

For one effective bond length shorter than the other four, R = 0.8 (Fig. 4.7), it is the trigonal bipyramids T$_0$ and T$_2$ that are stabilized, with the unique ligand in the equatorial plane. These trigonal bipyramids are connected by the square

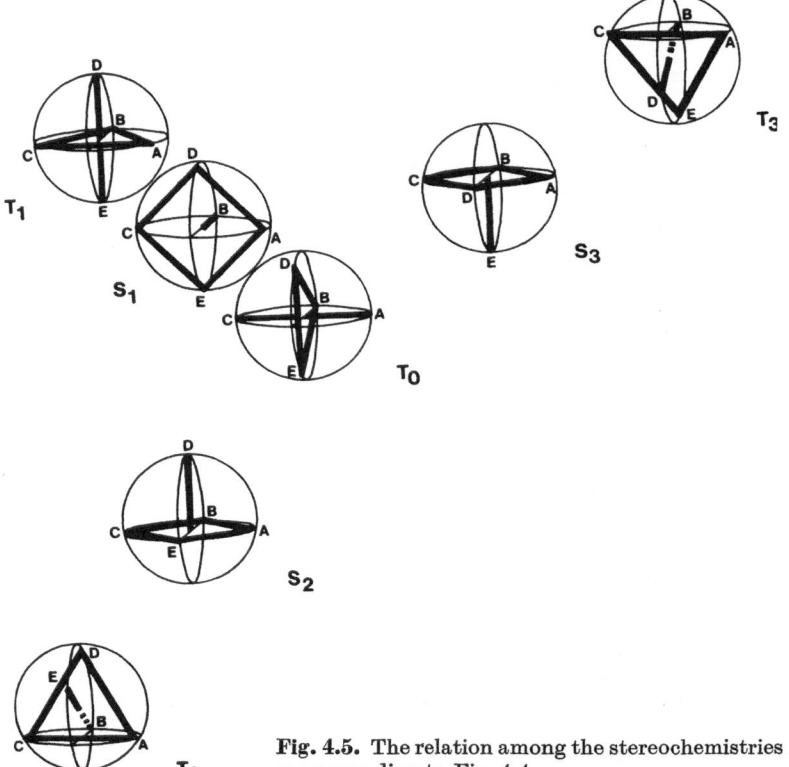

Fig. 4.5. The relation among the stereochemistries corresponding to Fig. 4.4

pyramid S_2, in which the unique ligand occupies the apical site. As R is further decreased, the minima corresponding to T_0 and T_2 move closer together until at R = 0.7 they coalesce to form the square pyramid S_2 as the sole minimum on the potential energy surface. On reduction to R = 0.4, this minimum again divides into two minima corresponding to a return to T_0 and T_2, which move further apart as R is decreased (Fig. 4.8).

These predicted stereochemical changes are summarized in Fig. 4.9.

Stereochemistries of complexes of the type [M(unidentate A)(unidentate B)$_4$] are given in Table 4.2. The observed distribution among the various stereochemistries is in general agreement with that predicted from the R(i/j) values for different metal-ligand bonds, and the above potential energy surfaces.

1. Axially Substituted Trigonal Bipyramid, T_1

The first two compounds in Table 4.2, [Sb(OMe)Ph$_4$] and [Sb(OH)Ph$_4$], have Ph—Sb—OX angles of 87.0°, which when fitted against calculations yield R(Ph/OX) = 1.15 (n = 6) [426], as expected for alkyl or aryl groups bonded to the heavier p-block elements, and the stabilisation of this isomer. The unique axial Sb—Ph bond is 4% longer than the three equatorial Sb—Ph bonds, which is

Table 4.2. Stereochemistries of [M(unidentate A)(unidentate B)$_4$] complexes

Complex	Stereochemistry	Ref.
[Sb(OMe)Ph$_4$]		[983]
[Sb(OH)Ph$_4$]		[85]
[Os{N(CH$_2$CH$_2$)$_3$CH}O$_4$]		[506]
[Fe(SC$_3$Ph$_2$)(CO)$_4$]		[315]
[Fe(NC$_5$H$_5$)(CO)$_4$]		[269]
[Fe(NC$_4$H$_4$N)(CO)$_4$]		[269]
[Fe(N$_2$C$_{13}$H$_{14}$O$_4$)(CO)$_4$]		[677]
[Fe(NC$_{10}$H$_{11}$)(CO)$_4$]		[832]
[Fe(PHPh$_2$)(CO)$_4$]	axially subs. trig. bipy., T$_1$	[647]
[Fe(PBu$_3$)(CO)$_4$]		[874]
[(CO)$_4$Fe(PMe$_2$PMe$_2$)Fe(CO)$_4$]		[614]
[Fe(PPh$_3$)(CO)$_4$]		[934]
[Fe(AsMe$_3$)(CO)$_4$]		[705]
[Fe(SbMe$_3$)(CO)$_4$]		[705]
[Fe(SbPh$_3$)(CO)$_4$]		[167]
[Fe(C$_5$H$_8$N$_2$)(CO)$_4$]		[600]
[(Ph$_3$P$_2$)$_2$N][Fe(CN)(CO)$_4$]		[486]
[(Ph$_3$P$_2$)$_2$N][FePr(CO)$_4$]		[601]
(Et$_4$N)[Fe(SiCl$_3$)(CO)$_4$]		[611]
[Co(SiF$_3$)(CO)$_4$]		[402]
[Co(SiCl$_3$)(CO)$_4$]		[938]
[Co(GeCl$_3$)(CO)$_4$]		[99]
[Co(GeCo$_3$(CO)$_9$)(CO)$_4$]		[121]
[Co{GeMePh(C$_{10}$H$_7$)}(CO)$_4$]		[290]
[Co(AuPPh$_3$)(CO)$_4$]		[646]
(Ph$_4$P)[Mn(PPh$_3$)(CO)$_4$]		[934]

Complex	Stereochemistry	Ref.
[P(C$_5$H$_6$N)F$_4$]		[981]
(Ph$_4$As)[SnMeCl$_4$]		[1083]
[NiBr(PMe$_3$)$_4$](BF$_4$)	equat. subs. trig. bipy., T$_0$	[300]
[NiBr{P(OMe)$_3$}$_4$](BF$_4$)		[789]
[Mn(NO)(CO)$_4$]		[452]
[Ru(PPh$_3$)(CNBu)$_4$]		[80]
[Ru(SbPh$_3$)(CO)$_4$]		[445]
(Et$_4$N)$_2$[Fe(NO)(CO)$_4$]		[673]
[Co(ClO$_4$)(Ph$_2$MeAsO)$_4$](ClO$_4$)		[588, 858]
[MnI(Ph$_3$PO)$_4$][MnI$_2$(CO)$_4$]		[238]
[ReOCl$_4$]		[380]
(Ph$_4$As)[CrOCl$_4$]		[458]
(Ph$_4$As)[MoOCl$_4$]		[461, 658]
[(Ph$_3$P$_2$)N][TcOCl$_4$]	axially subs. square py., S$_2$	[263]
(Ph$_4$As)[ReOCl$_4$]		[731]
(Ph$_4$As)[MoO(SPh)$_4$]		[144]
(Ph$_4$As)[MoNCl$_4$]		[658]
(Ph$_4$As)[RuNCl$_4$]		[871]
(Ph$_4$As)[OsNCl$_4$]		[872]
[WSCl$_4$]		[349]
[WSBr$_4$]		[349]
Rb[Bi(lone pair)(SCN)$_4$]	equat. subs. trig. bipy., T$_0$	[459]

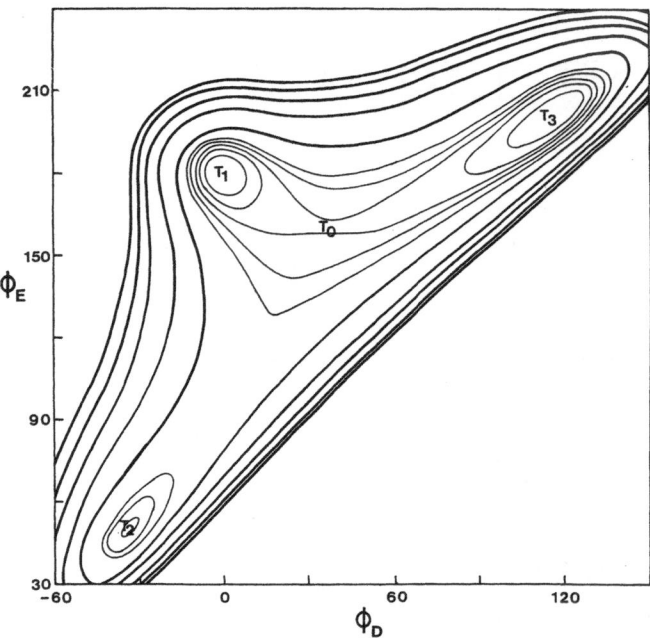

Fig. 4.6. Projection of the potential energy surface for [M(unidentate A)(unidentate B)$_4$] onto the $\phi_D - \phi_E$ plane (in degrees). The five *faint contour lines* are for successive 0.01 increments above the minima and the five *heavy contour lines* are for successive 0.1 increments above the minima, at T$_1$ and T$_3$. R = 1.2, n = 6. The positions of the trigonal bipyramids (T) are shown

again in agreement with the expectation that these type α compounds will follow predictions from simple repulsion theory.

The osmium tetroxide-quinuclidine complex, [Os{N(CH$_2$CH$_2$)$_3$CH}O$_4$] has three very small N—Os—O angles of 79.4°, showing R(N/O^{2-}) = 1.6, or R(O^{2-}/N) = 0.6 (n = 6).

A large group of compounds with this stereochemistry are the monosubstituted carbonyls of metal ions with the d^8 electron configuration, manganese(−I), iron(0), and cobalt(I). For compounds containing uncharged ligands, [MnL(CO)$_4$]$^-$ and [FeL(CO)$_4$], the L—M—CO bond angles of $\sim 90°$ imply R(L/CO) ~ 1.0. This result is significantly different from compounds containing charged ligands, [FeX(CO)$_4$]$^-$ and [CoX(CO)$_4$], where X—M—CO $\sim 85°$ and R(X$^-$/CO) ~ 1.2 (n = 6). As was noted for complexes containing five equivalent ligands, these type β molecules do not have the axial M—CO bond longer than the equatorial M—CO bonds. The ^{13}C NMR spectrum of [Fe(NC$_5$H$_5$)(CO)$_4$] shows this molecule to be fluxional, even down to −90 °C, as expected for R(L/CO) ~ 1.0 [150].

The metal hydrides [(Ph$_3$P)$_2$N][FeH(CO)$_4$] [996], [CoH(PF$_3$)$_4$] [453], and [RhH(PPh$_3$)$_4$]¹/$_2$C$_6$H$_6$ [66] also appear to be trigonal bipyramidal with the hydride in an axial site. They are more rigid in solution, as expected for large differences in ligand type [299, 320, 412].

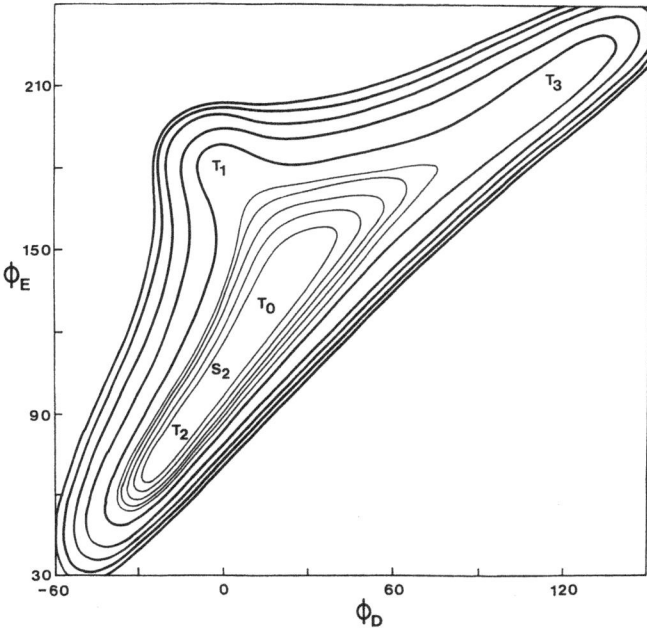

Fig. 4.7. Projection of the potential energy surface for [M(unidentate A)(unidentate B)$_4$] onto the $\phi_D - \phi_E$ plane (in degrees). The five *faint contour lines* are for successive 0.01 increments above the minima, and the five *heavy contour lines* are for successive 0.1 increments above the minima, at T$_0$ and T$_2$. R = 0.8, n = 6. The positions of the trigonal bipyramids (T) and square pyramid (S) are shown

2. Equatorially Substituted Trigonal Bipyramid, T$_0$

The alkyl group in [P(C$_5$H$_6$N)F$_4$] is found in one of the less hindered equatorial sites of a trigonal bipyramid, as expected from the low R(C/F) value for fluoro complexes. This point is confirmed by the RPF angles of 91.9° and 126.0°, showing greater repulsion from the phosphorus-alkyl bond than from the phosphorus-fluorine bonds. The axial P—F bond are 4% longer than the equatorial P—F bonds, which is again in agreement with the expectation that these type α compounds will follow the predictions from simple repulsion theory.

Similarly the low R(C/Cl) value expected for alkyl compounds of the heavier p-block elements leads to this T$_0$ stereochemistry for [SnMeCl$_4$]$^-$.

The dynamic behavior of complexes of the type [MRF$_4$], where M is a p-block element, has been extensively studied. The ^{19}F NMR spectra of [PRF$_4$] show that all fluorine atoms are equivalent, where R is Cl or a range of alkyl groups [217, 817]. However, when R is NMe$_2$ or SMe, rigid structures are formed at about −100 °C, with rapid rearrangements at room temperature [816, 860, 1040, 1089]. These compounds have also been studied by ^{31}P NMR, which shows that the rate of pseudorotation increases along the series NMe$_2$ < SMe, H < Cl < Me, F [391]. Similar work on related tetrafluoro compounds [AsPhF$_4$] [816] and [SiRF$_4$]$^-$

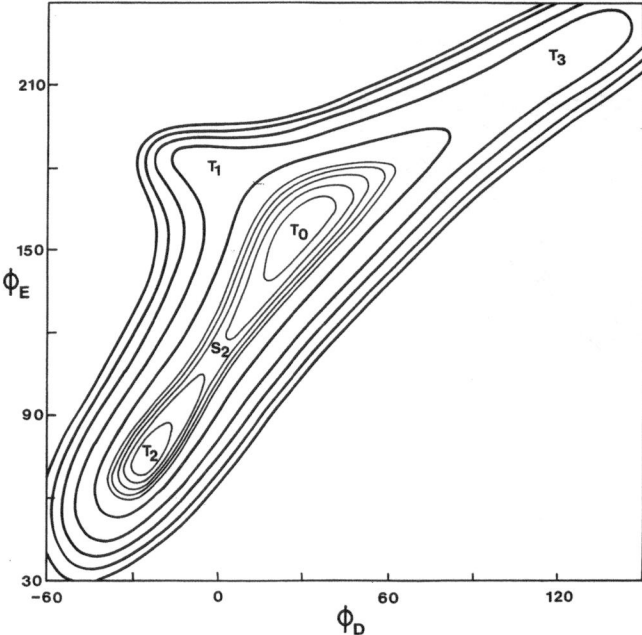

Fig. 4.8. Projection of the potential energy surface for [M(unidentate A)(unidentate B)$_4$] onto the $\phi_D - \phi_E$ plane (in degrees). The five *faint contour lines* are for successive 0.01 increments above the minima, and the five *heavy contour lines* are for successive 0.1 increments above the minima, at T$_0$ and T$_2$. R = 0.2, n = 6. The positions of the trigonal bipyramids (T) and square pyramid (S) are shown

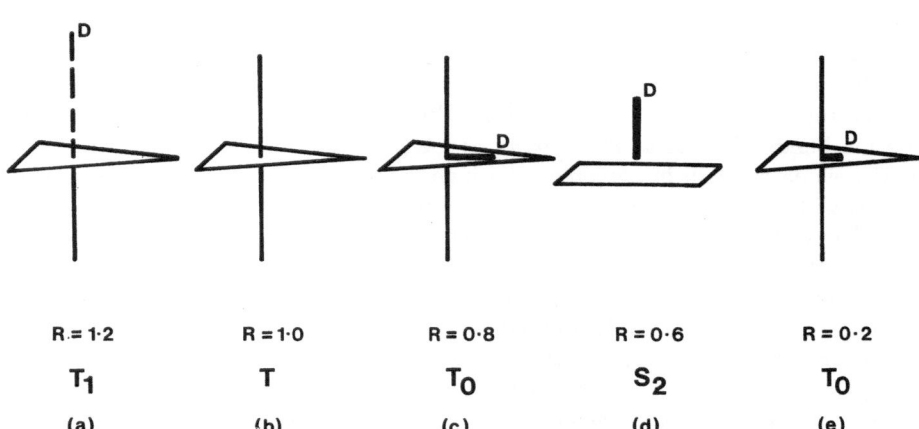

Fig. 4.9. a Axially substituted trigonal bipyramid. **b** Trigonal bipyramid. **c** Equatorially substituted trigonal bipyramid. **d** Axially substituted square pyramid. **e** Equatorially substituted trigonal bipyramid

[652] shows that all fluorine atoms are equivalent. The rapid fluorine exchange observed for [S(lone pair)F$_4$] appears to be largely intermolecular rather than intramolecular [819].

3. Axially Substituted Square Pyramid, S$_2$

Stereochemistry S$_1$, the square pyramid with the unique ligand in the axial site, is favored for complexes of the type [MOX$_4$]$^{x-}$ and [MNX$_4$]$^{x-}$. The OMX and NMX angles of about 104° imply R(O^{2-}/X$^-$) $\sim$ 0.8.

The observation that the two nitrosyl complexes [Mn(NO)(CO)$_4$] and (Et$_4$N)$_2$[Fe(NO)(CO)$_4$] have stereochemistries T$_0$ or S$_2$ implies greater repulsion from the nitrosyl ligand than from the other four unidentate ligands. This result is in agreement with the observed structures of other nitrosyl complexes [426].

4. [M(lone pair)(unidentate)$_4$], Stereochemistry T$_0$

The structure of Rb[Bi(lone pair)(SCN)$_4$] can be considered to be trigonal bipyramidal if it is assumed that the lone pair of electrons occupies an equatorial site. The SBiS angles of 92.6° and 158.2° show a considerable reduction from the regular trigonal bipyramidal angles of 120° and 180°, and imply R(:/SCN$^-$) $\sim$ 0.1$_5$ (n = 6). The Bi—S$_{axial}$ bonds are 6% longer than the Bi—S$_{equatorial}$ bonds.

Compounds containing a lone pair of electrons and four identical unidentate logands are relatively rare. However, there are many molecules known with a mixture of unidentate ligands. Examples include [As^{III}(lone pair) X$_3$(ligand)] [1099], [Te^{IV}(lone pair) X$_3$(R)] [655, 746], and [Te^{IV}(lone pair) X$_2$R$_2$] [900], (where X = Cl, Br, I; R = alkyl, aryl).

C. [M(unidentate A)$_2$ (unidentate B)$_3$]

The description of the general stereochemistry is the same as that used for [M(unidentate)$_5$] and [M(unidentate A)(unidentate B)$_4$] (Fig. 4.3). Two ligands lie on the mirror plane with M—D = M—E = R, the remaining effective bond lengths being defined as unity.

The potential energy surface calculated for R = 1.2 is shown in Fig. 4.10 and should be compared with Fig. 4.4 calculated for R = 1.0, and Fig. 4.5. It can be seen that stereochemistry T$_1$ is stabilized, with the two extended bonds at the apices of a trigonal bipyramid. Conversely as R is decreased to 0.8 (Fig. 4.11), it is the trigonal bipyramid T$_0$ that is stabilized, with the two shorter effective bond lengths in equatorial sites. As R is decreased even further, the repulsion between these two contracted bonds becomes the dominant term, and the minimum deepens and shifts so that there is a continuous change in stereochemistry from T$_0$, through the square pyramid S$_1$, to T$_1$ in which the pair of ligands again occupy the two axial sites [426].

Compounds of the type [M(unidentate A)$_2$(unidentate B)$_3$] are listed in Table 4.3. For compounds with R $\sim$ 1.0, there is considerable scattering along the long

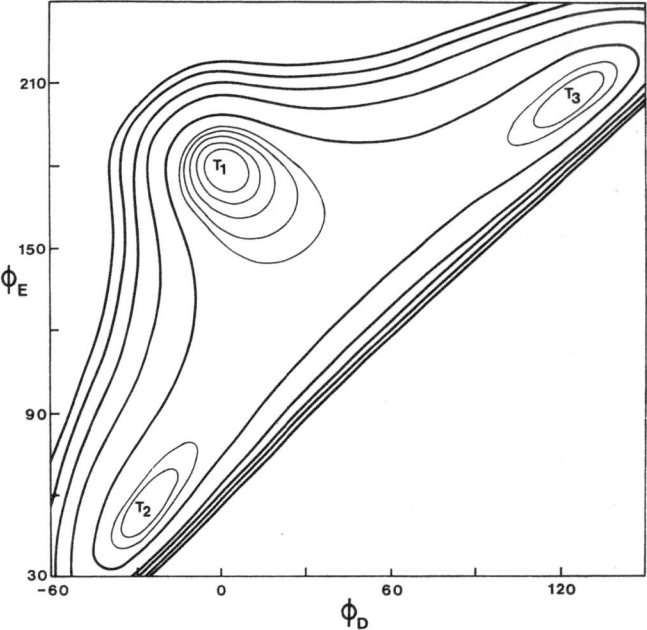

Fig. 4.10. Projection of the potential energy surface for [M(unidentate A)$_2$(unidentate B)$_3$] onto the $\phi_D - \phi_E$ plane (in degrees). The five *faint contour lines* are for successive 0.01 increments above the minimum, and the five *heavy contour lines* are for successive 0.1 increments above the minimum, at T$_1$. R = 1.2, n = 6. The position of the trigonal bipyramids (T) are shown

trough in the potential energy surface, and it is sometimes necessary to make some more-or-less arbitrary divisions if the molecules are to be grouped as belonging to a particular stereochemistry [426]. There is apparently a wide diversity of structure types, particularly among the transition metal complexes. However, nearly all molecules can be adequately described as trigonal bipyramidal T$_0$ or T$_1$, with an exceptional group of square pyramidal S$_2$ stereochemistry.

As would be expected from a comparison of the potential energy surfaces for [M(unidentate)$_5$] and [M(unidentate A)$_2$(unidentate B)$_3$], compounds of the type PX$_2$F$_3$ are more rigid than PF$_5$. For example, the intramolecular exchange barriers for PCl$_2$F$_3$ [753], PH$_2$F$_3$[472], and PMe$_2$F$_3$ [805] are 30, 43, and 75 kjoule/ mole, respectively. On the other hand, AsPh$_2$F$_3$ shows rapid exchange of fluorine atoms [733].

1. Bi(Axially Substituted) Trigonal Bipyramid, T$_1$

Atoms with two bonds more extended than the other three will adopt the trigonal bipyramidal structure, with the two extended bonds in the more crowded axial sites. Thus in [PF$_2$(C$_6$F$_5$)$_3$] the R(F/C) value of approximately 1.2 leads to a trigonal bipyramid with the fluorine atoms in axial sites and perfluorophenyl groups in equatorial sites.

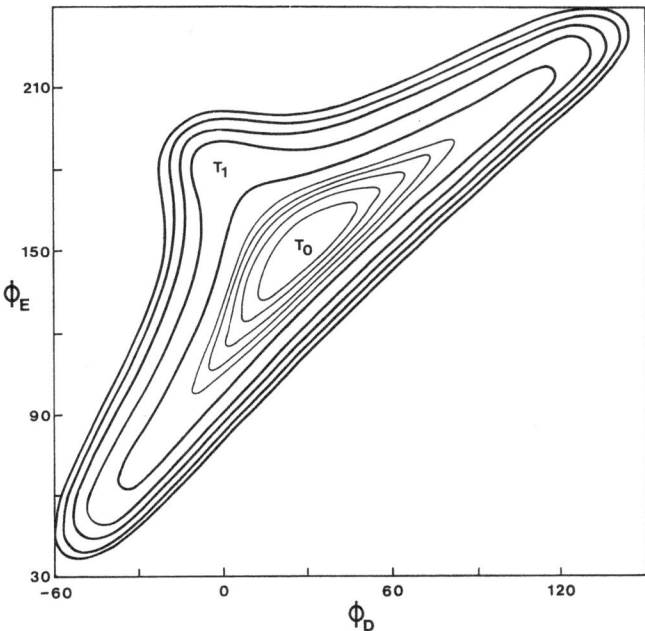

Fig. 4.11. Projection of the potential energy surface for [M(unidentate A)$_2$(unidentate B)$_3$] onto the $\phi_D - \phi_E$ plane (in degrees). The five *faint contour lines* are for successive 0.01 increments above the minimum, and the five *heavy contour lines* are for successive 0.1 increments above the minimum, at T_0. R = 0.8, n = 6. The positions of the trigonal bipyramids (T) are shown

Similarly in metal complexes [M^{III}(ligand)$_2$Cl$_3$] and [M^{II}(ligand)$_2$Cl$_3$]$^-$ (Table 4.3), R(ligand/Cl$^-$) = 1.2, and the isomer found has the uncharged ligands in the axial sites and the chlorine atoms in the equatorial sites. The same factors also lead to this structure for [Ru(OH)$_2$O$_3$]$^{2-}$, where R(OH$^-$/O^{2-}) > 1.0.

The substituted carbonyls [M(ligand)$_2$(CO)$_3$]$^{x\pm}$ (M = Fe^0, Os^0, Ir^I) are also of this stereochemistry (Table 4.3), in agreement with R(ligand/CO) > 1.0 usually observed for metals in low oxidation states.

2. Bi(Equatorially Substituted) Trigonal Bipyramid, T_0

Atoms with two bonds of shorter effective bond length than the other three will also adopt the trigonal bipyramidal structure, but now this pair of atoms will be in equatorial sites.

Transition metal complexes [M^{II}X$_2$(ligand)$_3$] (X = Cl, Br, I) are therefore trigonal bipyramidal, with the two halogen atoms in equatorial sites (Table 4.3). The complex [Mn(NO)$_2${PPh(OMe)$_2$}$_3$]$^+$ also has this structure, again indicating that nitrosyl groups correspond to short effective bond lengths.

The trifluoro complexes ClF$_3$ and XeF$_3^+$ (Table 4.3) are T-shaped, but can be considered to be of stereochemistry T_0 if lone pairs of electrons occupy two of the

Table 4.3. Stereochemistries of [M(unidentate A)₂(unidentate B)₃] complexes

Complexes	Stereochemistry	Ref.
[PF₂(C₆H₅)₃]		[982]
[SbF₂Me₃]		[974]
[Sb(OH)₂Ph₃]		[983]
[AsMe₂Cl₃]		[594]
[BiCl₂Ph₃]		[544]
[Sb(NCO)₂Ph₃]		[434]
[Fe[P(OMe)₃]₂(CO)₃]		[477]
[Os(PPh₃)₂(CO)₃]		[1007]
[Ir(PPhMe₂)₂(CO)₃](ClO₄)		[901]
[Al(Me₂NH)₂Cl₃]		[6]
[Al(Me₃N)₂H₃]	bi(axially subs.), trig. bipy., T	[554]
[Tl(ONC₅H₄CN)₂Cl₃]		[509]
[Tl(OH₂)₂Cl₃]		[480]
[Tl(OH₂)₂Br₃]		[949]
[T(Me₃N)₂Br₃]		[502]
[V(Me₃N)₂Cl₃]		[448]
[Cr(Me₃N)₂Cl₃]		[403]
[Co(PEt₃)₂Cl₃]		[946]
[Ni{N(CH₂CH₂)₃NMe}₂Cl₃](ClO₄)		[946]
[Cu{N(CH₂CH₂)₃NMe}₂Cl₃](ClO₄)		[894]
[Ni(CN)₂[P(C₆H₄C₆H₄)Et]₃]		[894]
[Ni(CN)₂[P(C₆H₄C₆H₄)Me]₃]MeOH		[495]
[Co(PPh(OEt)₂)₂(CNC₆H₄NO₂)₃](ClO₄)		[737]
[Co{P(OMe)₃}₂(CNC₆H₄F)₃]		[634]
Ba[Ru(OH)₂O₃]		[1067]
[Mo₃(C₅H₅)₃S₄][SnCl₂Me₃]		[531]
(Ph₃PCH₂Ph)[SnCl₂Bu₃]		[531]
(Ph₃AsCH₂COPh)[SnCl₂Ph₃]		

Complex	Stereochemistry	Ref.
[RuCl₂(PPh₃)₃]	bi(basal subs.) square py., S₁	[692]
[Cu(PhOCH₂COO)₂(H₂O)₃]		[483]
[Ni(CN)₂{P(C₆H₄C₆H₄Me)₃}		[894]
[CuCl₂(N₂C₃H₂Me₂)₃]	intermediate, T₀—S₁	[589]
[SnMe₂Cl(terpy)][SnMe₂Cl₃]		[388]
(C₉H₈N)[SnMe₂Cl₃]		[201]
[CoBr₂(Ph₂PH)₃]		[108]
[NiI₂(Ph₂PH)₃]		[108]
[NiI₂{P(OMe)₃}₃]		[1064]
[NiBr₂(PMe₃)₃]		[305]
[NiBr₂{P(C₆H₄C₆H₄)Et}₃]2CHCl₃	bi(equat. subs.) trig. bipy., T₀	[893]
[NiCl₂{P(C₆H₄C₆H₄)Et}₃]		[892]
[CoBr₂(PF₂Ph)₃]		[1011]
[Ta{N(SiMe₃)₂}₂Cl₃]		[145]
[SbPh₂Br₃]		[138]
[Xe(lone pair)₂F₃](SbF₆)		[126]
[Xe(lone pair)₂F₃](Sb₂F₁₁)		[750]
[Xe(lone pair)₂F₃](BiF₆)		[474]
[Cl(lone pair)₂F₃]		[181]
[Mn(NO)₂{PPh(OMe)₂}₃](BF₄)		[685]
[PdBr₂{P(C₆H₄C₆H₄)Et}₃]PhCl		[236]
[PtBr₂{P(C₆H₄C₆H₄)Et}₃]PhBr		[236]
[PtBr₂{P(C₆H₄C₆H₄)Me}₃]PhBr	square pyramid, S₂	[236]
[PdBr₂{PPh(CH₂)₂C₆H₄}₃]Me₂CO		[236]
[PdBr₂{PPh(CH₂)₂C₆H₄}₃]		[236]
[PdCl₂(PMe₂Ph)₃]CH₂Cl₂		[739]
[MnCl₂(N₂C₃H₃Me)₃]	trigonal bipyramid, T₂	[869]

equatorial sites completing a five-coordinate trigonal bipyramidal arrangement of electron pairs. The FClF and FXeF angles of 87° and $\sim 83°$, respectively, imply R(:/F) values of ~ 0.9 and $0.5-0.8$ respectively (n = 6), these values being considerably larger than those obtained for complexes containing only one non-bonding pair of electrons. Closely similar structures have been deduced from early microwave studies on ClF$_3$ (FClF = 87.5°) [994] and BrF$_3$ (FBrF = 86.2°) [752].

Stereochemistry T_0 has two different types of (unidentate B) ligands, the two in the axial A and C sites experiencing greater repulsion than the one in the equatorial B site. This results in the two outer M—F bonds being significantly longer than the central M—F bond:

$$\frac{Cl-F(A)}{(Cl-F(B)} = 1.06, \qquad \frac{Xe-F(A)}{Xe-F(B)} = 1.05.$$

The linear trihalide ions such as $(I-I-I)^-$ can likewise be regarded as being of stereochemistry T_1 with three nonbonding pairs of electrons in the three equatorial sites [838, 899, 948].

3. Square Pyramid, S$_2$

Palladium(II) and platinum(II) complexes of the type [MX$_2$(phosphine)$_3$] (X = Cl, Br), are exceptional as all are square pyramidal, with one halogen atom in an axial site and one in a basal site (Table 4.3).

These molecules can alternatively be regarded as four-coordinate and square-planar [MX(phosphine)$_3$]$^+$, with a more weakly bonded halide ion completing the pyramid:

$$\frac{M-X(axial)}{M-X(planar)} = 1.15-1.24 \text{ (average 1.19)}.$$

The structures of these complexes are quite different from the nickel(II) and the cobalt(II) complexes [MX$_2$(phosphine)$_3$], which are of stereochemistry T_0 in accord with the expected R(X$^-$/ligand) ~ 0.8.

CHAPTER 5

Five-Coordinate Compounds Containing Chelate Groups

A. [M(bidentate)(unidentate)$_3$]

The interconversions between the trigonal bipyramid and square pyramid described in the previous chapter for [M(unidentate)$_5$] complexes involve relatively large changes in the length of the polyhedral edges, the limits being $3^{1/2}r = 1.73$ r for the edges linking the equatorial sites of a trigonal bipyramid, and 1.39 r (for n = 6) for the edges linking the basal sites of a square pyramid. The introduction of a chelating ligand of fixed normalized bite must therefore substantially alter the form of the potential energy surface, and a more rigid and well-defined structure might be expected. The results of the calculations outlined below show that this expectation is not realized for bidentate ligands of low normalized bite where the potential energy surfaces are even flatter than those calculated for [M(unidentate)$_5$], and is only partially realized for those of high normalized bite.

1. The Theoretical Stereochemistries

The axes for the general stereochemistry are defined by placing the bidentate ligand symmetrically across the "North Pole" at $\phi = 0$, with $\theta_A = 0$ and $\theta_B = 180°$ (Fig. 5.1). The angles ϕ_i and θ_i are defined by Fig. 5.1. The total repulsion energy is a function of ϕ_C, θ_C, ϕ_D, θ_D, ϕ_E and θ_E. Potential energy surfaces are shown projected onto the $\theta_C - \theta_D$ plane, which most clearly separates the different stereochemistries. To prevent the overlap of several minima on these projections, due merely to an interchange of labels on the donor atoms, it is also convenient to impose the conditions $\phi_E \geq \phi_C$, $\phi_E \geq \phi_D$, and $\theta_D \leq \theta_E \leq (360° + \theta_C)$.

For values of the normalized bite b less than 1.3, the single minimum consists of an extraordinarily long and level valley encompassing stereochemistries I and II (Fig. 5.2). It should be noted that these surfaces are even flatter than those calculated for [M(unidentate)$_5$].

In both stereochemistries I and II, $\phi_C \sim \phi_D \sim \phi_E \sim 120°$, and $|\theta_C - \theta_D| \sim |\theta_D - \theta_E| \sim |\theta_E - \theta_C| \sim 120°$, and the structures may be considered to consist of the bidentate ligand lying above and parallel to an approximately equilateral triangular arrangement of the three unidentate ligands (Fig. 5.3). The interconversion between I and II can be readily envisaged as free spinning of the bidentate ligand above the triangular array of unidentate ligands.

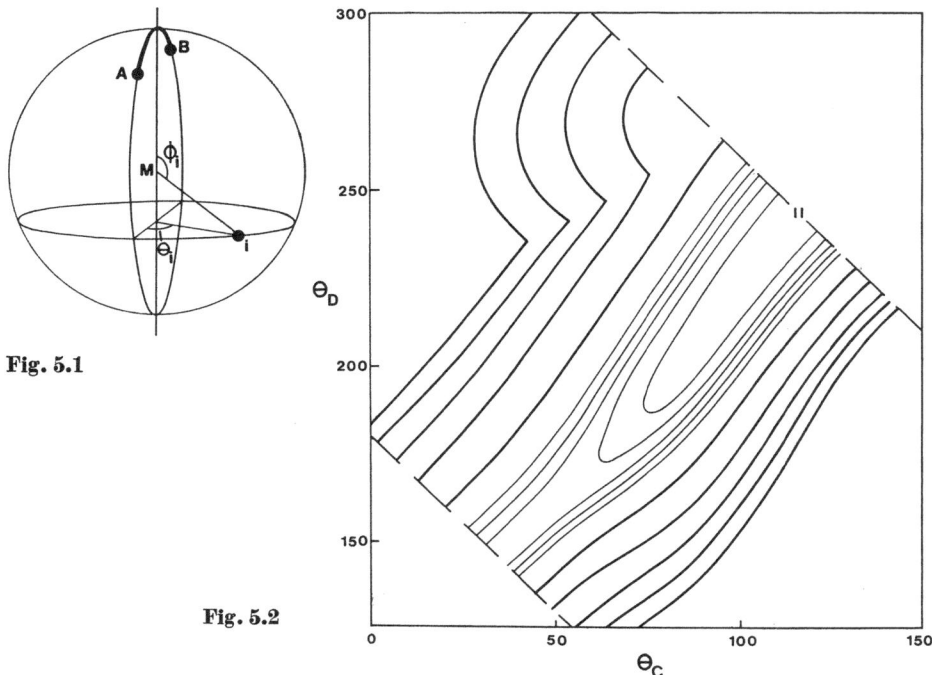

Fig. 5.1. General stereochemistry of [M(bidentate)(unidentate)$_3$]

Fig. 5.2. Projection of the potential energy surface of [M(bidentate)(unidentate)$_3$] onto the $\theta_C - \theta_D$ plane (in degrees). The five *faint contour lines* are for successive 0.01 increments above the minimum, and the five *heavy contour lines* are for successive 0.1 increments above the minimum, at II. b = 1.2, n = 6. The position of stereochemistries I and II are shown

In stereochemistry I, the bidentate ligand AB is parallel to the edge DE, and this stereochemistry may alternatively be pictured as a distorted square pyramid with the bidentate ligand spanning one of the basal edges (Fig. 5.4a). A regular pyramid is attained, that is, ABDE form a square, when the normalized bite is equal to the basal edge length of the square pyramid calculated for [M(unidentate)$_5$] (b = 1.39 for n = 6). As expected from the results for [M(unidentate)$_5$], the unidentate ligand at the apical C site is subjected to less repulsion than are the unidentate ligands at the D and E sites.

In stereochemistry II, the projection of the bidentate ligand AB is normal to the edge CD, and this stereochemistry may be pictured as an irregular trigonal bipyramid (Fig. 5.4b). As b is increased to $2^{1/2} = 1.414$, the bipyramid becomes regular. As expected, the unidentate ligands C and D are less repelled than is E, and the A end of the bidentate ligand is less repelled than the B end.

As the normalized bite is further increased, stereochemistry II becomes increasingly stable with respect to stereochemistry I, a typical potential energy surface being shown in Fig. 5.5. There is also a large and continuous change in

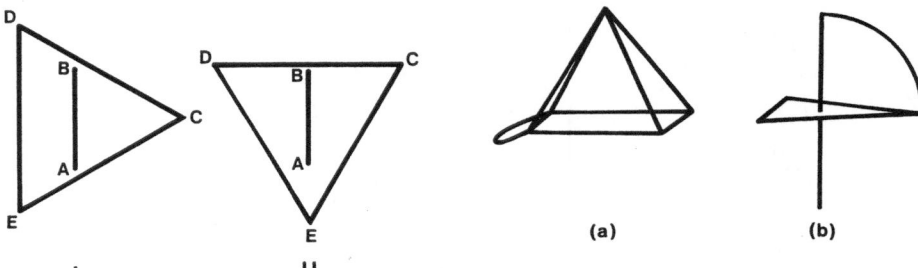

Fig. 5.3. Stereochemistries I and II for [M(bidentate)(unidentate)$_3$] at low b

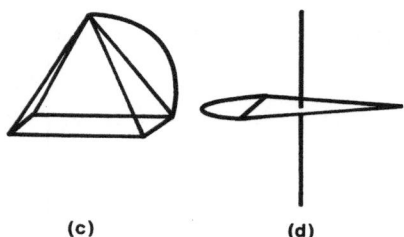

Fig. 5.4a—d. Stereochemistries for [M(bidentate) (unidentate)$_3$].
a Stereochemistry I, b = 1.39. **b** Stereochemistry II, b = 1.414. **c** Stereochemistry II, b = 1.55. **d** Stereochemistry III, b = 1.732

angular parameters, until a square pyramid with the bidentate ligand spanning the axial and one basal site is attained at b = 1.55 (for n = 6) (Fig. 5.4 c), with the same dimensions as was calculated for [M(unidentate)$_5$]. At even higher values of the normalized bite, the stereochemistry continues to change until a regular trigonal bipyramid is formed at b = $3^{1/2}$ = 1.732 with the bidentate ligand spanning two equatorial sites (Fig. 5.4 d). This limit has been labelled stereochemistry III.

In these stereochemistries the ϕ angular parameters decrease as the effective bond length ratio R(unidentate/bidentate) decreases, whereas the θ angular parameters remain relatively constant. Isomers II and III become increasingly stable relative to isomer I as R(unidentate/bidentate) decreases.

2. Comparison with Experiment

The difference in energies between stereochemistries I and II is small, and it can only be predicted that the observed stereochemistry will lie somewhere in the valley between these extremes. This is in agreement with the known structures listed in Table 5.1.

The pattern of relative bond lengths is in general agreement with that expected. For example in the square pyramidal compounds [Ru(S$_2$CHPMe$_2$Ph)(PMe$_2$Ph)$_3$] · (PF$_6$)   and   [Sb(O$_2$C$_6$H$_4$)Ph$_3$][Sb(O$_2$C$_6$H$_4$)Ph$_3$(H$_2$O)],   (Ru—P$_{ax}$)/(Ru—P$_{basal}$) = 0.96 and (Sb—Ph$_{ax}$)/(Sb—Ph$_{basal}$) = 0.98. Similarly the trigonal bipyramidal compounds of stereochemistry   II, [Sn(PhCOCHCOPh)Ph$_3$] and [P(C$_{14}$H$_8$O$_2$)

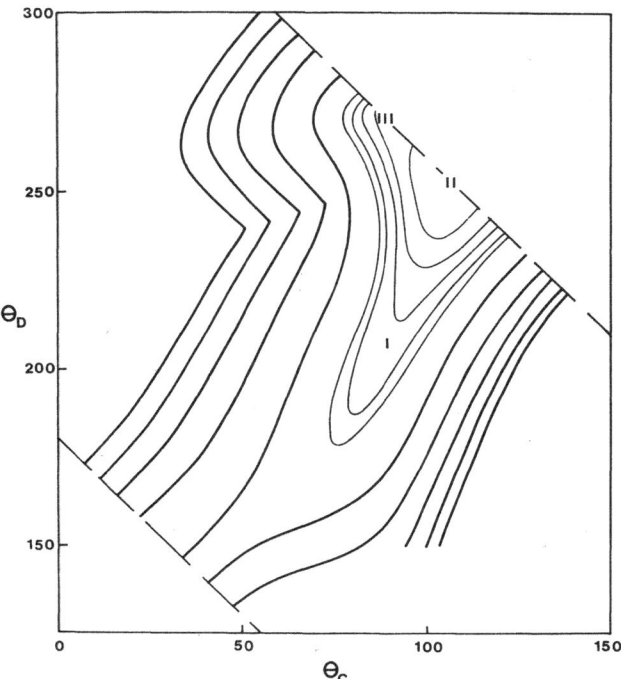

Fig. 5.5. Projection of the potential energy surface for [M(bidentate)(unidentate)$_3$] onto the $\theta_C - \theta_D$ plane (in degrees). The five *faint contour lines* are for successive 0.01 increments above the minimum, and the five *heavy contour lines* are for successive 0.1 increments above the minimum, at II. b = 1.5, n = 6. The positions of stereochemistries I, II, and III are shown

· (OPr)$_3$], (M-unidentate$_{eq}$)/(M-unidentate$_{ax}$) = 0.99 and 0.97 respectively, with the bidentate ligands being bonded even more unsymmetrically (M-eq)/(M-ax) = 0.92 and 0.93 respectively. The reverse behaviour of the Fe^0—CO bonds in

Table 5.1. Structural parameters for [M(bidentate)(unidentate)$_3$] complexes

Complex	b	Stereo-chemistry	ϕ_C	θ_C	ϕ_D	θ_D	ϕ_E	θ_E	Ref.
[Sn(NO$_3$)(SnPh$_3$)$_3$]	0.72	I—II	106°	109°	101°	231°	106°	347°	[828]
[Ru(S$_2$CHPMe$_2$Ph)(PMe$_2$Ph)$_3$](PF$_6$)		I							[54]
[Fe(Ph$_2$PCH$_2$PPh$_2$)(CO)$_3$]	1.20	I—II	108°	98°	116°	230°	128°	344°	[266]
[Sn(PhCOCHCOPh)Ph$_3$]	1.26	II	100°	111°	105°	241°	125°	355°	[68]
[Sb(O$_2$C$_6$H$_4$)Ph$_3$][Sb(O$_2$C$_6$H$_4$)Ph$_3$(H$_2$O)]	1.28	I	103°	90°	122°	204°	121°	324°	[518]
[Fe{C$_6$H$_4$(AsMe$_2$)$_2$}(CO)$_3$]	1.34	II	108°	113°	108°	247°	134°	0°	[160]
[Fe{(C$_6$H$_{11}$)$_2$PC$_2$(CF$_2$)$_2$P(C$_6$H$_{11}$)$_2$}(CO)$_3$]	1.39	I—II	115°	107°	114°	239°	131°	348°	[386]
[P(C$_{14}$H$_8$O$_2$)(OPr)$_3$]	1.41	II							[522, 1003]

[Fe{C$_6$H$_4$ · (AsMe$_2$)$_2$}(CO)$_3$], (Fe—C$_{eq}$)/(Fe—C$_{ax}$) = 1.08, is again the normal observation for a type β complex. The d^8 gold(III) complexes with substituted bipyridyl or 1,10-phenanthroline ligands, [Au(bidentate)X$_3$] (X=Cl, Br), have one Au—N bond 25% longer than the other, and they may be considered as square-planar complexes with four unidentate ligands [841, 939].

The angular parameters in Table 5.1 can be fitted to the calculated angular parameters to obtain a value for the effective bond length ratio R(unidentate/ bidentate), for example:

[Sn(PhCOCHCOPh)Ph$_3$],   R(Ph/O) = 0.7 (n = 6) ;

[Fe{C$_6$H$_4$(AsMe$_2$)$_2$}(CO)$_3$],   R(CO/As) = 1.0 (n = 6).

The very flat floor in the potential energy surface connecting stereochemistries I and II agrees with the observed nonrigid behavior of a number of substituted iron carbonyls, [Fe(bidentate)(CO)$_3$] [8, 266, 691]. However, more rigid behavior is expected at higher normalized bites, and it is therefore relevant that the ^{19}F-NMR spectrum of (CH$_2$)$_4$PF$_3$ is temperature dependent, the molecule becoming rigid below −70 °C. On the other hand, (CH$_2$)$_5$PF$_3$, in which the normalized bite will be larger, shows no evidence of an intramolecular exchange process up to ∼ 100° [817].

Molecules of more complicated stoichiometry have stereochemistries that can be deduced from the general principles outlined above. In addition, unsymmetric bidentate ligands are expected to favor stereochemistry II, in which the two ends of the bidentate are attached to different types of vertex, rather than stereochemistries I or III, in which both ends of the bidentate ligands are stereochemically equivalent. Thus stereochemistry II is observed for [Sn{OC(Ph)N(Ph)O}Ph$_3$] [527], [Sn{C$_6$H$_3$Me · C(C$_6$H$_4$Me)NH}Cl$_3$] [440], and [Sn{CH$_2$CH$_2$C(OMe)O}Cl$_3$] [528].

One interesting example of compounds with a more complex stoichiometry is the tellurium(IV) complexes [Te(bidentate)(unidentate)$_2$(lone pair)] [573, 654, 656, 657, 744, 745, 757, 758, 902]. These compounds are of stereochemistry III if it is assumed that a lone pair of electrons occupies the least crowded equatorial site E. This stabilization of stereochemistry III can be simulated by repeating the above calculations with the metal-lone pair effective bond length being much shorter than the other four effective bond lengths (Fig. 5.6).

B. [M(bidentate)$_2$(unidentate)]

1. The Theoretical Stereochemistries

The calculations described here are similar to those described in Chap. 4 for [M(unidentate)$_5$]. The problem is first simplified by imposing a twofold axis along the metal-unidentate ligand bond, which generates potential energy surfaces similar to those calculated for [M(unidentate)$_5$] in which C$_{2v}$ symmetry was enforced. Second, potential energy surfaces are calculated in which no symmetry is enforced, and are comparable with those calculated for [M(unidentate)$_5$] in which only a mirror plane was enforced.

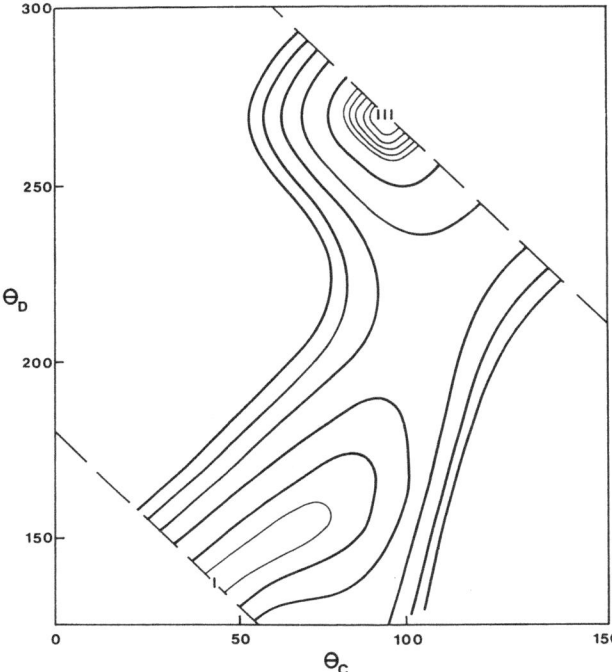

Fig. 5.6. Projection of the potential energy surface for [M(bidentate)(unidentate A)₂ (lone pair)] onto the θ_C — θ_D plane (in degrees). The five *faint contour lines* are for successive 0.01 increments above the minimum, and the five *heavy contour lines* are for successive 0.1 increments above the minimum, at III. The effective bond length ratio for the lone pair is R = 0.5, all others being equal to unity. b = 1.5, n = 6. The positions of stereochemistries I and III are shown

a) Symmetrical Structures

Figure 5.7 defines the general stereochemistry, the bidentate ligands spanning the AB and CD edges. Two representative potential energy surfaces are shown in Figs. 5.8 and 5.9, and should be compared with Fig. 4.2, calculated for five unidentate ligands.

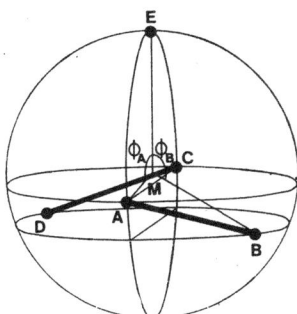

Fig. 5.7. General stereochemistry for [M(bidentate)₂-(unidentate)] with C₂ symmetry

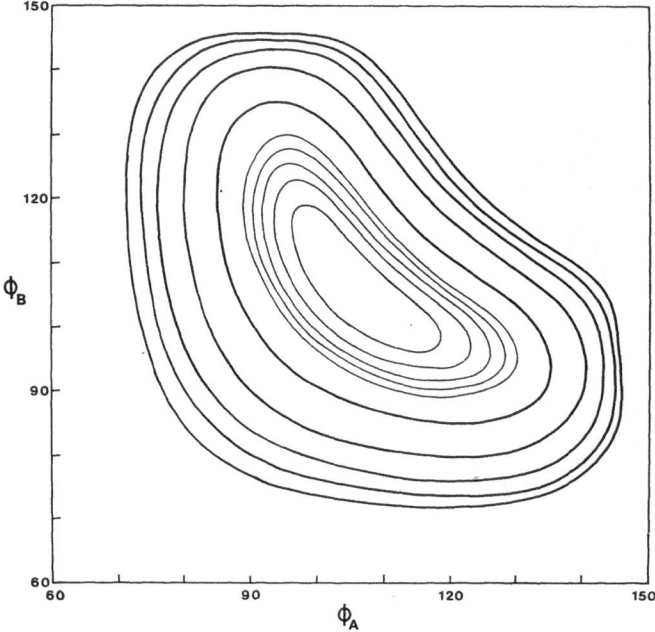

Fig. 5.8. Projection of the potential energy surface for [M(bidentate)$_2$(unidentate)] onto the $\phi_A - \phi_B$ plane (in degrees). The five *faint contour lines* are for successive 0.01 increments above the minimum, and the five *heavy contour lines* are for successive 0.1 increments above the minimum. b = 1.2, n = 6

For small values of the normalized bite (Fig. 5.8), a single minimum on the potential energy surface occurs at $\phi_A = \phi_B$, corresponding to the square pyramid (or more correctly, a rectangular pyramid). As the normalized bite is progressively increased to b = 1.28, the minimum symmetrically splits into two minima corresponding to two equivalent irregular bipyramids. These minima progressively deepen and move further apart as the normalized bite is further increased (Fig. 5.9).

The splitting of the rectangular pyramidal minimum into the two trigonal bipyramidal minima occurs at the higher value of b = 1.37 as the effective bond length ratio R(unidentate/bidentate) is decreased from 1.0 to 0.6 (Fig. 5.10). However, this trend is reversed at still lower values of R(unidentate/bidentate) (Fig. 5.10). The stabilization of the square pyramid relative to the trigonal bipyramid for R < 1.0 is similar to the behaviour observed for [M(unidentate A)-(unidentate B)$_4$] (Chap. 4).

b) Unsymmetrical Structures

The general stereochemistry for [M(bidentate)$_2$(unidentate)] with no assumed symmetry is shown in Fig. 5.11, and may be compared with Fig. 4.3 which defines [M(unidentate)$_5$]. The axes are again chosen so that $\phi_A = \phi_B = \phi_C$ and $\theta_B = 180°$, and the structure is completely defined by ϕ_A, θ_A, θ_C, ϕ_D, and ϕ_E, the remaining variables θ_D and θ_E being calculated from the normalized bite.

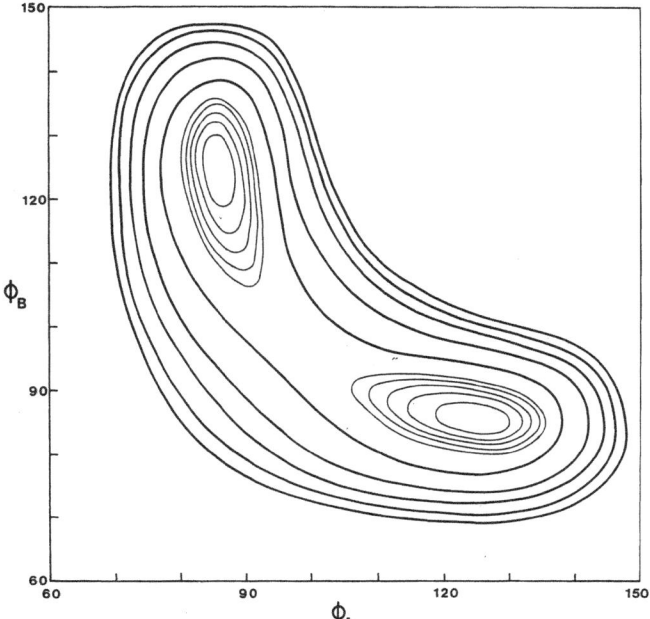

Fig. 5.9. Projection of the potential energy surface for [M(bidentate)₂(unidentate)] onto the $\phi_A - \phi_B$ plane (in degrees). The *five faint contour lines* are for successive 0.01 increments above the minima, and the five *heavy contour lines* are for succesive 0.1 increments above the minima. b = 1.5, n = 6

A typical potential energy surface, projected onto the $\phi_D - \phi_E$ plane, is shown in Fig. 5.12. The stereochemistries corresponding to the points marked A, B and C on Fig. 5.12 are shown in Fig. 5.13. The minimum at A at $\phi_D = 180 - \phi_E = 20.3°$ and $\phi_A = 90°$ is the same as that shown in Fig. 5.8. Movement along the line $\phi_D = 180° - \phi_E$ in Fig. 5.12 maintains $\phi_A = 90°$ and the twofold axis through B, and corresponds to movement along the reaction coordinate in Fig. 5.8.

The potential energy surface in Fig. 5.12 shows that significant distortion with loss of the twofold axis is possible, particularly in the direction toward the distorted square pyramid at B, in which the unidentate ligand now occupies a basal site.

This interchange between the square pyramid isomers A and B can alternatively by pictured as rotation of one of the bidentate ligands above the triangular plane formed by the unidentate ligand and the other bidentate ligand (Fig. 5.14).

The unsymmetrical square pyramid at B becomes increasingly possible as the effective bond length ratio R(unidentate/bidentate) is increased, a typical potential energy surface being shown in Fig. 5.15.

2. Comparison with Experiment

Compounds of the type [M(bidentate)₂(unidentate)] are given in Table 5.2. The four unidentate-metal-bidentate bond angles, that is, ϕ_A, ϕ_B, ϕ_C, and ϕ_D (Fig. 5.7) are listed, the lowest value being defined as ϕ_A.

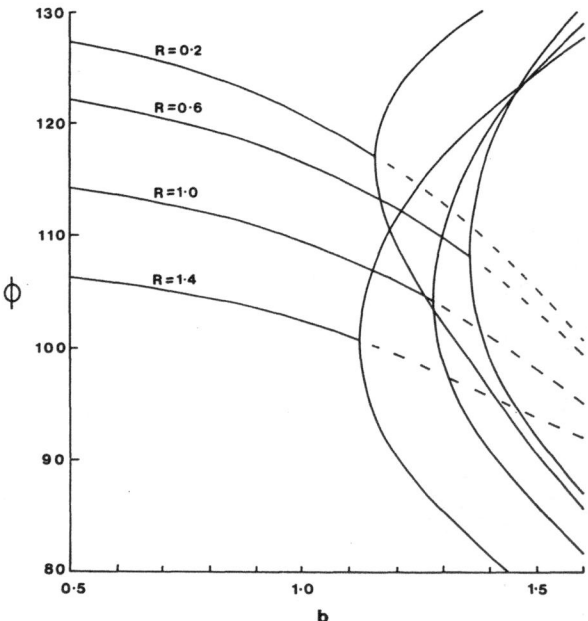

Fig. 5.10. Angular coordinates (degrees) for [M(bidentate)$_2$(unidentate)] as a function of normalized bite b and effective bond length ratio R(unidentate/bidentate). n = 6

The first group of compounds are those of phosphorus(V) and arsenic(V). All have a reasonable twofold axis, that is $\phi_A \sim \phi_C$ and $\phi_B \sim \phi_D$, and the stereochemistries range from near square pyramidal ($\phi_A = \phi_B$) to near trigonal bipyramidal ($\phi_A = 90, \phi_B = 120°$):

$$[P(O_2C_6Cl_{4/2})_2Ph]: \qquad \phi_{A,C} = 103.2°, \qquad \phi_{B,D} = 105.0°$$

$$[As(OCMe_2CMe_2O)_2Ph]: \quad \phi_{A,C} = 95.1°, \qquad \phi_{B,D} = 117.7°.$$

These bond angles can be fitted against those calculated as a function of normalized bite and effective bond length ratio to yield R(unidentate/bidentate) = 1.0 for [P(O$_2$C$_6$H$_4$)$_2$F] and R(unidentate/bidentate) $\sim$ 0.8 (n = 6) for the other complexes, as expected. Also as expected for the trigonal bipyramidal molecules, the

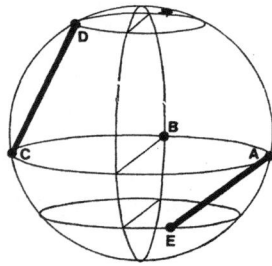

Fig. 5.11. General stereochemistry for [M(bidentate)$_2$(unidentate)] with no assumed symmetry

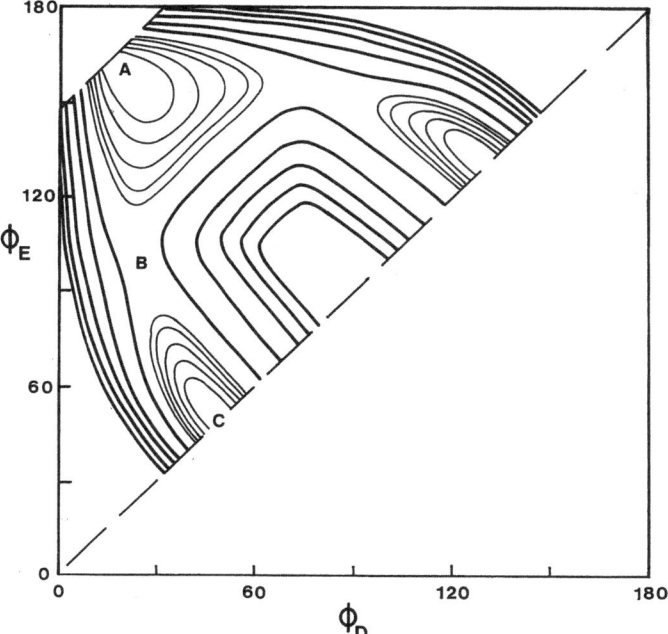

Fig. 5.12. Projection of the potential energy surface for [M(bidentate)$_2$(unidentate)] onto the ϕ_D — ϕ_E plane (in degrees). The five *faint contour lines* are for successive 0.01 increments above the minima, and the five *heavy contour lines* are for successive 0.1 increments above the minima. b = 1.2, n = 6.

apical P(As)—O$_{A,C}$ bonds are about 2% longer than the equatorial P(As)—O$_{B,D}$ bonds.

The next five compounds in Table 5.2 are transition metal complexes containing oxide or nitride as the unidentate ligand. All have close to square pyramidal stereochemistry as expected for O^{2-} and N^{3-} with short effective bond lengths, and the bond angles correspond to R(unidentate/bidentate) $\sim$ 0.8 (n = 6) (Fig. 5.10).

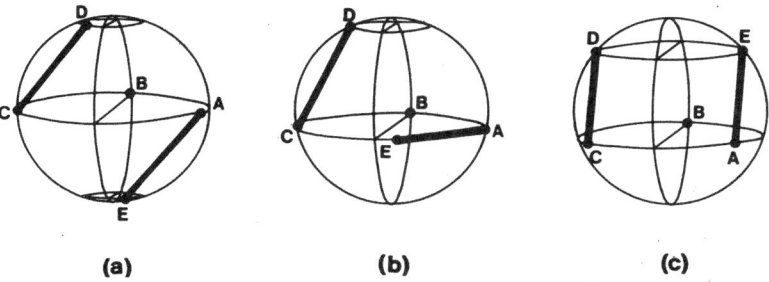

Fig. 5.13a—c. Stereochemistries corresponding to points marked on Fig. 5.12

Table 5.2. Stereochemical parameters for [M(bidentate)$_2$(unidentate)] complexes.

Complex	b	ϕ_A	ϕ_B	ϕ_C	ϕ_D	Ref.
[P(O$_2$C$_6$H$_4$)$_2$F]	1.43	95.8°	106.6°	96.0°	107.3°	[1108]
[P(O$_2$C$_6$H$_4$)$_2$Cl]	1.42	98.8°	105.3°	98.3°	105.0°	[166]
[P(O$_2$C$_6$H$_4$)$_2$Ph]	1.42	99.9°	106.1°	100.0°	108.7°	[174, 1105]
[P(O$_2$C$_6$H$_4$)$_2$Me]	1.42	101.2°	106.1°	102.2°	106.0°	[1104]
[P(O$_2$C$_6$H$_4$)$_2$(CMe$_3$)]	1.40	102.5°	106.4°	102.7°	106.6°	[1107]
[P(O$_2$C$_6$Cl$_4$)$_2$Ph]	1.42	103.1°	104.9°	103.4°	105.1°	[242]
[P(OCPhCPhO)$_2$Ph]	1.41	99.8°	105.9°	103.2°	105.6°	[241]
[As(O$_2$C$_6$H$_4$)$_2$Me]	1.40	100.2°	108.6°	101.1°	108.3°	[1106]
[As(OCH$_2$CH$_2$O)$_2$(OH)]	1.40	96.3°	109.1°	99.4°	110.5°	[489]
[As(OCMe$_2$CMe$_2$O)$_2$Ph]	1.39	94.5°	117.4°	95.7°	118.0°	[489]
[P(S$_2$C$_6$H$_4$)$_2$Me]		101.0°	108.3°			[392]
[V(MeCOCHCOMe)$_2$O]	1.38	104.8°	106.2°	105.6°	108.4°	[331, 571]
[Os(OCH$_2$CH$_2$O)$_2$O]	1.33	107.3°	112.8°	107.3°	112.8°	[870]
[V(S$_2$CNEt$_2$)$_2$O]	1.21	107.4°	108.3°	107.4°	110.2°	[555]
[Mo(S$_2$CNPr$_2$)$_2$O]	1.18	108.3°	109.6°	111.7°	110.3°	[926]
[Re(S$_2$CNEt$_2$)$_2$N]	1.19	106.8°	107.1°	107.9°	109.1°	[441]
[Fe{S$_2$C$_2$(CF$_3$)$_2$}$_2$(AsPh$_3$)]	1.41	92.3°	101.0°	98.2°	93.5°	[413]
[Fe(S$_2$C$_2$Ph$_2$)$_2${P(OMe)$_3$}]	1.40	90.3°	103.7°	106.9°	95.2°	[800]
[Fe(MeCOCHCOMe)$_2$Cl]		~ 105°	~ 105°	~ 105°	~ 105°	[725]
[Fe{C$_6$H$_4$(AsMe$_2$)$_2$}$_2$(NO)](ClO$_4$)$_2$	1.34	95.9°	98.8°	99.1°	97.1°	[405]
[Fe(S$_2$CNEt$_2$)$_2$Cl]	1.24	105.1°	105.5°	106.7°	105.4°	[581]
[Fe(S$_2$CNEt$_2$)$_2$Br]	1.23	104.7°	105.5°	105.7°	106.0°	[226]
[Fe(S$_2$CNEt$_2$)$_2$I]	1.23	99.4°	107.8°	103.1°	109.9°	[550]
[Fe(S$_2$CNC$_4$H$_8$)$_2$I]·^{1}/$_2$I$_2$	1.24	99.6°	102.9°	105.8°	106.8°	[642]
[Fe(S$_2$CNMe$_2$)$_2$(NO)]	1.23	103.1°	104.3°	108.3°	107.5°	[301]
[Fe(S$_2$CNPr$_2$)$_2$Cl]CHCl$_3$	1.23	104.2°	104.6°	109.5°	107.7°	[797]
[Fe(S$_2$CNEt$_2$)$_2$(NCS)]	1.24	102.7°	103.1°	105.6°	107.0°	[903]
[Fe(S$_2$CNEt$_2$)$_2$(NO)]	1.23	~ 106°	~ 106°	~ 106°	~ 106°	[253]
[Co{C$_6$H$_4$(AsMe$_2$)$_2$}$_2$(NO)](ClO$_4$)$_2$	1.36	90.3°	135.5°	92.1°	125.2°	[406]
[Co(Ph$_2$PCH$_2$CH$_2$PPh$_2$)$_2$Cl](SnCl$_3$)	1.32	90.0°	95.8°	94.5°	94.0°	[1006]
[Co(Ph$_2$PCH$_2$CH$_2$PPh$_2$)$_2$Cl](SnCl$_3$)PhCl	1.30	91.7°	128.1°	92.1°	126.1°	[1006]
[Co{NH(CH$_2$CH$_2$CH$_2$NH}$_2$Cl]Cl·EtOH						[1008]
[Co(S$_2$CNMe$_2$)$_2$(NO)]	1.24	102.2°	103.6°	103.5°	103.5°	[404]
[Co(S$_2$PPh$_2$)$_2$(NC$_9$H$_7$)]	1.32	85.2°	131.9°	85.2°	131.9°	[1068]
[Ni(S$_2$PEt$_2$)$_2$(NC$_9$H$_7$)]	1.32	99.7°	100.7°	98.9°	110.3°	[985]
[Cu(NH$_2$CH$_2$CH$_2$CH$_2$NH$_2$)$_2$(NCS)](ClO$_4$)	1.42	89.5°	116.8°	90.2°	109.0°	[213]
[Cu(bipy)$_2$(NO$_3$)](NO$_3$)H$_2$O	1.31	85.4°	91.9°	85.9°	127.7°	[826]
[Cu(bipy)$_2$I]I	1.31	89°	122°	92°	124°	[71]
[Cu(bipy)$_2$Cl]Cl·6H$_2$O	1.28	90.9°	118.6°	90.9°	118.7°	[1013]
[Cu(bipy)$_2$(NH$_3$)](BF$_4$)$_2$	1.28	91.6°	122.3°	92.7°	129.5°	[1012]
[Cu(bipy)$_2${SC(NH$_2$)$_2$}](ClO$_4$)	1.28	92.3°	124.4°	93.4°		[435]
[Cu(bipy)$_2$(H$_2$O)](S$_5$O$_6$)	1.29	89.0°	143.6°	91.4°	104.9°	[532]
[Cu(bipy)$_2$Cl]$_2$(S$_5$O$_6$)6H$_2$O	1.28	92.0°	130.7°	93.3°	122.0°	[532]
[Cu{(C$_5$H$_4$N)$_2$NH}$_2$I]I(ClO$_4$)	1.41	87.3°	124.7°	87.8°	135.5°	[621]
	1.36	88.0°	137.9°	88.4°	119.8°	[621]
[Cu(phen)$_2$(H$_2$O)](NO$_3$)$_2$	1.33	85.5°	110.0°	85.5°	110.0°	[824]
[Cu(phen)$_2$(H$_2$O)](BF$_4$)$_2$	1.32	86.4°	111.7°	86.4°	111.7°	[825]
[Cu(phen)$_2$I]I·S$_8$	1.29	92.3°	125.3°	92.3°	125.3°	[521]
[Cu(phen)$_2$(CN)](NO$_3$)H$_2$O	1.29	91.8°	129.0°	93.9°	132.4°	[44]
[Zn(MeCOCHCOMe)$_2$(H$_2$O)]	1.39	97.5°	104.9°	100.3°	104.9°	[802]
[Zn(S$_2$CNEt$_2$)$_2$(C$_5$H$_5$N)]·^{1}/$_2$C$_6$H$_6$	1.19	93.9°	116.8°	95.9°	116.1°	[451]

Complex	b	ϕ_A	ϕ_B	ϕ_C	ϕ_D	Ref.
[Zn(S₂COEt)₂(C₅H₅N)]	1.17	103.8°	115.4°	103.8°	115.4°	[916]
[Mo(Ph₂PCH₂CH₂PPh₂)₂(CO)]	1.28	90.4°	92.1°	95.0°	95.0°	[960]
[Ru(Ph₂PCH₂CH₂PPh₂)₂(NO)](Ph₄B)Me₂CO	1.30	90.1°	126.1°	93.1°	134.5°	[878]
[Ir(Ph₂PCH₂CH₂PPh₂)₂(CNMe)](ClO₄)	1.34	87.2°	106.9°	89.3°	132.7°	[484]
[Ir(Ph₂PCH₂CH₂PPh₂)₂(CO)]Cl	1.33	91°	109°	91°	143°	[613]
(Et₄N)[Cd(S₂COEt)₂(SCSOEt)]	1.09	111.8°	111.9°	119.1°	112.7°	[578]

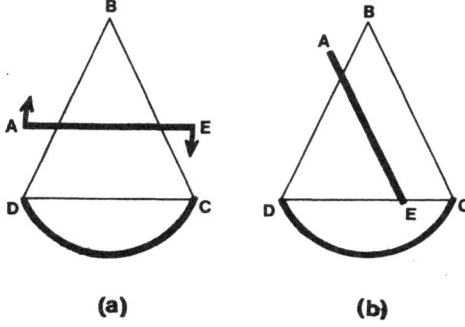

(a)   **(b)**

Fig. 5.14 a, b. Alternative view of stereochemistries corresponding to points marked on Fig. 5.12

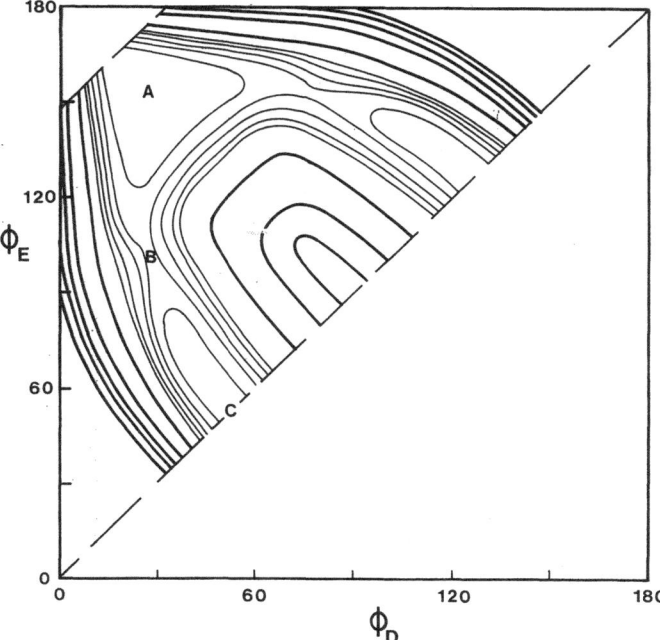

Fig. 5.15. Projection of the potential energy surface for [M(bidentate)₂(unidentate)] onto the $\phi_D - \phi_E$ plane (in degrees). The five *faint contour lines* are for successive 0.01 increments above the minima, and the five *heavy contour lines* are for successive 0.1 increments above the minima. R = 1.4, b = 1.2, n = 6

The stereochemistries of many of the remaining transition metal complexes are more complicated, and are similar to many of the [M(unidentate)$_5$] complexes which were labelled "type β" (Chap. 4). A number show marked distortions from twofold symmetry, that is $(\phi_A - \phi_C) \neq (\phi_B - \phi_D)$. The compounds showing the most extreme departure from twofold symmetry are listed below:

	$\phi_A - \phi_C$	$\phi_B - \phi_D$
[Fe(S$_2$C$_2$Ph$_2$)$_2${P(OMe)$_3$}]	−17°	8°
[Cu(bipy)$_2$(NO$_3$)](NO$_3$)H$_2$O	0°	−36°
[Cu(bipy)$_2$(H$_2$O)](S$_5$O$_6$)	− 2°	39°
[Ir(Ph$_2$PCH$_2$CH$_2$PPh$_2$)$_2$(CNMe)](ClO$_4$)	− 2°	−26°
[Ir(Ph$_2$PCH$_2$CH$_2$PPh$_2$)$_2$(CO)]Cl	0°	−34°

This distortion is toward the unsymmetrical square pyramid, with the unidentate ligand in a basal site (Figs. 5.13—5.15).

The stereochemistries of compounds containing two unsymmetrical bidentate ligands are consistent with the general principles described above. For example, in the following compounds of the p-block elements, it is the more electronegative ends of the bidentate ligands that are found in the more hindered axial sites of a trigonal bipyramid:

[P(OCHPhCHMeNMe)$_2$H] [833]   [P(OC$_6$H$_4$NH)$_2$H]   [786]

[P(OOCC$_6$H$_4$)$_2$Ph] [1061]   [S{OC(CF$_3$)$_2$C$_6$H$_3$Bu}$_2$O] [863]

[P(OCH$_2$CH$_2$NH)$_2$H] [786]   [Ga(OC$_9$H$_5$MeN)$_2$Cl] [374]

Table 5.3. Stereochemical parameters for [M(bidentate)$_2$(lone pair)] complexes

Complex	b	ϕ_A	ϕ_B	$\dfrac{M-A}{M-B}$	Ref.
Symmetric Bidentate Ligands					
[Te(O$_2$C$_6$H$_4$)$_2$(lone pair)]	1.30	102.9°	130.9°	1.04	[726]
[Sb(S$_2$CNBu$_2$)$_2$(lone pair)]	1.15	107.5°	132.7°	1.07	[1065]
[Sn(S$_2$CNEt$_2$)$_2$(lone pair)]	1.10	110.2°	131.9°	1.08	[579, 891]
[Sn(S$_2$COMe)$_2$(lone pair)]	1.09	109.6°	130.4°	1.07	[420]
Na$_2$[Sn(C$_2$O$_4$)$_2$(lone pair)]	1.17	107.3°	140.7°	1.05	[333]
[Sn(BuCOCHCOBu)$_2$(lone pair)]	1.26	106.6°	137.1°	1.07	[1058]
[Sn(PhCOCHCOPh)$_2$(lone pair)]	1.30	104.8°	132.7°	1.07	[419]
[Pb(S$_2$COBu)$_2$(lone pair)]	1.0	109°	134°	1.04	[512]
[Pb(S$_2$COEt)$_2$(lone pair)]	1.02	111.4°	130.9°	1.05	[513]
[Pb(S$_2$CNEt$_2$)$_2$(lone pair)]	1.04	113.4°	131.9°	1.05	[609]
[Pb(S$_2$CNPr$_2$)$_2$(lone pair)]	1.06	110.7°	131.1°	1.06	[607]
[Pb{S$_2$P(OEt)$_2$}$_2$(lone pair)]	1.16	110.9°	133.0°	1.09	[608]
Unsymmetric Bidentate Ligands					
[S(C$_6$H$_4$COO)$_2$(lone pair)]	1.40	90.8°	126.1°		[629]
[S{C$_6$H$_3$Bu·C(CF$_3$)$_2$O}$_2$(lone pair)]	1.37	91.2°	126.1°		[863]
[Se(C$_6$H$_4$COO)$_2$(lone pair)]	1.34	93.8°	129.5°		[291]
[Se(CH$_2$CH$_2$COO)$_2$(lone pair)]	1.37	93.8°	129.3°		[292]
[Te(SC$_6$H$_4$O)$_2$(lone pair)]	1.32	99.9°	129.1°		[1075]

A number of compounds of the p-block elements can be considered to have a trigonal bipyramidal arrangement of electron pairs, with a nonbonding electron pair in one of the equatorial sites (Table 5.3). The high values of ϕ_A and ϕ_B, compared with 90.0 and 120.0° for a regular trigonal bipyramid, imply R(lone pair/bidentate) values of 0.6 for sulfur(IV) compounds, decreasing to ~ 0.4 for selenium(IV) and ~ 0.2 for tellurium(IV), antimony(III), tin(II), and lead(II).

As expected from the calculated potential energy surfaces the barriers to intramolecular rearrangement are generally higher for [M(bidentate)$_2$(unidentate)] complexes, than for [M(unidentate)$_5$] complexes [488, 570, 792, 863].

C. [M(tridentate)(unidentate)$_2$]

1. The Theoretical Stereochemistries

The locations of the symmetric tridentate ligand ABC, the unidentate ligands D and E, and the axes are defined by specifying that $\phi_A = \phi_B = \phi_C$, $\theta_B = 180.0°$, and $\theta_C = -\theta_A$ (Fig. 5.16). The general structure is similar to that of [M(unidentate)$_5$] containing a mirror plane (Fig. 4.3). The geometry of the tridentate ligand is given by two variables, the normalized bite b of each chelate ring, and the tridentate angle ABC which is equal to θ_A.

Representative potential energy surfaces projected onto the $\phi_D - \phi_E$ plane, calculated assuming a flexible tridentate ligand, are shown in Figs. 5.17 and 5.18. They should be compared with Fig. 4.4, calculated for [M(unidentate)$_5$].

At low values of the normalized bite (Fig. 5.17), the only minimum is at the center of the potential energy surface. This stereochemistry always contains two mirror planes, through MABC and MBDE, but there is a smooth and continuous transformation from one coordination polyhedron into another as the normalized bite is increased (Fig. 5.19).

A second minimum develops at higher normalized bites (b > 1.32 for n = 6) (Fig. 5.18). At b = $2^{1/2}$, a trigonal bipyramid is attained with the tridentate ligand spanning the equatorial-axial-equatorial sites (Fig. 5.20).

An important point to note from the potential energy surfaces is that the long trough is almost at 45° to the ϕ_D and ϕ_E axes. That is, for b < ~ 1.4:

$$\phi_D - \phi_E = DME \sim \text{constant} \sim 100°.$$

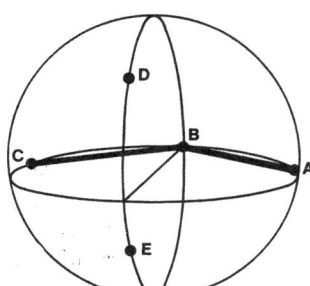

Fig. 5.16. General stereochemistry for [M(tridentate)(unidentate)$_2$]

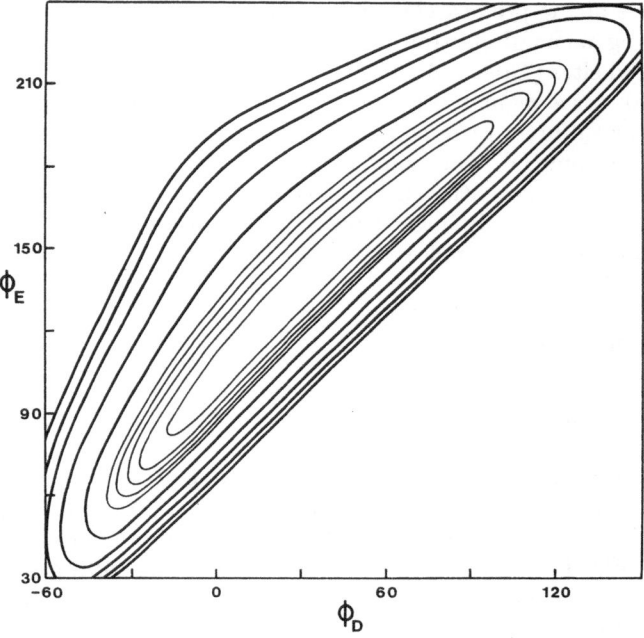

Fig. 5.17. Projection of the potential energy surface for [M(flexible tridentate)(unidentate)$_2$] onto the $\phi_D - \phi_E$ plane (in degrees). The five *faint contour lines* are for successive 0.01 increments above the minimum, and the five *heavy contour lines* are for successive 0.1 increments above the minimum. b = 1.2, n = 6

The stereochemical changes that occur on movement along this trough are shown in Fig. 5.21. These changes involve changes in ϕ_A, and hence θ_A, which is equal to the tridentate angle ABC. No tridentate ligand can be completely regarded as being freely hinged at atom B, and according to the design of the particular ligand some preferred value of ABC, and hence θ_A and ϕ_A, will be favored. This is expected to largely determine the position of the molecule along the trough, the stereochemistry at the center of the trough having higher values of ABC and θ_A.

Imposing rigidity on the tridentate ligand, conversely, fixes ϕ_A and allows only a much more restricted variation in ϕ_D and ϕ_E. This is shown by the typical potential energy surface in Fig. 5.22.

2. Comparison with Experiment

Molecules containing a symmetrical tridentate ligand and two identical unidentate ligands are listed in Table 5.4. The variation of ϕ_D and ϕ_E is also displayed in Fig. 5.23, and should be compared with the calculated potential energy surfaces.

The complexes are divided into two groups. In the first group, denoted by open circles in Fig. 5.23, the angle between the two metal-unidentate ligand bonds, DME $= \phi_E - \phi_D \sim 100°$ as predicted. In these cases the tridentate ligands are based on flexible aliphatic chains with ABC $= \theta_A = 84—97°$, and are of the type RA(CH$_2$CH$_2$AR$_2$)$_2$ (where A is N or P) or X(CH$_2$CH$_2$X$^-$)$_2$ (where X is O or S).

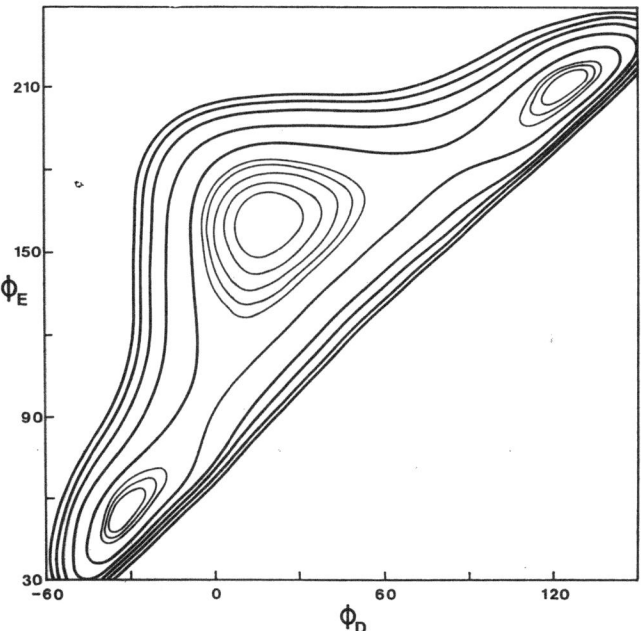

Fig. 5.18. Projection of the potential energy surface for [M(flexible tridentate)(unidentate)$_2$] onto the $\phi_D - \phi_E$ plane (in degrees). The five *faint contour lines* are for successive 0.01 increments above the minimum, and the five *heavy contour lines* are for successive 0.1 increments above the minimum. b = 1.5, n = 6

The second group of complexes, denoted as filled circles in Fig. 5.23, contain much more rigid tridentate ligands in which the ligand design enforces a larger tridentate angle ABC, and hence stereochemistries closer to the center of the potential energy surface. These complexes have significantly greater angles between the two metal-unidentate ligand bonds, $\phi_D - \phi_E = $ DME, which can be attributed to two effects. Firstly, bulky unidentate ligands will increase this angle, the most extreme example being [Cd(terpy){Mn(CO)$_5$}$_2$]. Secondly, the presence of bulky substituents in the metal-tridentate ligand plane forces the unidentate ligands apart, as in [Ni{HN(CH$_2$C$_5$H$_3$MeN)$_2$}Br$_2$] and [Cd{NC$_5$H$_3$

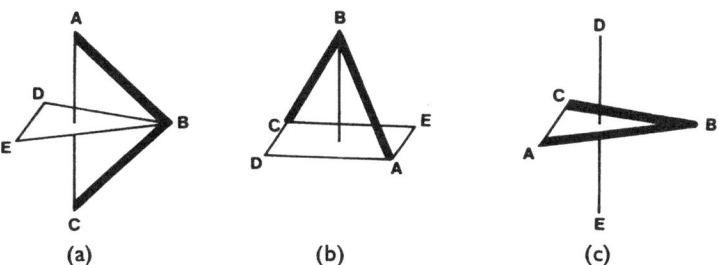

Fig. 5.19a—c. Stereochemistries of [M(tridentate)(unidentate)$_2$].
a b = 1.414;   **b** b = 1.55;   **c** b = 1.732

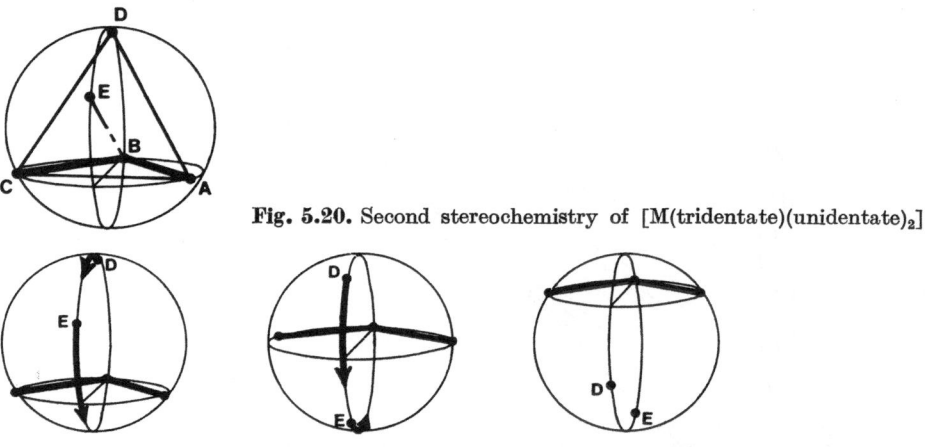

Fig. 5.20. Second stereochemistry of [M(tridentate)(unidentate)$_2$]

Fig. 5.21. Interconversion of stereochemistries of [M(flexible tridentate) (unidentate)$_2$]

· (CMe : NC$_6$H$_4$SMe)$_2$}I$_2$] (Fig. 5.24). A similar but smaller effect attributable to the hydrogen atoms is present in the terpyridyl complexes, $\phi_D - \phi_E = DME \sim 110°$. In contrast, the tridentate ligands in the first group of compounds have no additional groups attached to the donor atoms, or have tetrahedrally coordinated donor atoms with no in-plane interactions of this type.

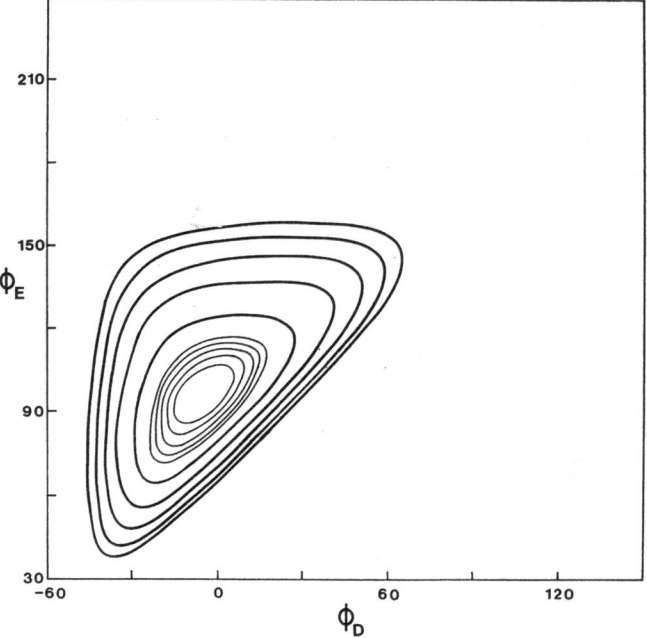

Fig. 5.22. Projection of the potential energy surface for [M(rigid tridentate)(unidentate)$_2$] onto the $\phi_D - \phi_E$ plane (in degrees). The five *faint contour lines* are for successive 0.01 increments above the minimum, and the five *heavy contour lines* are for successive 0.1 increments above the minimum. ABC = 100°, $\phi_A = 111.0°$, b = 1.2, n = 6

Table 5.4. Stereochemical parameters for [M(tridentate)(unidentate)$_2$] complexes

Complex	b	ABC	ϕ_A	ϕ_D	ϕ_E	DME	Ref.
[Co{(C$_3$H$_7$S)N(CH$_2$CH$_2$NEt$_2$)$_2$}(NCS)$_2$]	1.22	86°	118°	−27°	72°	99°	[296]
[Co{MeN(CH$_2$CH$_2$NMe$_2$)$_2$}Cl$_2$]	1.28	93°	112°	−18°	86°	104°	[328]
[Co{HN(CH$_2$CH$_2$NEt$_2$)$_2$}Cl$_2$]	1.29	84°	119°	−36°	67°	102°	[336]
[Ni{(C$_{14}$H$_{14}$As)N(CH$_2$CH$_2$NEt$_2$)$_2$}(NCS)$_2$]	1.32	96°	99°	− 1°	100°	100°	[327]
[Cd{MeN(CH$_2$CH$_2$NMe$_2$)$_2$}(NCS)$_2$]	1.25	97°	110°	− 7°	97°	104°	[212]
[Ge{S(CH$_2$CH$_2$S)$_2$}Cl$_2$]	1.29	85°	119°	−38°	60°	98°	[337]
[Ge{O(CH$_2$CH$_2$S)$_2$}Cl$_2$]	1.30	88°	115°	−29°	74°	102°	[338]
[Sn{S(CH$_2$CH$_2$S)$_2$}Cl$_2$]^a	1.33	84°	116°	−31°	65°	96°	[339]
[Sn{O(CH$_2$CH$_2$S)$_2$}Cl$_2$]	1.29	93°	111°	−22°	78°	100°	[340]
[Cd(terpy){Mn(CO)$_5$}$_2$]	1.10	114°	90°	22°	155°	132°	[246]
[Zn(terpy)Cl$_2$]	1.20	106°	98°	7°	119°	112°	[260, 387]
[Co(terpy)Cl$_2$]	1.22	104°	99°	0°	111°	111°	[487]
[Co(terpy)(NCO)$_2$]	1.23	104°	90°	35°	145°	111°	[640]
[Ni{HN(CH$_2$C$_5$H$_3$MeN)$_2$}Br$_2$]	1.31	98°	93°	6°	155°	149°	[942]
[Cd{NC$_5$H$_3$(CMe:NC$_6$H$_4$SMe)$_2$}I$_2$]	1.11	112°	90°	23°	143°	119°	[347]
Cs[V{NC$_5$H$_3$(CO$_2$)$_2$}O$_2$]H$_2$O	1.22	104°	95°	27°	136°	110°	[840]

^{a} Angles obtained from atomic coordinates, rather than those quoted in reference.

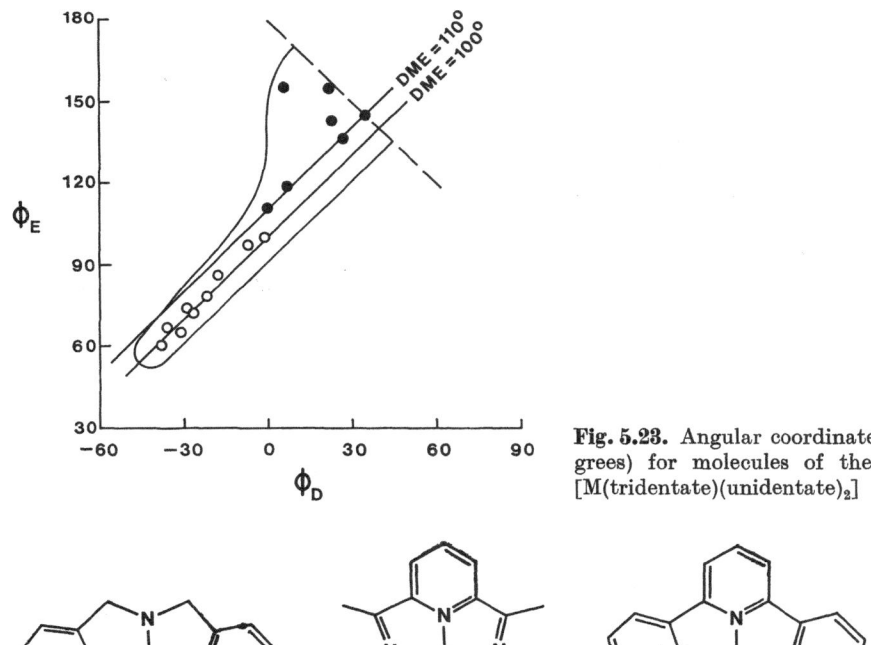

Fig. 5.23. Angular coordinates (degrees) for molecules of the type [M(tridentate)(unidentate)$_2$]

Fig. 5.24. Chelate geometry in [Ni{HN(CH$_2$C$_5$H$_3$MeN)$_2$}Br$_2$], [Cd{NC$_5$H$_3$(CMe:NC$_6$H$_4$SMe)$_2$}I$_2$] and [M(terpyridyl)L$_2$]

Six-Coordinate Compounds Containing only Unidentate Ligands

A. [M(unidentate A)(unidentate B)₅]

The general stereochemistry for octahedral molecules containing one ligand different to the other five is shown in Fig. 6.1. The effective bond length to the unique ligand A lying on the fourfold axis is R, the other five metal-ligand bond lengths being defined as unity. The repulsion energy calculations confirm that as R is decreased from 1.0, AMB increases as expected (Fig. 6.2).

As R is decreased from 1.0, the increase in AMB and decrease in BMF increases the repulsion experienced by atom F relative to atom B (Fig. 6.3). It is therefore expected that the bond *trans-* to the unique ligand will be longer than the four *cis-*

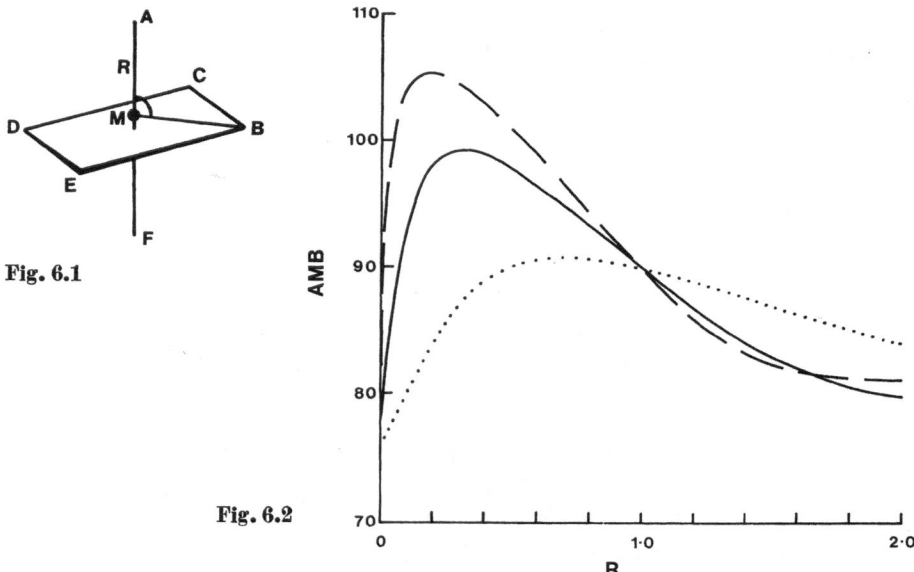

Fig. 6.1

Fig. 6.2

Fig. 6.1. General stereochemistry for [M(unidentate A)(unidentate B)₅]

Fig. 6.2. Bond angles (degrees) for [M(unidentate A)(unidentate B)₅], as a function of effective bond length ratio R. *Dotted line*, n = 1; *full line*, n = 6; *broken line*, n = 12

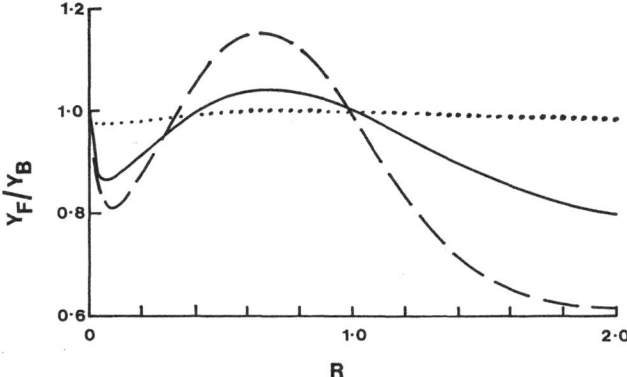

Fig. 6.3. The *trans*-influence, Y_F/Y_B, as a function of R for [M(unidentate A)(unidentate B)$_5$]. *Dotted line*, n = 1; *full line*, n = 6; *broken line*, n = 12

bonds. However, it is important to note that at much lower values, R < 0.4, the opposite *trans*-effect is expected, Y_F/Y_B < 1.0 (Fig. 6.3).

Experimental bond angles AMB, and the length of the *trans*-metal-ligand bond relative to the length of the *cis*-metal-ligand bonds, (M—F)/(M—B), for a selection of compounds are collected in Table 6.1. The experimental bond angles for any molecule can be fitted against the calculated bond angles (Fig. 6.2), and a value of R(i/j) obtained for any pair of ligands i and j (Table 6.1). As was observed in Chap. 3 for tetrahedral complexes [M(unidentate A)(unidentate B)$_3$], it is again important to note that for reasonable distortions from a regular polyhedron, that is for AMB > 91°, it is not possible to obtain a fit between the experimental data and calculations which are based on using n = 1 in the repulsion law.

For nitrido complexes [MNX$_5$]$^{x-}$ and oxo complexes [MOX$_5$]$^{x-}$, the NMX and OMX bond angles of about 95° correspond to R(N^{3-}/X$^-$) and R(O^{2-}/X$^-$) values of about 0.7 (n = 6). As expected, the nitrido complexes appear more distorted with lower R values than the oxo complexes, and the fluoro complexes are similarly more distorted than the chloro or thiocyanato complexes. The two oxo actinoid complexes, (Et$_4$N)$_2$[PaOCl$_5$] and (Ph$_4$P)[UOCl$_5$], appear exceptional with OMCl angles closer to 90°.

In these nitrido and oxo complexes the predicted large weakening of the bond *trans*- to the N^{3-} or O^{2-} group is observed, this bond being about 10% longer than the four *cis*-metal-ligand bonds (Table 6.1).

Electron diffraction and microwave studies [74] on [IOF$_5$] yield structural parameters similar to those obtained for the above transition metal complexes:

$$\text{OIF} = 98.0°, \ (\text{I—F})_{trans}/(\text{I—F})_{cis} = 1.03, \ \text{R(O/F)} = 0.50 \ (n = 6).$$

The nitroprusside ion [Fe(CN)$_5$(NO)]$^{2-}$ has ON—Fe—CN angles of 95°, corresponding to R(NO$^+$/CN$^-$) = 0.7 (n = 6) (Table 6.1). This low R value is typical for the nitrosyl ligand.

Table 6.1. Stereochemical parameters for octahedral [M(unidentate A)(unidentate B)$_5$] complexes. R(A/B) values obtained for n = 6

Complex	AMB	$\dfrac{M-F}{M-B}$	R(A/B)	Ref.
K$_2$[OsNCl$_5$]	96.2°	1.10	0.63	[150]
(Ph$_4$As)$_2$[ReN(NCS)$_5$]	96.0°	1.14	0.64	[216]
(C$_2$H$_{10}$N$_2$)[VOF$_5$]	98.2°	1.17	0.48	[933]
(N$_2$H$_6$)[NbOF$_5$]H$_2$O	98.0°	1.14	0.50	[958]
K$_2$[NbOF$_5$]	94.5°	1.12	0.74	[884]
(Ph$_4$As)$_2$[NbO(NCS)$_5$]	96°	1.09	0.64	[631]
K$_2$[MoOCl$_5$]	95.6°	1.08	0.67	[1049]
(Et$_4$N)$_2$[PaOCl$_5$]	91.6°	0.92	0.91	[159]
(Ph$_4$P)[UOCl$_5$]	89.7°	0.96	1.02	[321]
Na$_2$[Fe(CN)$_5$(NO)]2 H$_2$O	95.4°	0.99	0.69	[140]
Ba[Fe(CN)$_5$(NO)]2 H$_2$O	96°	1.00	0.64	[690]
Sr[Fe(CN)$_5$(NO)]4 H$_2$O	94.6°	1.02	0.73	[221]
(Ph$_4$As)$_2$[Fe(CN)$_5$(NO)]	94°	1.03	0.76	[222]
[Co(NHCOMe)(NH$_3$)$_5$](ClO$_4$)$_2$	89.6°	1.02	1.03	[971]
[Co(SCN)(NH$_3$)$_5$]Cl$_2$·H$_2$O	89.8°	0.99	1.01	[999]
[Co(S$_2$O$_3$)(NH$_3$)$_5$]Cl, H$_2$O	90.3°	1.01	0.98	[924]
[Co{S(O)$_2$Ph}(NH$_3$)$_5$][SnCl$_3$(ClO$_4$)]	90.6°	1.03	0.97	[395]
[Co(SO$_3$)(NH$_3$)$_5$]Cl, H$_2$O	90.0°	1.05	1.00	[396]
[Co{S(O)$_2$C$_6$H$_4$Me}(NH$_3$)$_5$](ClO$_4$)$_2$·H$_2$O	90.2°	1.03	0.99	[396]
[As{N(Me)S(O)F$_2$}F$_5$]	87.7°	0.99	1.14	[91]
[Cr(NH$_3$)$_6$][Mn(H$_2$O)Cl$_5$]	87.6°	0.98	1.15	[244]
(NH$_4$)$_2$[Fe(H$_2$O)Cl$_5$]	87.6°	0.98	1.15	[437]
[U(Ph$_3$PO)Cl$_5$]	89.0°	0.99	1.06	[129]

Much smaller distortions are expected for [M(X$^-$)(uncharged ligand)$_5$]. For example the cobalt(III) pentaammine complexes listed in Table 6.1 have X—Co—NH$_3$ angles of 90.2° and R(X/NH$_3$) = 0.99 (n = 6), with only a small *trans*-effect, (Co—NH$_3$)$_{trans}$/(Co—NH$_3$)$_{cis}$ = 1.02.

As expected, AMB angles of less than 90° are observed for [M(uncharged ligand)X$_5$] (X = F, Cl) (Table 6.1). Angles of 88° correspond to R(ligand/X$^-$) = 1.1 (n = 6). As predicted, the *trans*-M—X bond is now shorter than the four *cis*-M—X bonds, (M—X)$_{trans}$/(M—X)$_{cis}$ = 0.98.

A particularly important type of complex is [M(lone pair)(unidentate)$_5$]$^{x\pm}$. The series [SbX$_5$(lone pair)]$^{2-}$ (X = F [952], Cl [1082, 1102], Br [215]), [TeF$_5$(lone pair)]$^-$ [626, 769], [IF$_5$(lone pair)] [182, 623], and [XeF$_5$(lone pair)]$^+$ [77, 78, 79 701] have the expected square pyramidal structure, with a nonbonding pair of electrons presumed to occupy the sixth octahedral site. The structures are grossly distorted, the (lone pair)-M—F angles of ~100° corresponding to R(:/F) ~ 0.3 (n = 6). It is important to note that, in contrast to the nitrido and oxo complexes above, the *trans*-M—F bond is now *shorter* than the *cis*-M—F bond, (M—F)/(M—B) = 0.96, and that this behavior is in complete agreement with the predictions which were made above. Electron diffraction and microwave data on IF$_5$ [553] and BrF$_5$ [552] are in accord with the X-ray results.

B. [M(unidentate A)$_2$(unidentate B)$_4$]

1. The Theoretical Stereochemistries

The general stereochemistry for complexes of the type [M(unidentate A)$_2$ (unidentate B)$_4$] [154] is shown in Fig. 6.4. The M—A, M—B, M—C, and M—D effective bond lengths are defined as unity, and the M—E and M—F effective bond lengths are defined as R, the axes being defined so that $\phi_E = \phi_F$, $\theta_E = 90°$, and $\theta_F = -90°$.

The potential energy surface is shown projected onto the $\theta_A - \theta_B$ plane in Fig. 6.5, calculated for six equal metal-ligand bonds, R = 1.0. The deep minima at $\theta_A = 0$, $\theta_B = -90°$, and $\theta_A = 90$, $\theta_B = 0°$, correspond to the *cis*-octahedron whereas the equally deep minimum at $\theta_A = 0$, $\theta_B = 0°$, corresponds to the *trans*-octahedron. These three minima are connected via high saddles corresponding to two of the three possible disubstituted trigonal prisms.

The path between the *cis*-octahedron at $\theta_A = 0°$, $\theta_B = -90°$, and the *trans*-octahedron at $\theta_A = 0$, $\theta_B = 0°$, is shown in Fig. 6.6, and corresponds to trigonal twisting of the ABF face relative to the CDE face. The alternative reaction path, between the *cis*-octahedron at $\theta_A = 0°$, $\theta_B = -90°$, and the alternative *cis*-octahedron at $\theta_A = 90°$, $\theta_B = 0°$, is illustrated in Fig. 6.7, and corresponds to trigonal twisting of the ACE face relative to the BDF face. This second pathway interchanges the AD and BC atoms, while retaining the *cis*-arrangement of E and F.

Reducing R to 0.7 results in the *trans*-octahedral structure existing as an even deeper minimum (Fig. 6.8). Conversely increasing R to 1.5 stabilises the *cis*-structure, and in addition results in a much flatter potential energy surface (Fig. 6.9). It is now important to note that a *cis*-to-*trans* interconversion, or a *cis*-to-*trans*-to-*cis* interconversion, is possible, but not a direct *cis*-to-*cis* interconversion.

2. Comparison with Experiment

For R ~ 0.8 —1.4, the *cis*- and *trans*-structures are of comparable stability, and both structures would be expected. Indeed the ability to form both these octahedral isomers has been recognized since the beginning of coordination chemistry [1084].

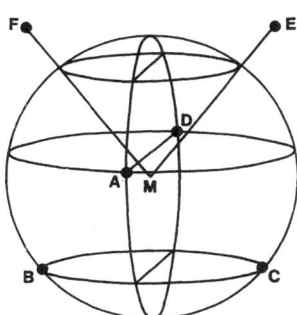

Fig. 6.4. General stereochemistry for [M(unidentate A)$_2$(unidentate B)$_4$]

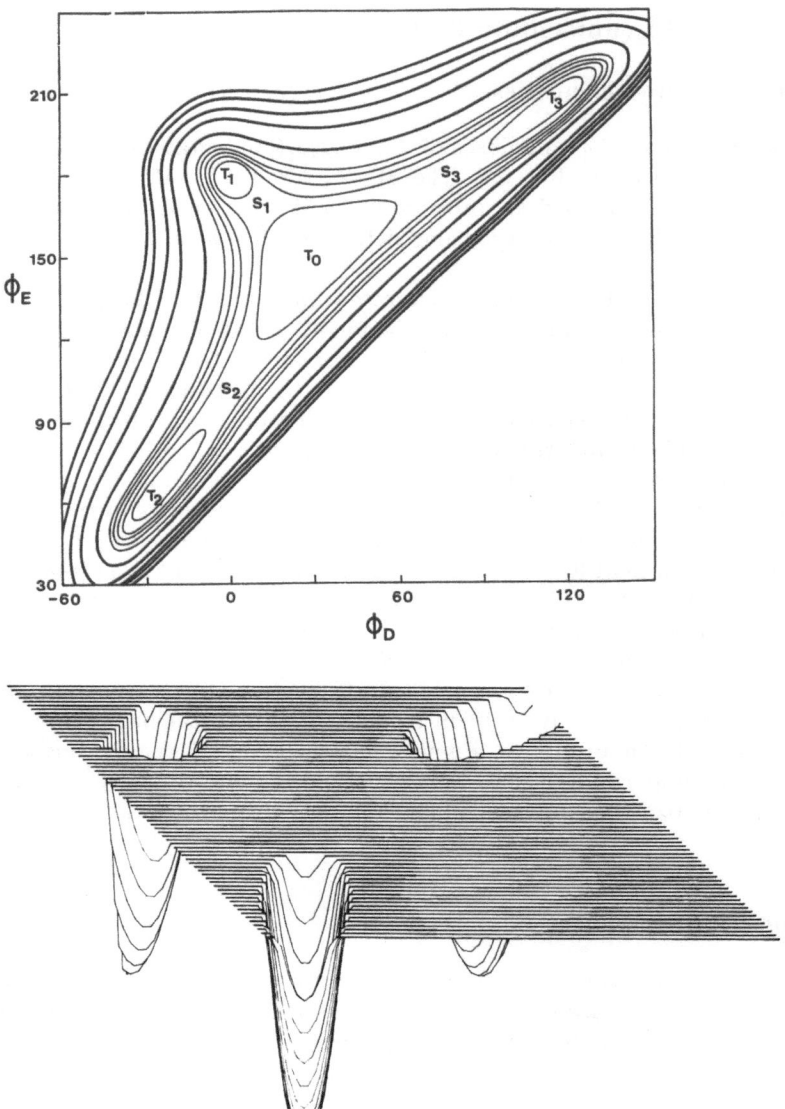

Fig. 6.5. a Projection of the potential energy surface for [M(unidentate A)$_2$(unidentate B)$_4$] onto the $\theta_A - \theta_B$ plane (in degrees). The five *faint contour lines* are for successive 0.02 increments above the minima, and the five *heavy contour lines* are for successive 0.2 increments above the minima, at C and T. R = 1.0, n = 6. The positions of the *cis*- (C) and *trans*-isomers (T) are shown. **b** As in **a**, but with truncation at X = 0.1

The *trans*-structure is expected for R < ∼ 0.7 (Fig. 6.8), the two important experimental cases being [M(lone pair)$_2$(unidentate)$_4$] and [MO$_2$(unidentate)$_4$]. Compounds containing two nonbonding pairs of electrons and four unidentate

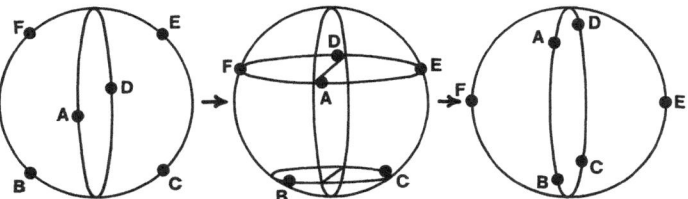

Fig. 6.6 Conversion of *cis*-[M(unidentate A)₂(unidentate B)₄] to the *trans*-isomer

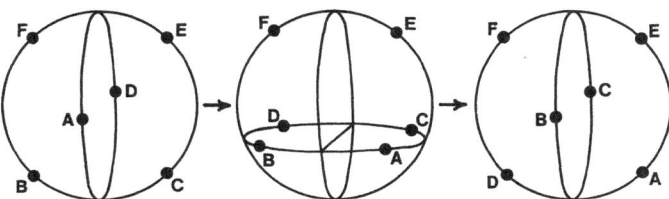

Fig. 6.7. Rearrangement of *cis*-[M(unidentate A)₂(unidentate B)₄]

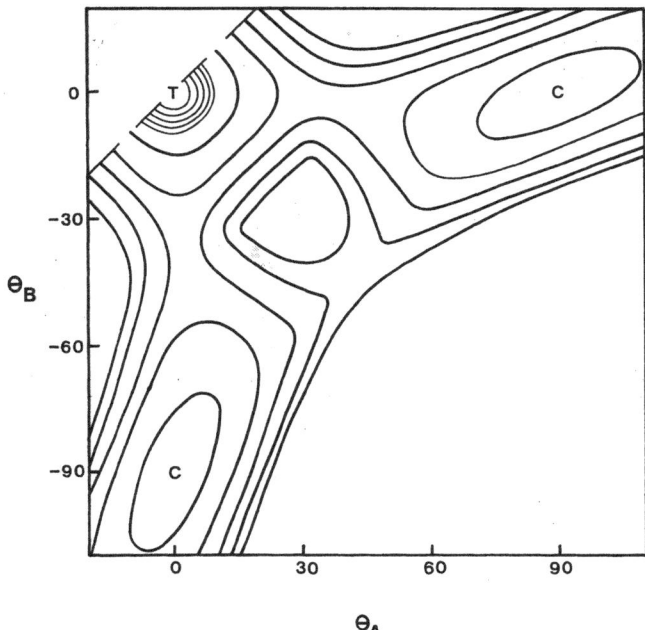

Fig. 6.8. Projection of the potential energy surface for [M(unidentate A)₂(unidentate B)₄] onto the $\theta_A - \theta_B$ plane (in degrees). The five *faint contour lines* are for successive 0.02 increments above the minimum, and the five *heavy contour lines* are for successive 0.2 increments above the minimum, at T. R = 0.7, n = 6. The position of the *cis*- (C) and *trans*-isomers (T) are shown

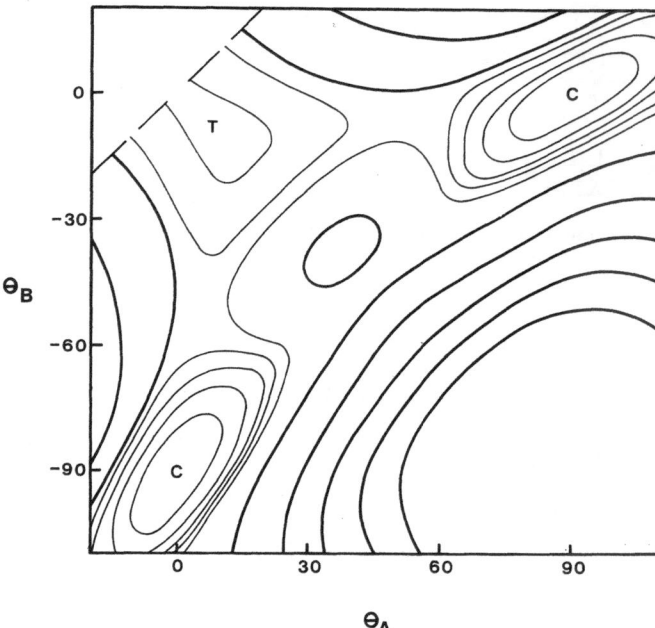

Fig. 6.9. Projection of the potential energy surface for [M(unidentate A)$_2$(unidentate B)$_4$] onto the $\theta_A - \theta_B$ plane (in degrees). The five *faint contour lines* are for successive 0.02 increments above the minima, and the five *heavy contour lines* are for successive 0.2 increments above the minima, at C. R = 1.5, n = 6. The positions of the *cis*- (C) and *trans*-isomers (T) are shown

ligands are invariably observed to be square planar, with a presumed *trans*-arrangement of the nonbonding electron pairs completing the octahedron. Examples include [Te^{II}(ligand)$_4$(lone pair)$_2$]$^{2+}$ [61, 398, 538, 675], [I^{III}Cl$_4$(lone pair)$_2$]$^-$ [381, 400], and [Xe^{IV}F$_4$(lone pair)$_2$] [186, 188, 604, 1041]. Uranyl complexes [UO$_2$X$_4$]$^{2-}$ have only been observed as the *trans*-isomer [123, 326, 497], other examples of *trans*-dioxo complexes being [ReO$_2$(C$_5$H$_5$N)$_4$]Cl·2H$_2$O [205, 736], K$_3$[ReO$_2$(CN)$_4$] [432, 823], and NaK$_3$[MoO$_2$(CN)$_4$]6H$_2$O [306]. However complexes with more complex stoichiometry, that is containing two different unidentate ligands in addition to O^{2-}, are known as both *cis*- and *trans*-isomers.

Experimental bond angles for *cis*-[M(unidentate A)$_2$(unidentate B)$_4$] can be fitted against calculated values and R(i/j) obtained as before. For example, for complexes of the type [M(uncharged ligand)$_2$Cl$_4$] [128, 1121], R(ligand/Cl$^-$) $\sim$ 1.2 (n = 6) in agreement with expectation. It should be noted that fits between experimental data and calculations based on using n = 1 in the repulsion law can again not be obtained.

In agreement with predictions made from the potential energy surfaces, intramolecular rearrangements are restricted to compounds which have two long effective bond lengths, and four short effective bond lengths.

The intramolecular rearrangement of *cis*-[M(ER$_3$)$_2$(CO)$_4$] (where M=Fe, Ru, Os; E=Si, Ge, Sn, Pb; R = alkyl, Ph, Cl) has been studied in some detail [886,

887, 1062, 1063]. Detailed ^{13}C-nmr line shape analysis shows that the CO scrambling proceeds by a *cis*-to-*trans*-to-*cis* isomerisation process, the results not being consistent with a *cis*-to-*cis* process, in agreement with the potential energy surface shown in Fig. 6.9. The intramolecular isomerisation of *cis*-[Mo(PBu$_3$)$_2$(CO)$_4$] in tetrachloroethylene at 65—80 °C has been examined by studying the C—O stretching vibration in the infra-red [298]. It was noted that there was no exchange between the two pairs of carbonyl groups prior to formation of the *trans*-isomer, again showing the absence of the direct *cis*-to-*cis* process.

3. Copper(II) Complexes

Octahedral compounds with an odd number of electrons in the e$_g$ orbitals, for example manganese(III) (t$_{2g}^3$e$_g^1$) and copper(II) (t$_{2g}^6$e$_g^3$), are often observed to be distorted.

Complexes of the type [Cu(ligand)$_6$]$^{x\pm}$ are relatively rare, being restricted to those where the ligand is H$_2$O, OH$^-$, NH$_3$, NO$_2^-$, and a few organic molecules. The imidazole complex [Cu(C$_3$H$_4$N$_2$)$_6$](NO$_3$)$_2$ [748] is tetragonally elongated, two *trans*-copper-nitrogen bonds being 28% longer than the other four. This is in contrast to the cobalt [898], nickel [957], and mercury [788] analogues which have regular octahedral structures. The 2-pyridone complex [Cu(C$_5$H$_5$NO)$_6$](ClO$_4$)$_2$ similarly has two *trans*-copper-oxygen bonds 31% longer than the other four [1033]. In these cases the two *trans*-ligands have clearly been "pushed back" by some non-bonding electron density each side of the plane, that is, in the copper d$_{z^2}$ orbital.

In contrast, the pyridine-1-oxide complexes [Cu(C$_5$H$_5$NO)$_6$](ClO$_4$)$_2$ [842, 1034] and [Cu(C$_5$H$_5$NO)$_6$](BF$_4$)$_2$ [842] have regular octahedral structures, with six equal copper-oxygen bond lengths. However these X-ray results are still consistent with the idea that each static [Cu(C$_5$H$_5$NO)$_6$]$^{2+}$ ion is tetragonally distorted, if it is assumed that there is either a rapid oscillation between the three mutually perpendicular tetragonal distortions at each cation site, or a random orientation of the distorted cations over all sites in the crystal.

The stereochemistry of [Cu(NO$_2$)$_6$]$^{4-}$ has been extensively studied, and may be tetragonally compressed, tetragonally elongated, or regular octahedral, depending upon the choice of cations and the temperature [285, 617, 820, 1026, 1027, 1029 to 1032]. For example K$_2$Pb[Cu(NO$_2$)$_6$] has a small tetragonal compression at 3 °C (two *trans*-Cu-N bonds 5% shorter than the other four), but is regular octahedral above 7 °C. On the other hand the cobalt [107] and nickel [1027, 1028, 1031] analogues are regular octahedral.

Associated with these distortions is the tendency of copper(II) to form complexes of stoichiometry [Cu(unidentate A)$_2$(unidentate B)$_4$]$^{x\pm}$. These complexes typically have a large tetragonal elongation, and there is a continuous series from *trans*-octahedral complexes to square planar [Cu(unidentate B)$_4$] complexes. It must be remembered that in any square planar molecule in the solid or in solution, the very open spaces outside the square faces inevitably allows the approach of other atoms, for example from anions, other square planar molecules, or solvent molecules, which can sometimes be regarded as forming one (as in five-coordinate copper(II) complexes) or two additional weak bonds. It may be noted that repul-

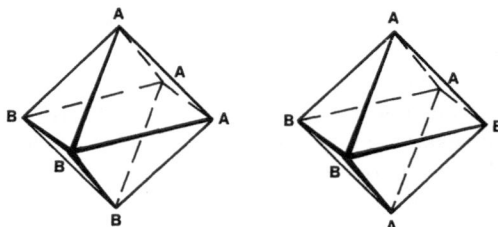

Fig. 6.10. *Facial-* and *meridional-*isomers of [M(unidentate A)$_3$(unidentate B)$_3$]

sion energy calculations show that the two unidentate A ligands need only approach to R(A/B) $\leq$ 1.4 (n = 6) to enforce a square planar structure. However even genuine four-coordinate copper(II) complexes are not necessarily tetrahedral. For example [CuCl$_4$]$^{2-}$ is most commonly a squashed tetrahedron, although it has also been observed as a regular tetrahedron and as a square plane. In Sect. 3.C it was noted that the squashing of a tetrahedron into a square requires only a small amount of electron density on each side of the metal atom, for example from an electron pair in the d$_{z^2}$ orbital.

C. [M(unidentate A)$_3$(unidentate B)$_3$]

Complexes of the type [M(unidentate A)$_3$(unidentate B$_3$)] can exist as *facial-* or *meridional-*isomers (Fig. 6.10). The difference in repulsion energy is very small, for example it is less than one tenth of the difference in repulsion energy between the *cis-* and *trans-*isomers of [M(unidentate A)$_2$(unidentate B)$_4$].

As predicted, structurally characterized complexes are approximately equally distributed between *fac-* and *mer-*isomers.

CHAPTER 7

Six-Coordinate Compounds [M(Bidentate)$_2$(Unidentate)$_2$]

A. Relative Stability of cis- and trans-Octahedral Complexes

The two stereochemistries usually envisaged for complexes of the type [M(biden-tate)$_2$(unidentate$_2$)] are the *cis*- and *trans*-octahedral structures (Fig. 7.1 a, b). It will be shown in more detail in later sections that both these structures become significantly distorted for bidentate ligands of small normalized bite. It is suffi-cient to note at this stage that the *trans*-octahedral structure distorts to form the skewtrapezoidal bipyramidal structure (Fig. 7.1 c, d).

The general stereochemistry is shown in Fig. 7.2, where AB and CD are the two bidentate ligands, and E and F the two unidentate ligands. A twofold axis

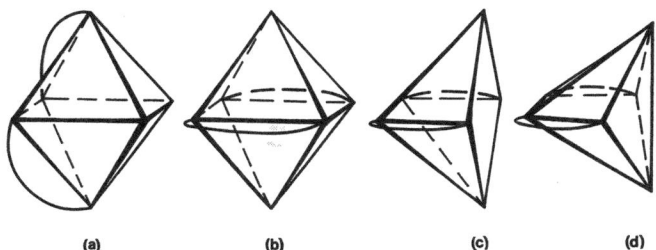

(a)   (b)   (c)   (d)

Fig. 7.1 a—d. Stereochemistries of [M(bidentate)$_2$(unidentate)$_2$].
a *cis*-octahedral; **b** *trans*-octahedral; **c, d** skew-trapezoidal bipyramidal

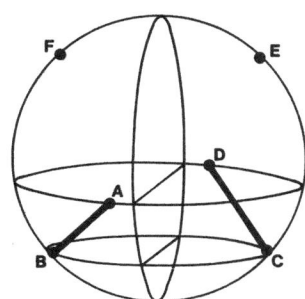

Fig. 7.2. General stereochemistry for *cis*-[M(bidentate)$_2$(unidentate)$_2$]

passes through $\phi = 0$ and $\phi = 180°$, the θ coordinates being defined relative to $\theta_E = 90°$ and $\theta_F = 180°$.

Typical potential energy surfaces projected onto the $\theta_A - \theta_B$ plane are shown in Figs. 7.3—7.5, and should be compared with the similar surface calculated for [M(unidentate A)₂(unidentate B)₄] (Fig. 6.5). The angular distortion of the *cis*-isomer as the normalized bite is reduced is evident in these figures, and is discussed in more detail in the next section. At $b = 2^{1/2}$, both the *cis*- and the *trans*-structures have the same energy, but as the normalized bite is reduced the *cis*-structure becomes significantly more stable than the *trans*-structure (Fig. 7.4), and below $b = 0.90$ ($n = 6$) the *trans*-structure ceases to remain as a discrete minimum (Fig. 7.5). However the *trans*-structure remains possible at low b when the effective bond length ratio, R(unidentate/bidentate) > 1.0 [637].

All compounds of known crystal structure having two identical and symmetrical bidentate ligands, and two identical unidentate ligands, are listed in Table 7.1 in order of increasing normalized bite. As predicted, a marked preference is seen for compounds with small normalized bite ligands to form the *cis*-octahedral structure. Copper(II) complexes are excluded from Table 7.1, as typically they contain a planar arrangement of two bidentate ligands with a tetragonally distorted octahedron completed by two weakly bonded *trans*-ligands (compare with Sect. 6. B. 3).

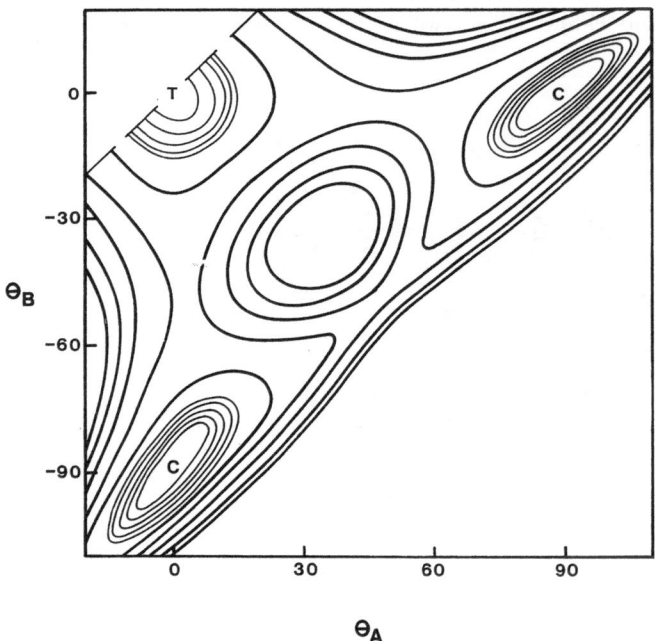

Fig. 7.3. Projection of the potential energy surface for [M(bidentate)₂(unidentate)₂] onto the $\theta_A - \theta_B$ plane (in degrees). The five *faint contour lines* are for successive 0.02 increments above the minima, and the five *heavy contour lines* are for successive 0.2 increments above the minima, at C. $b = 1.4$, $n = 6$. The positions of the *cis*- (C) and *trans*-isomers (T) are shown

Table 7.1. Structures of $[M(bidentate)_2(unidentate)_2]^{x\pm}$

Complex	b	Structure	Ref.
$[Sn(NO_3)_2Me_2]$	0.92	skew-trap. bipy.	[559]
$[Co(NO_3)_2(Me_3PO)_2]$	0.96	*cis*	[268]
$(Ph_4As)_2[Ni(NO_3)_4]CH_2Cl_2$	1.02	*trans*	[89]
$[Ni(O_2CPh)_2(NC_5H_4Me)_2]$	1.03	*cis*	[593]
$[Ni(O_2CPh)_2(NC_9H_7)_2]$	1.04	*trans*	[593]
$[Sn(S_2CNMe_2)_2Me_2]$	1.06	skew-trap. bipy.	[648]
$[Sn(S_2CNEt_2)_2Ph_2]$	1.11	*cis*	[724]
$[Hg(phen)_2(SCN)_2]$	1.11	*cis*	[86]
$[Mo(S_2CNPr_2)_2O_2]$	1.12	*cis*	[926]
$[Mo(S_2CNEt_2)_2O_2]$	1.13	*cis*	[674]
$[Mo(S_2CNEt_2)_2(NPh)_2]$	1.13	*cis*	[546]
$[Sn(S_2CNMe_2)_2(SCSNMe_2)_2]$	1.14	*cis*	[890]
$[Sn\{S_2P(OEt)_2\}_2Ph_2]$	1.14	skew-trap.bipy.	[722]
$[Sn(S_2COEt)_2(SCSOEt)_2]$	1.14	*cis*	[904]
$[Sn(S_2COEt)_2I_2]$	1.14	*cis*	[1086]
$[Sn(S_2COEt)_2Br_2]$	1.16	*cis*	[1086]
$[Sn(S_2CNEt_2)_2(SCSNEt_2)_2]$	1.16	*cis*	[526]
$[Ni(S_2CCH_2Ph)_2(C_5H_5N)_2]$	1.18	*cis*	[134]
$K[Rh(S_2CO)_2(PMe_2Ph)_2]3\,H_2O$	1.20	*trans*	[491]
$[Fe\{S_2CN(CH_2)_5\}_2(CO)_2]$	1.21	*cis*	[927]
$[Cd(NH_2CH_2CH_2NH_2)_2(NCS)_2]$	1.23	*trans*	[987]
$(NH_4)_3[V(C_2O_4)_2O_2]2\,H_2O$	1.23	*cis*	[968]
$K_3[V(C_2O_4)_2O_2]3\,H_2O$	1.23	*cis*	[369]
$[Ti(MeOCH_2CH_2OMe)_2Br_2][TiBr_4(MeOCH_2CH_2OMe)]$	1.24	*cis*	[449]
$[Cr(bipy)_2(C_6H_4OMe)_2]$	1.24	*cis*	[294]
$[Cr(bipy)_2Ph_2]$	1.25	*cis*	[295]
$[Ga(bipy)_2Cl_2](GaCl_4)$	1.25	*cis*	[925]
$[Al(bipy)_2Cl_2]Cl\cdot MeCN$	1.28	*cis*	[92]
$[Re(NH_2CH_2CH_2NH_2)_2O_2]Cl$	1.28	*trans*	[736]
$[Mo(Ph_2PCH_2CH_2PPh_2)_2(N_2)_2]$	1.28	*trans*	[1059]
$[Si(bipy)_2(OH)_2]I_2\cdot 2H_2O$	1.29	*cis*	[964]
$[Mo(MeCOCHCOMe)_2O_2]$	1.29	*cis*	[630]
$[Mo(PhCOCHCOPh)_2O_2]$	1.29	*cis*	[665]
$[Ni(biimidazole)_2(H_2O)_2](NO_3)_2$	1.29	*trans*	[787]
$[Os(NH_2CH_2CH_2NH_2)_2O_2](HSO_4)_2$	1.29	*trans*	[755]
$[Co(bipy)_2Cl_2]_2(CoCl_4)$	1.30	*cis*	[561]
$[Co(bipy)_2Me_2](AlEt_4)$	1.31	*cis*	[667]
$[Ni(S_2PPh_2)_2(C_5H_5N)_2]$	1.31	*trans*	[889]
$[Ni\{S_2P(OEt)_2\}_2(C_5H_5N)_2]$	1.31	*trans*	[845]
$[Ni(NH_2CH_2CH_2NH_2)_2(NCS)_2]$	1.31	*trans*	[155]
$[Ni\{C_6H_4(NH_2)_2\}_2(NH_2C_6H_4NH_2)_2]Cl_2\cdot 2\,C_6H_4(NH_2)_2$	1.31	*trans*	[397]
$[Ti(MeCOCHCOMe)_2(OC_{12}H_{17})_2]$	1.32	*cis*	[113]
$[Co(Me_2NCH_2CH_2NMe_2)_2(OOCCF_3)_2](CF_3CO_2)$	1.32	*trans*	[95]
$[Pt(NH_2CH_2CH_2NH_2)_2Cl_2][CuCl_4]H_2O$	1.32	*trans*	[694]
$[Pt(NH_2C_6H_{10}NH_2)_2Cl_2][Pt(NH_2C_6H_{10}NH_2)_2]Cl_4$	1.33	*trans*	[695]
$[Tc\{C_6H_4(AsMe_2)_2\}_2Cl_2](ClO_4)$	1.33	*trans*	[481]
$[Mg(MeCONHCOMe)_2(H_2O)_2](ClO_4)_2$	1.33	*trans*	[469]
$[Ni(NH_2CH_2CH_2NH_2)_2(NO_2)_2]$	1.34	*trans*	[795]
$[Pt(NH_2CH_2CH_2NH_2)_2Cl_2]Cl_2$	1.34	*cis*	[734]
$[Zn(CF_3COCHCOCF_3)_2(C_5H_5N)_2]$	1.34	*cis*	[896]
$[Ni(MeNHCH_2CH_2NHMe)_2(ONO)_2]$	1.34	*trans*	[346]

(Continued)

Table 7.1 (continued)

Complex	b	Structure	Ref.
[Ni(MeNHCH$_2$CH$_2$NHMe)$_2$Br$_2$]	1.35	*trans*	[852]
[Cr(NH$_2$CH$_2$CH$_2$NH$_2$)$_2$Cl$_2$]Cl · HCl · 2 H$_2$O	1.35	*trans*	[846]
[Co(NH$_2$CH$_2$CH$_2$NH$_2$)$_2$(NO$_2$)$_2$][Co(C$_2$O$_4$)(NO$_2$)$_2$(NH$_3$)$_2$] · H$_2$O	1.35	*cis*	[986]
[Co(NH$_2$CH$_2$CH$_2$NH$_2$)$_2$Cl$_2$](NO$_3$)	1.36	*trans*	[848]
[Co(NH$_2$CH$_2$CH$_2$NH$_2$)$_2$(NO$_2$)$_2$](NO$_3$)	1.36	*cis*	[139]
[Co(NH$_2$CH$_2$CH$_2$NH$_2$)$_2$(NO$_2$)$_2$](NO$_3$)	1.36	*trans*	[139]
Na$_2$[Co(NH$_2$CH$_2$CH$_2$NH$_2$)$_2$(SO$_3$)$_2$](ClO$_4$)3 H$_2$O	1.36	*cis*	[919]
[Sn(MeCOCHCOMe)$_2$Me$_2$]	1.36	*trans*	[790]
[Ru(Ph$_2$PCH$_2$CH$_2$PPh$_2$)$_2$H$_2$]C$_6$H$_6$	1.36	*cis*	[866]
[Ir(Ph$_2$PCH$_2$CH$_2$PPh$_2$)$_2$H$_2$](B$_9$H$_{14}$)	1.37	*cis*	[503]
[Mn(MeCOCHCOMe)$_2$(H$_2$O)$_2$]	1.37	*trans*	[804, 844]
[Ni(MeCOCH$_2$COMe)$_2$Br$_2$]	1.37	*trans*	[663]
[Co{C$_6$H$_4$(AsMe$_2$)$_2$}$_2$Cl$_2$]Cl	1.37	*trans*	[103]
[Ni{C$_6$H$_4$(AsMe$_2$)$_2$}$_2$Cl$_2$]Cl	1.37	*trans*	[103]
[Mg(PhCOCHCOPh)$_2$(OCHNMe$_2$)$_2$]	1.37	*cis*	[569]
[Co(NH$_2$CH$_2$CH$_2$NH$_2$)$_2$Br$_2$]Br · 2 H$_2$O	1.37	*trans*	[847]
[Co(NH$_2$CH$_2$CH$_2$NH$_2$)$_2$Cl$_2$]$_2$(S$_5$O$_6$)H$_2$O	1.37	*trans*	[763]
[Ni(NH$_2$CH$_2$CH$_2$NH$_2$)$_2$(H$_2$O)$_2$](ClO$_4$)$_2$	1.37	*trans*	[794]
[Co(NH$_2$CH$_2$CH$_2$NH$_2$)$_2$(H$_2$O)$_2$)][Co(CN)$_6$]3 H$_2$O	1.38	*cis*	[632]
[Sn(MeCOCHCOMe)$_2$Cl$_2$]	1.38	*cis*	[791]
[Ni{C$_6$H$_4$(AsMe$_2$)$_2$}$_2$I$_2$]	1.38	*trans*	[1014]
[Co{C$_6$H$_4$(AsMe$_2$)$_2$}$_2$Cl$_2$](ClO$_4$)	1.39	*trans*	[857]
[Co(NH$_2$CH$_2$CH$_2$NH$_2$)$_2$(N$_3$)$_2$](NO$_3$)	1.39	*cis*	[850]
[Co(NH$_2$CH$_2$CH$_2$CH$_2$NH$_2$)$_2$(NO$_3$)$_2$](NO$_3$)	1.39	*trans*	[1110]
[Ni(NH$_2$CH$_2$CH$_2$CH$_2$NH$_2$)$_2$(H$_2$O)$_2$](ClO$_4$)$_2$	1.39	*trans*	[853]
[Ni(MeCOCH$_2$COMe)$_2$(H$_2$O)$_2$](ClO$_4$)$_2$	1.39	*trans*	[277]
[Mg(MeCOCHCOMe)$_2$(H$_2$O)$_2$]	1.40	*trans*	[809]
[Re(MeCOCHCOMe)$_2$Cl$_2$]	1.41	*trans*	[163]
[Co(MeCOCHCOMe)$_2$(C$_5$H$_5$N)$_2$]	1.41	*trans*	[393]
[Co(MeCOCHCOMe)$_2$(C$_{10}$H$_9$N)$_2$]	1.41	*trans*	[591]
[Ni(MeCOCHCOMe)$_2$(C$_5$H$_5$NO)$_2$]	1.42	*cis*	[575]
(Ph$_4$As)$_2$[Re(MeCOCHCOMe)$_2$Cl$_2$]	1.42	*trans*	[735]
[Pt(NHCMeCH$_2$CMeNH)$_2$(NH$_3$)$_2$](ClO$_4$)$_2$	1.42	*trans*	[147]
[Co(MeCOCHCOMe)$_2$(H$_2$O)$_2$]	1.44	*trans*	[177]
[Ni(MeCOCHCOMe)$_2$(C$_5$H$_5$N)$_2$]	1.44	*trans*	[394]
[Ni(MeCOCHCOMe)$_2$(H$_2$O)$_2$]	1.44	*trans*	[803]
[Co(NH$_2$CH$_2$CH$_2$CH$_2$NH$_2$)$_2$(NO$_2$)$_2$]Cl · H$_2$O	1.44	*cis*	[1005]
[Ni(NH$_2$CH$_2$CH$_2$CH$_2$NH$_2$)$_2$(H$_2$O)$_2$](NO$_3$)$_2$	1.45	*trans*	[851]
[Co(NH$_2$CH$_2$CH$_2$CH$_2$NH$_2$)$_2$Cl$_2$]Cl · HCl · 2 H$_2$O	1.47	*trans*	[773]
[Pt(MeCOCHCOMe)$_2$I$_2$]	1.48	*trans*	[258]
[Co(MeCOCHCOMe)$_2$(NO$_2$)(C$_7$H$_{10}$N$_5$)]	1.48	*trans*	[228]
[Co(MeCOCHCOMe)$_2$(NO$_2$)$_2$]			

B. Angular Distortions in cis-Octahedral Complexes

The angular coordinates of the *cis*-isomer, are shown in Fig. 7.6, axes being defined by Fig. 7.2. The variation with normalized bite is not as may have been intuitively expected. If decreasing b from the regular octahedral value of 1.414 merely

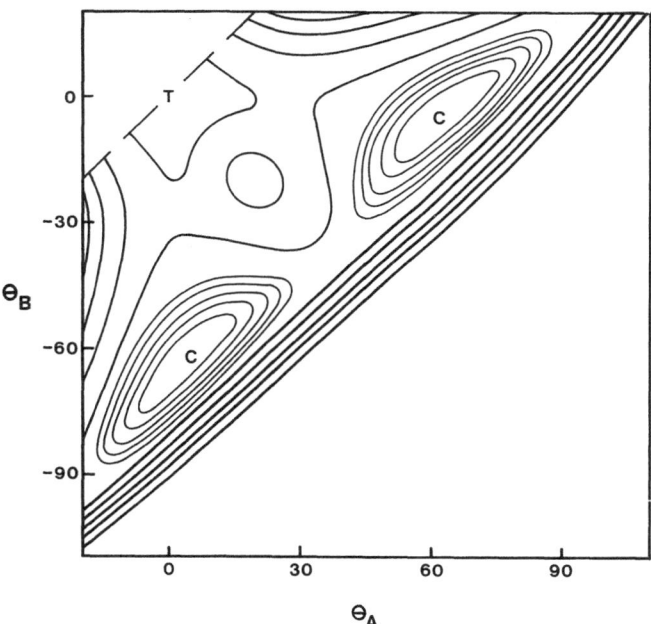

Fig. 7.4. Projection of the potential energy surface for $[M(\text{bidentate})_2(\text{unidentate})_2]$ onto the $\theta_A - \theta_B$ plane (in degrees). The five *faint contour lines* are for successive 0.02 increments above the minima, and the five *heavy contour lines* are for successive 0.2 increments above the minima, at C. b = 1.1, n = 6. The position of the *cis*- (C) and *trans*-isomers (T) are shown

resulted in atoms A and B moving uniformly toward one another, then θ_A would move to negative values, and θ_B would increase above $-90°$. However, Fig. 7.6 shows that θ_A is predicted to *increase*, with a corresponding very large increase in θ_B. This distortion corresponds to a rotation of both bidentate ligands about the twofold axis to higher values of θ (Fig. 7.7).

This increase in θ_A and θ_B is experimentally observed, as is shown in Table 7.2, which is restricted to molecules with bidentate ligands that form four-membered chelate rings, with small normalized bites. Exact agreement is not expected with Fig. 7.6, which was calculated for R(unidentate/bidentate) = 1.0. Nevertheless, in all cases the largest distortion from the parameters of the regular octahedron is the increase in θ_B, whereas θ_A slightly *increases*.

C. Bond-length Distortions in cis-Octahedral Complexes

As the normalized bite of the bidentate ligand is reduced from 1.414, the A end of each ligand is associated with less repulsion than is the B end (Fig. 7.8). It is therefore predicted that the bidentate ligand will be unsymmetrically coordinated, with the M—A bond being shorter than the M—B bond, the latter being *trans*-

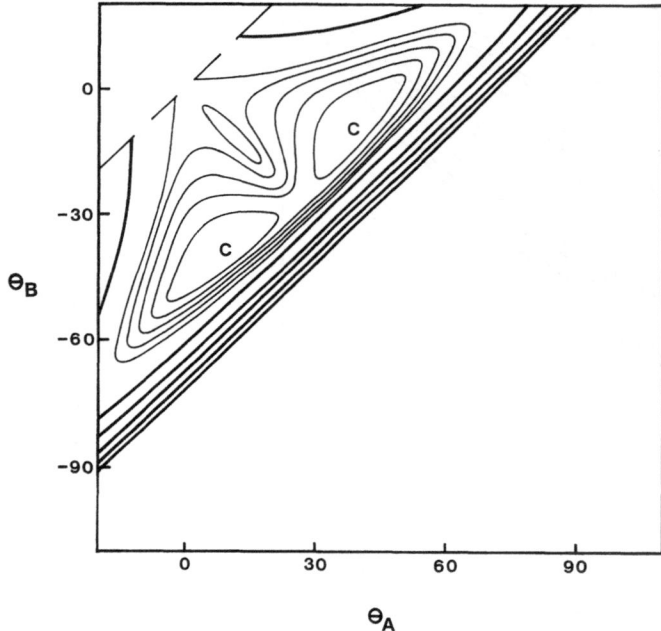

Fig. 7.5. Projection of the potential energy surface for [M(bidentate)₂(unidentate)₂] onto the θ$_A$ — θ$_B$ plane (in degrees). The five *faint contour lines* are for successive 0.02 increments above the minima, and the five *heavy contour lines* are for successive 0.2 increments above the minima, corresponding to the *cis*-isomer (C). b = 0.8, n = 6

to a unidentate ligand. This effect increases as R(unidentate/bidentate) decreases (Fig. 7.8).

This predicted behavior is experimentally observed, and is shown most clearly in compounds in which the unidentate ligand is O^{2-} or NPh^{2-} (Table 7.3). The M—A bond, which is *cis* to both unidentate ligands, is about 9% shorter than the

Table 7.2. Angular parameters (°) of *cis*-octahedral [M(bidentate)₂(unidentate)₂] containing four-membered chelate rings

Complex	b	ϕ_A	ϕ_B	ϕ_E	θ_A	θ_B
Regular octahedron	1.41	90°	135°	45°	0°	−90°
[Co(NO₃)₂(Me₃PO)₂]	0.96	101°	138°	53°	−1°	−53°
[Sn(S₂CNEt₂)₂Ph₂]	1.11	103°	140°	51°	5°	−65°
[Mo(S₂CNPr₂)₂O₂]	1.12	104°	139°	53°	14°	−59°
[Mo(S₂CNEt₂)₂O₂]	1.13	106°	140°	57°	15°	−60°
[Mo(S₂CNEt₂)₂(NPh)₂]	1.13	104°	140°	52°	6°	−64°
[Sn(S₂COEt)₂(SCSOEt)₂]	1.14	103°	136°	46°	0°	−78°
[Sn(S₂COEt)₂I₂]	1.14	102°	139°	49°	5°	−68°
[Sn(S₂COEt)₂Br₂]	1.16	99°	136°	48°	2°	−69°
[Sn(S₂CNEt₂)₂(SCSNEt₂)₂]	1.16	102°	131°	41°	1°	−73°
[Fe{S₂CN(CH₂)₅}₂(CO)₂]	1.21	99°	135°	46°	1°	−77°

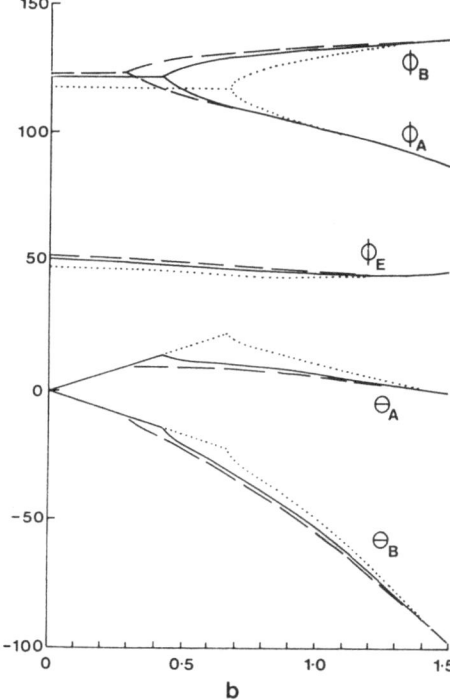

Fig. 7.6. Angular coordinates (degrees) for *cis*-[M(bidentate)₂(unidentate)₂] as a function of normalized bite b. *Dotted lines,* n = 1; *full lines,* n = 6; *broken lines,* n = 12

M—B bond. As expected, the effect is much smaller for unidentate ligands of lower charge [637], a typical example being the unsymmetrical bonding of the nitrate group in [Co(NO₃)₂(Me₃PO)₂], (M—A)/M—B) = 0.98. This relatively small effect must not be confused with that observed in complexes containing nitrate groups which approach unidentate behavior [683, 684, 879].

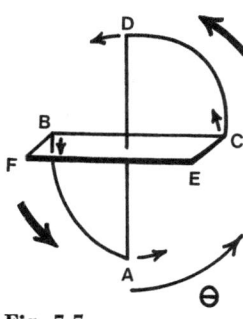

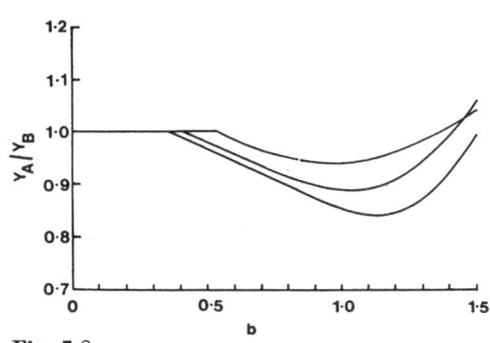

Fig. 7.7   Fig. 7.8

Fig. 7.7. Angular distortions in *cis*-[M(bidentate)₂(unidentate)₂]
Fig. 7.8. Ratio of individual atom-repulsion coefficients for the two ends of each bidentate ligand in *cis*-[M(bidentate)₂(unidentate)₂] as a function of normalized bite b. n = 6.
Upper curve, effective bond length ratio R = 1.5; *middle curve,* R = 1.0; *lower curve* R = 0.75

Table 7.3. Bond-length ratios for *cis*-[M(bidentate)$_2$(unidentate)$_2$] containing O^{2-} or NPh^{2-} ligands

Complex	b	(M—A)/M—B)
[Mo(S$_2$CNPr$_2$)$_2$O$_2$]	1.12	0.92
[Mo(S$_2$CNEt$_2$)$_2$O$_2$]	1.13	0.93
[Mo(S$_2$CNEt$_2$)$_2$(NPh)$_2$]	1.13	0.92
(NH$_4$)$_3$[V(C$_2$O$_4$)$_2$O$_2$]2H$_2$O	1.23	0.90
K$_3$[V(C$_2$O$_4$)$_2$O$_2$]3H$_2$O	1.23	0.91
[Mo(MeCOCHCOMe)$_2$O$_2$]	1.29	0.90
[Mo(PhCOCHCOPh)$_2$O$_2$]	1.29	0.92

D. Distortion of the trans-Octahedral Structure to the Skew-Trapezoidal Bipyramidal Structure

Bidentate ligands that form five- or six-membered chelate rings and have relatively large normalized bites (b ∼ 1.2—1.5) normally form undistorted *trans*-structures.

As the normalized bite is further decreased, however, repulsion energy calculations show that the rectangle formed by the two coplanar bidentate ligands becomes distorted and forms a planar trapezoid. The unidentate ligands are

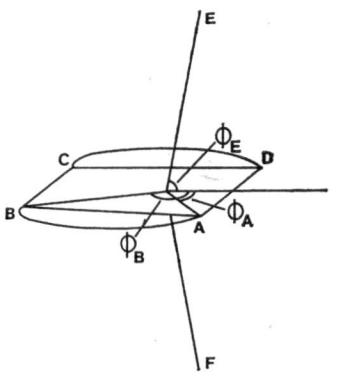

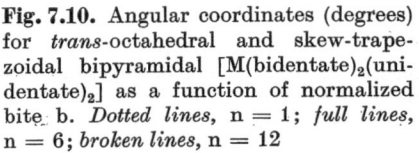

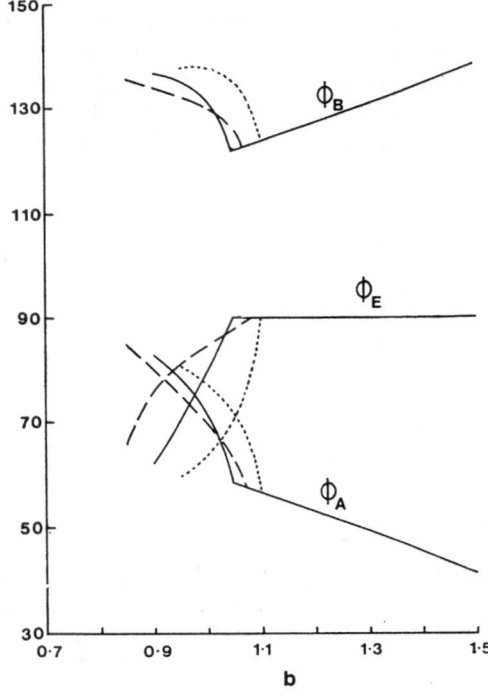

Fig. 7.9. General stereochemistry for *trans*-octahedral and skew-trapezoidal bipyramidal [M(bidentate)$_2$(unidentate)$_2$]

Fig. 7.10. Angular coordinates (degrees) for *trans*-octahedral and skew-trapezoidal bipyramidal [M(bidentate)$_2$(unidentate)$_2$] as a function of normalized bite. b. *Dotted lines*, n = 1; *full lines*, n = 6; *broken lines*, n = 12

simultaneously skewed toward (Fig. 7.1c), or even past (Fig. 7.1d), the long edge of the trapezoid. Two mirror planes are retained throughout this distortion, and the stereochemistry is completely described by three angular parameters, the angles the metal-ligand bonds make with the twofold axis (Fig. 7.9). The dependence of these parameters, ϕ_A, ϕ_B, and ϕ_E on the normalized bite is shown in Fig. 7.10. This structure does not remain as a discrete minimum on the potential energy surface at very low values of the normalized bite, as indicated by the extent of the lines in Fig. 7.10. The repulsion energy calculations also show that the distortion to the skew-trapezoidal bipyramidal structure commences at larger values of the normalized bite as R(unidentate/bidentate) decreases.

The only examples of skew-trapezoidal bipyramidal structures found with two equivalent unidentate ligands and two symmetrical bidentate ligands are three dimethyl- and diphenyl-bis(bidentate)tin complexes. The detailed structural parameters are given in Table 7.4, and the trapezoids illustrated in Fig. 7.11, together with [Sn(MeCOCHCOMe)$_2$Me$_2$] which has a larger normalized bite and an undistorted *trans* structure, and two closely related skew-trapezoidal bipyramidal molecules with unsymmetrical bidentate ligands [529, 530].

A notable feature of these complexes in the very unsymmetrical bonding of the bidentate ligands, the tin-ligand(A) bonds being 10—30% longer than the tin-ligand(B) bonds (Table 7.4 and Fig. 7.11). This arises from the considerably

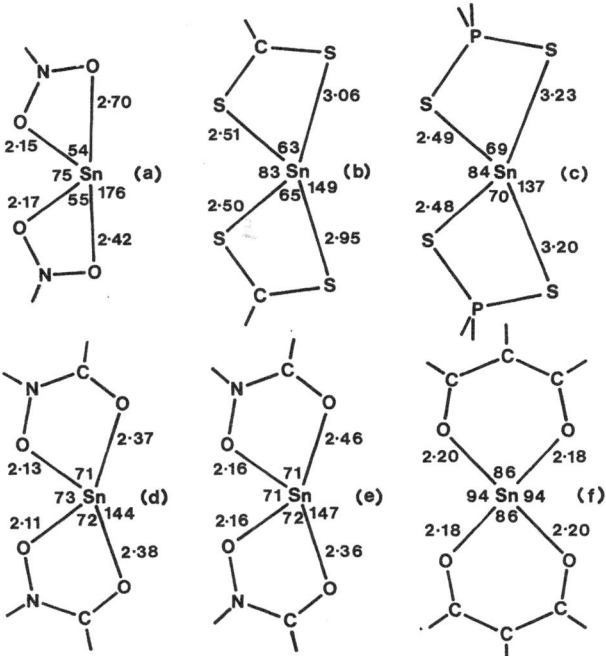

Fig. 7.11a—f. Geometries, degrees and Å, of the Sn(bidentate)$_2$ planes in: a [Sn(NO$_3$)$_2$Me$_2$]; b [Sn(S$_2$CNMe$_2$)$_2$Me$_2$]; c [Sn{S$_2$P(OEt)$_2$}Ph$_2$]; d [Sn(ONMeCMeO)$_2$Me$_2$]; e [Sn(ONHCMeO)$_2$ · Me$_2$]H$_2$O; f [Sn(MeCOCHCOMe)$_2$Me$_2$]

Table 7.4. Stereochemical parameters of skew-trapezoidal bipyramidal [Sn(bidentate)₂R₂] complexes (R=Me, Ph)

Complex	b	ϕ_A	ϕ_B	ϕ_E	(M—A)/(M—B)
[Sn(NO₃)₂Me₂]	0.92	88.1°	142.6°	71.8°	1.19
[Sn(S₂CNMe₂)₂Me₂]	1.06	74.5°	138.6°	68°	1.20
[Sn{S₂P(OEt)₂}₂Ph₂]	1.14	68.4°	137.8°	67.5°	1.29
[Sn(ONMeCMeO)₂Me₂]	1.17	72.0°	143.4°	72.9°	1.12
[Sn(ONHCMeO)₂Me₂]	1.17	73.3°	144.6°	78.4°	1.12

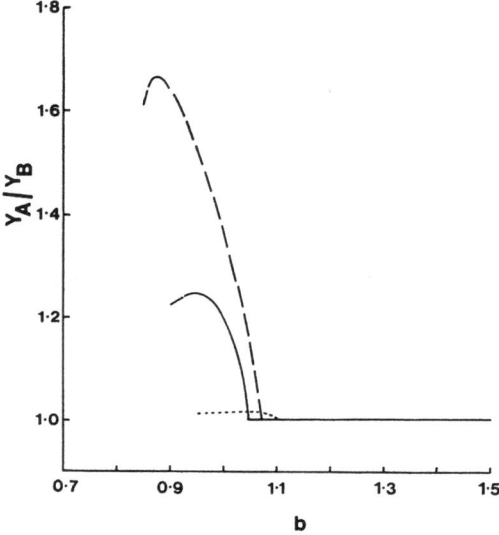

Fig. 7.12. Ratio of individual atom-repulsion coefficients for the two ends of each bidentate ligand in *trans*-octahedral and skew-trapezoidal bipyramidal [M(bidentate)₂(unidentate)₂] as a function of normalized bite b. *Dotted line,* n = 1; *full line,* n = 6; *broken line,* n = 12

greater repulsion energy associated with the A end of the bidentate ligand than with the B end (Fig. 7.12). Repulsion energy calculations show that the extent of this distortion increases as R(unidentate/bidentate) decreases.

E. Bis(dithiochelate) Complexes of Selenium(II) and Tellurium(II)

The thiourea complexes [Te^{II}{SC(NH₂)₂}₄]Cl₂ and [Te^{II}{SC(NH₂)₂}₄]Cl₂·2H₂O have a square planar arrangement of sulfur atoms about the tellurium atom (Sect. 6.B).

Selenium(II) and tellurium(II) xanthates, dithiocarbamates, and diselenocarbamates, on the other hand, have a planar trapezoid structure (see Fig. 7.13 for typical dimensions). These compounds have bidentate ligands with normalized

Table 7.5. Stereochemical parameters of bis(dithiochelate) complexes of selenium(II) and tellurium(II)

Complex	b	ϕ_A	ϕ_B	(M − A)/(M − B)	Ref.
[Te(S$_2$COMe)$_2$]	1.09	71.1°	137.4°	1.14	[494]
[Te(S$_2$COEt)$_2$]	1.09	72.2°	138.2°	1.16	[595]
[Te{S$_2$CN(C$_4$H$_8$O)}$_2$]	1.10	72.8°	139.5°	1.13	[596]
[Se{S$_2$CN(C$_4$H$_8$O)}$_2$]	1.15	67.5°	137.7°	1.21	[41]
[Se(S$_2$CNEt$_2$)$_2$]	1.15	65.7°	136.2°	1.18	[598]
[Se(Se$_2$CNEt$_2$)$_2$]	1.20	61.6°	135.5°	1.14	[492]

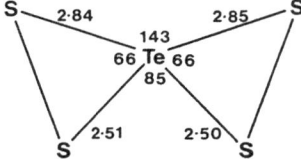

Fig. 7.13. Geometries, degrees and Å, of [Te(S$_2$COMe)$_2$

bites of b = 1.0 to 1.2 and can clearly be regarded as having skew-trapezoidal bipyramidal stereochemistry, with the *trans*-apical sites being occupied by the two nonbonding pairs of electrons. It is important to note that one central atom-donor atom bond to each chelate is 10—20% longer than the other, as predicted in the previous section.

Detailed structural parameters are given in Table 7.5.

F. Unsymmetrical Bidentate Ligands

For bidentate ligands in which the two ends are chemically dissimilar, there are three *cis*-octahedral isomers and two *trans*-octahedral isomers possible (Fig. 7.14).

The relative stabilities of each of these isomers can be calculated as before, a typical results being shown in Fig. 7.15, calculated for R(A/B) = R(D/C) = 0.8, and R(B/E) = 1.0. The energy difference is small, and it would be expected that all five isomers would be possible depending upon the choice of donor atoms, and indeed they have all been structurally characterized.

One general result is that for b < ∼ 1.3, the *cis*-B isomer is the most stable of the *cis*-isomers. That is, it is predicted that the metal-bidentate ligand bonds with the shortest effective bond lengths will be *trans* to each other, and *cis* to the uni-dentate ligands. A similar conclusion can be reached from Sect. 7.C, where it was found that these mutually *trans* sites were subject to less repulsion than were the other ends of the bidentate ligands.

These predictions are followed for compounds with relatively simple bonding. For example the oxinate and related compounds of titanium(IV) [113, 1022], molybdenum(IV) [1109], and tin(IV) [918, 970] have the negatively charged ends of the bidentate ligands mutually *trans*, with the uncharged ends being mutually *cis*. Similarly in the molybdenum(VI) complex [Mo(OCH$_2$CH$_2$OH)$_2$O$_2$], the negatively charged alkoxide ends of the bidentate ligands are mutually *trans*, and the alcohol ends are mutually *cis* [192, 973]. However in complexes of the later transition metals, the preference of certain donor atoms to selected sites is more complicated [637].

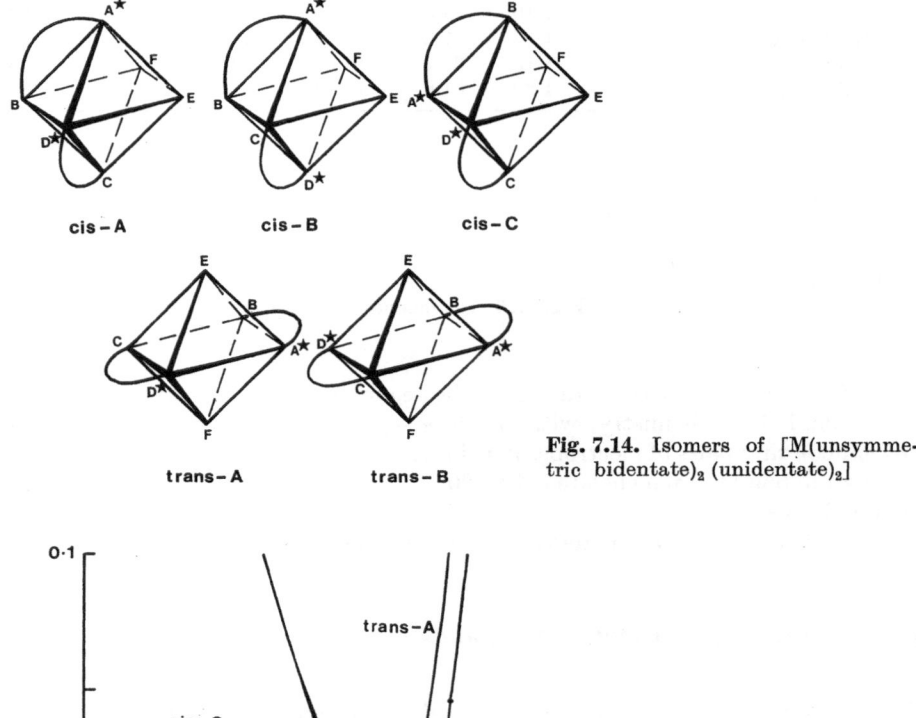

Fig. 7.14. Isomers of [M(unsymmetric bidentate)₂ (unidentate)₂]

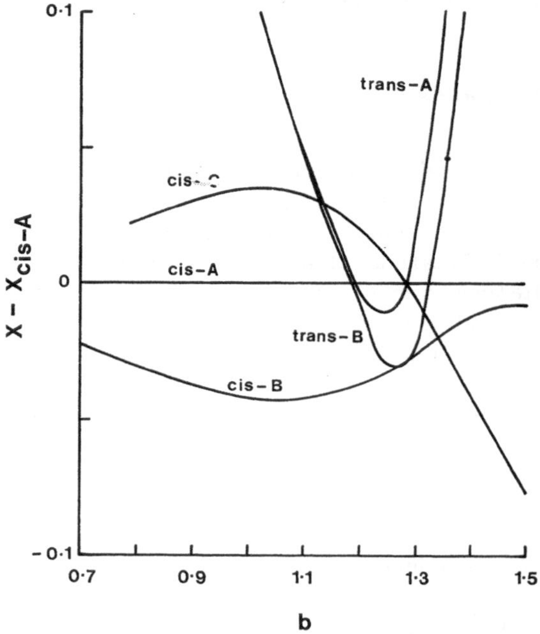

Fig. 7.15 Repulsion energy coefficient, above that corresponding to the *cis*-A isomer, for [M(unsymmetric bidentate)₂(unidentate)₂], as a function of b. n = 6

G. Pentagonal Pyramidal [M(bidentate)$_2$(unidentate A)(unidentate B)] Complexes

Compounds containing two bidentate ligands and two chemically similar unidentate ligands can exist as the usual *cis-* and *trans*-octahedral isomers, or as the skew-trapezoidal bipyramidal structure observed for bidentate ligands of small normalized bite. However a different stereochemistry, which can be represented as a pentagonal pyramid, is found under the combined conditions of small normalized bite and a large difference between the natures of the two unidentate ligands.

The stereochemistry is shown in Fig. 7.16. The two unidentate ligands lie on a mirror plane, the effective bond lengths being given by R_A and R_B respectively, the other four effective bond lengths being defined as unity. Repulsion energy calculations show that this structure is expected provided the normalized bite is less than about 0.9, and provided the difference $(R_B - R_A) > \sim 0.4$. In all cases it is found that the bond angle AMB is less than AMC and AMD, so that there is significant buckling of the pentagonal base of the pyramid. It should also be noted that both unidentate ligands lie on the same side of the CDEF plane, which distinguishes this stereochemistry from the skew-trapezoidal bipyramid where the two unidentate ligands lie on opposite sides of the trapezium formed by the two chelate rings.

The known pentagonal pyramidal molecules are [Cr(O$_2$)$_2$O(C$_5$H$_5$N)] [1018], (NH$_4$)[V(O$_2$)$_2$O(NH$_3$)] [368], and (NH$_4$)$_4$[{V(O$_2$)$_2$O}$_2$O] [1025], where two peroxide groups behave as bidentate ligands with b = 0.77, and the unidentate A ligand is O^{2-}. The remaining unidentate ligand is on the same side of the (O$_2$)$_2$ trapezium as the O^{2-} group, O=M-bidentate $\sim$ 106°, O=M-unidentate $\sim$ 98°.

These structures are in contrast to those of related complexes containing bidentate ligands of larger normalized bite such as (NH$_4$)$_2$[V(C$_2$O$_4$)$_2$O(H$_2$O)] [446] and [V(oxine)$_2$O(OPr)] [966] (b = 1.24), which have conventional *cis*-octahedral structures.

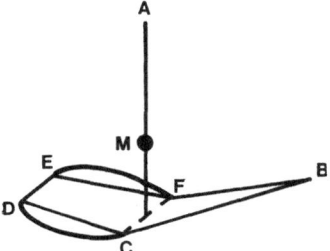

Fig. 7.16. Pentagonal pyramidal [M(bidentate)$_2$(unidentate A)(unidentate B)]

CHAPTER 8

Six-Coordinate Compounds [M(Bidentate)$_3$]

A. The Theoretical Stereochemistry

Repulsion energy calculations show that the stereochemistry corresponding to the single minimum on any potential energy surface always contains a threefold axis. The stereochemistry is, therefore, completely defined by the normalized bite of the bidentate ligand b and the angle of twist θ between the upper and lower triangular faces (Fig. 8.1). The regular octahedron corresponds to θ = 30°, and b = 2$^{1/2}$. The trigonal prism is the eclipsed arrangement with θ = 0°.

The repulsion energy coefficient X is shown as a function of the angle of twist θ and the normalized bite b in Fig. 8.2.

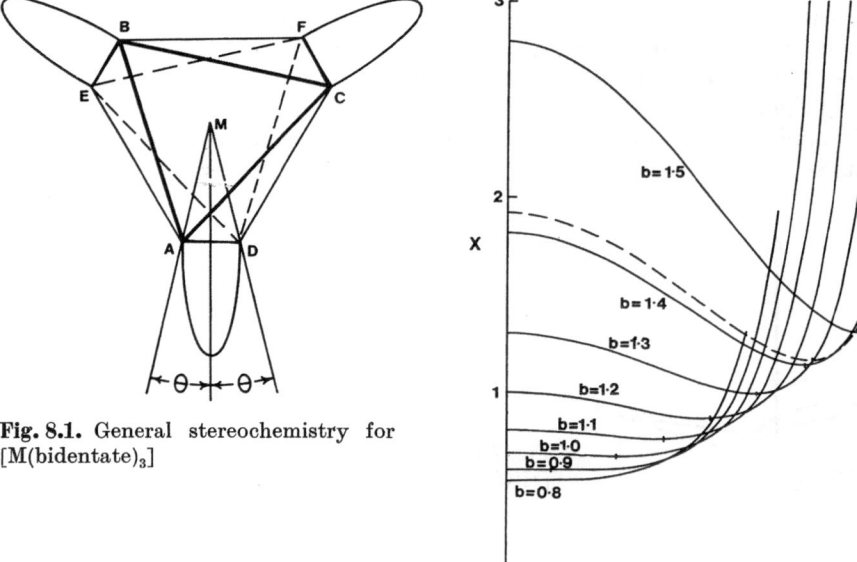

Fig. 8.1. General stereochemistry for [M(bidentate)$_3$]

Fig. 8.2. Repulsion-energy coefficient X for [M(bidentate)$_3$] as function of angle of twist θ (degrees) and normalized bite b. n = 6. *Broken line* corresponds to b = 2$^{1/2}$

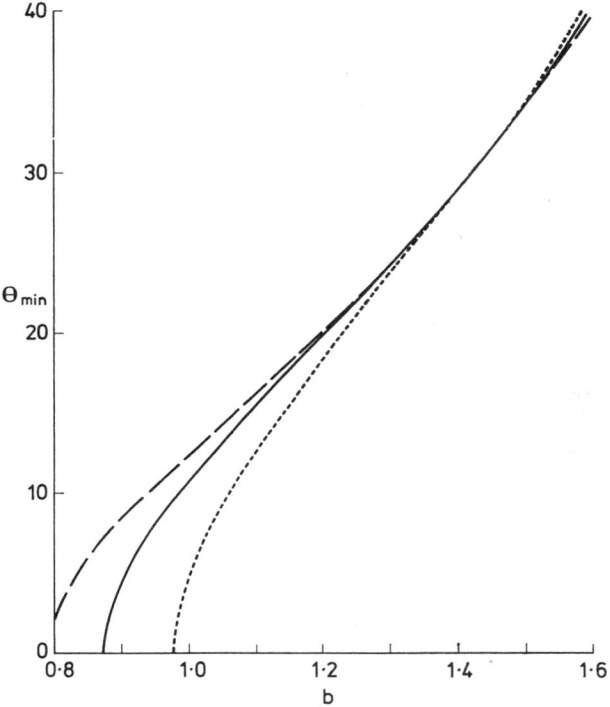

Fig. 8.3. Most stable stereochemistry (degrees) for [M(bidentate)$_3$] as function of normalized bite b. *Dotted line*, n = 1; *full line*, n = 6; *broken line*, n = 12

For normalized bites of b = $2^{1/2}$ = 1.414, the regular octahedron at θ = 30° exists as a deep minimum. The energy difference between this minimum and the maximum at θ = 0° is the activation energy for racemization of these optically active compounds by this simple twist mechanism.

As the normalized bite of the bidentate ligands is progressively decreased (Fig. 8.2), two important effects are observed. Firstly, the upper triangular face ABC of the octahedron is brought into a more eclipsed configuration relative to the lower triangular face DEF, and the energy minimum moves to lower values of θ. The location of this minimum, θ_{min}, is plotted as a function of b in Fig. 8.3.

Secondly, the decrease in b causes the potential energy surface to become much shallower, resulting in structures which are easier to distort (see Sect. 8.B and 8.C) and which have lower activation energies for racemization (see Sect. 8.H).

B. Crystal Structures of 158 Complexes

Those molecules which contain three identical symmetrical bidentate ligands are listed in Table 8.1. The structures of the majority conform reasonably well with the above predictions (Fig. 8.4).

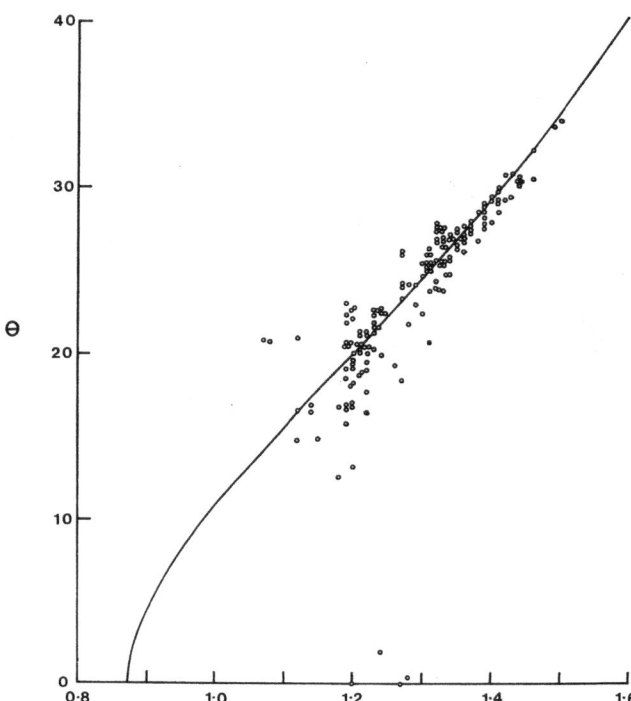

Fig. 8.4. Angle of twist θ (degrees) and normalized bite b for complexes of the type [M(bidentate)$_3$]; theoretical curve for n = 6 also shown

Table 8.1. Structural parameters for [M(bidentate)$_3$]$^{x\pm}$

Complex	b	θ	Ref.
[Co(MeCOCHCOMe)$_3$]	1.50	34.0°	[572]
	1.49	33.7°	[678]
[Ru(MeCOCHCOMe)$_3$]	1.46	32.2°	[225]
[Fe(MeCSCHCSMe)$_3$]	1.46	30.5°	[88]
[Ga(MeCOCHCOMe)$_3$]	1.44	30.6°	[375]
[Al(MeCOCHCOMe)$_3$]	1.44	30.4°	[572]
(NH$_4$)$_2$[Pt(S$_5$)$_3$]2 H$_2$O	1.44	30.3°	[624]
[Co(NH$_2$CHMeCH$_2$NH$_2$)$_3$][Cr(OOCCH$_2$COO)$_3$]3H$_2$O (anion)	1.44	30.1°	[199]
(Et$_3$NH)[P(O$_2$C$_6$H$_4$)$_3$]	1.43	29.4°	[37]
rac-[Cr(MeCOCHCOMe)$_3$]	1.43	30.8°	[808]
(−)-[Cr(MeCOCHCOMe)$_3$]	1.42	30.7°	[682]
[Al(PhCOPCOPh)$_3$]	1.42	29.2°	[87]
[Co{NHC(NH$_2$)NHC(NH$_2$)NH}$_3$]Cl$_3$ · H$_2$O	1.41	29.2°	[998]
K[V(N$_2$C$_3$H$_3$ · BH$_2$ · C$_3$H$_3$N$_2$)$_3$]EtOH	1.41	29.9°	[297]
[Cr(NH$_2$CH$_2$CH$_2$CH$_2$NH$_2$)$_3$][Ni(CN)$_5$]2 H$_2$O	1.41	29.0°	[628]
CaK[Co(S$_2$C$_2$O$_2$)$_3$]4 H$_2$O	1.41	28.5°	[198]

Complex	b	θ	Ref.
$[Co(NH_2CHMeCH_2CHMeNH_2)_3][Co(CN)_6]5\,H_2O$	1.41	29.8°	[963]
$[Co(NH_2CH_2CH_2CH_2NH_2)_3]Br_3$	1.40	27.9°	[961]
$[Co\{NHC(NH_2)NHC(NH_2)NH\}_3]Br_3 \cdot H_2O$	1.40	29.2°	[810]
$(C_5H_5NH)_2[Si(O_2C_6H_4)_3]$	1.40	29.4°	[444]
$[Mo(MeCOCHCOMe)_3]$	1.39	28.9°	[913]
$[Co\{OP(NMe_2)_2OP(NMe_2)_2O\}_3](ClO_4)_2$	1.39	28.8°	[616]
$[Co(NH_2CH_2CH_2NH_2)_3]Br_3 \cdot H_2O$	1.39	28.5°	[827]
$K_3[Mn(OOCCH_2COO)_3]2\,H_2O$	1.39	27.8°	[732]
$K[As(O_2C_6H_4)_3]1\frac{1}{2}H_2O$	1.39	27.5°	[660]
$[Fe(MeCOCHCOMe)_3]AgClO_4 \cdot H_2O$	1.39	28.2°	[830]
$[Fe(MeCOCHCOMe)_3]$	1.38	26.8°	[603]
$[Co\{NHC(NH_2)NC(NH_2)NH\}_3]2\,H_2O$	1.38	28.5°	[810]
$[Mg\{OP(NMe_2)_2OP(NMe_2)_2O\}_3](ClO_4)_2$	1.37	27.7°	[616]
$[Co(NH_2C_6H_{10}NH_2)_3]Cl_3 \cdot 5\,H_2O$	1.37	27.3°	[766]
$[Co(NH_2CH_2CH_2NH_2)_3](SnCl_3)Cl_2$	1.37	27.8°	[540]
$[Co(NH_2CH_2CH_2NH_2)_3]_2(Pb_2Cl_9)Cl \cdot 3\,H_2O$	1.37	27.4°	[539]
$[Co(NH_2CH_2CH_2NH_2)_3]Cl_3 \cdot 3\,H_2O$	1.36	27.7°	[1091]
$[Co(NH_2CH_2CH_2NH_2)_3](SCN)_3$	1.36	26.9°	[170]
$[Co(NH_2CH_2CH_2NH_2)_3](NO_3)_3$	1.36	27.1°	[1103]
$[Co(NH_2CH_2CH_2NH_2)_3](CdCl_6)Cl_2 \cdot 2\,H_2O$	1.36	26.7°	[1066]
$[Co(NH_2CH_2CH_2NH_2)_3](C_4H_4O_6)Cl \cdot 5\,H_2O$	1.36	26.9°	[1042]
$[In(MeCOCHCOMe)_3]$	1.36	26.1°	[943]
$[Co(NH_2CH_2CH_2NH_2)_3][Cr(CN)_5(NO)]2\,H_2O$	1.35	27.3°	[407]
$[Co(O_2C_7H_5)_3]$	1.35	27.5°	[330]
$(C_{14}H_{19}N_2)[Mg(CF_3COCHCOCF_3)_3]$	1.35	26.5°	[1057]
$[Co(NH_2CH_2CH_2NH_2)_3][Cr(NH_2CH_2CH_2NH_2)_3]Cl_6 \cdot 6H_2O$			
(Cr cation)	1.35	26.2°	[1090, 1092]
(Co cation)	1.34	27.2°	[1090, 1092]
$K[Ni(phen)_3][Co(C_2O_4)_3]2\,H_2O$ (anion)	1.34	27.0°	[197]
$[Cr\{NHC(NH_2)NC(NH_2)NH\}_3]H_2O$	1.34	25.5°	[250]
$[Co(NH_2C_6H_{10}NH_2)_3]Cl_3 \cdot 5\,H_2O$	1.34	26.8°	[962]
$[Co(NH_2C_6H_{10}NH_2)_3]Cl_3 \cdot H_2O$	1.34	25.5°	[661]
$[Ti(MeCOCHCOMe)_3](ClO_4)$	1.34	24.7°	[1045]
$[Co\{S_2P(OMe)_2\}_3]$	1.34	24.7°	[743]
$[Cr(NH_2CH_2CH_2NH_2)_3](NCS)_3$	1.33	26.7°	[7]
$[Fe(phen)_3]I_2 \cdot 2\,H_2O$	1.33	26.5°	[619]
$K_3[Cr(O_2C_6H_4)_3]1\frac{1}{2}H_2O$	1.33	25.2°	[922]
$K_3[Al(C_2O_4)_3]3\,H_2O$	1.33	23.7°	[1035]
$[Rh(NH_2C_6H_{10}NH_2)_3](NO_3)_3 \cdot 3\,H_2O$	1.33	25.3°	[799]
$[Co(NH_2CH_2CH_2NH_2)_3](Cu_2Cl_8)Cl_2 \cdot 2\,H_2O$	1.33	26.9°	[566]
$[Rh(NH_2CH_2CH_2NH_2)_3]Cl_3 \cdot 3\,H_2O$	1.33	27.6°	[1093]
$[Co(NH_2CH_2CH_2NH_2)_3][Cr(NH_2CH_2CH_2NH_2)_3](SCN)_6 \cdot nH_2O$			
(Co cation)	1.33	25.5°	[173]
(Cr cation)	1.33	26.4°	[173]
$[Cr(NH_2CH_2CH_2NH_2)_3][Rh(NH_2CH_2CH_2NH_2)_3]Cl_6 \cdot 6\,H_2O$			
(Rh cation)	1.33	27.6°	[1094]
(Cr cation)	1.32	26.8°	[1094]
$[Fe(phen)_3](ClO_4)_3 \cdot H_2O$	1.32	27.8°	[65]
$[Fe(phen)_3](SbC_4O_6H_2)$	1.32	27.6°	[1118]
$(bipyH)[Fe(bipy)_3](ClO_4)_4$	1.32	27.6°	[438]
$[Al(O_2C_7H_5)_3]$	1.32	24.3°	[814]
$K_3[Cr(C_2O_4)_3]3\,H_2O$	1.32	23.9°	[1035]

(Continued)

Table 8.1 (continued)

Complex	b	θ	Ref.
[Cr{S$_2$P(OEt)$_2$}$_3$]	1.32	23.8°	[972]
[Cr(NH$_2$CH$_2$CH$_2$NH$_2$)$_3$][Ni(CN)$_5$]1$\frac{1}{2}$H$_2$O	1.32	25.6°	[920]
[Cr(NH$_2$CH$_2$CH$_2$NH$_2$)$_3$]Cl$_3$·3H$_2$O	1.32	26.7°	[1091]
[Cr(NH$_2$CH$_2$CH$_2$NH$_2$)$_3$]Cl$_3$·2H$_2$O	1.31	26.3°	[1095, 1096]
[Cr(NH$_2$CH$_2$CH$_2$NH$_2$)$_3$][Co(CN)$_6$]6H$_2$O	1.31	25.2°	[921]
[Cr(NH$_2$CH$_2$CH$_2$NH$_2$)$_3$](SCN)$_3$	1.31	25.9°	[169]
[Cr(NH$_2$CH$_2$CH$_2$NH$_2$)$_3$](SCN)$_3$·O.75H$_2$O	1.31	25.1°	[171, 172]
[Cr(NH$_2$CH$_2$CH$_2$NH$_2$)$_3$]I$_3$·H$_2$O	1.31	25.3°	[28]
[Ru(NH$_2$CH$_2$CH$_2$NH$_2$)$_3$]Cl$_3$·3H$_2$O	1.31	26.0°	[861]
[Ni(NH$_2$CH$_2$CH$_2$NH$_2$)$_3$](SO$_4$)	1.31	25.2°	[776]
[Ni(NH$_2$CH$_2$CH$_2$NH$_2$)$_3$](CH$_3$COO)$_2$·2H$_2$O	1.31	25.2°	[280]
[Sc(MeCOCHCOMe)$_3$]	1.31	23.7°	[45]
[Cr(O$_2$C$_6$Cl$_4$)$_3$]CS$_2$·$\frac{1}{2}$C$_6$H$_6$	1.31	25.5°	[876]
[V{S$_2$P(OEt)$_2$}$_3$]	1.31	20.7°	[456]
[Ni(NH$_2$CH$_2$CH$_2$NH$_2$)$_3$](ClO$_4$)$_2$·H$_2$O	1.30	24.6°	[915]
[Ni(NH$_2$CH$_2$CH$_2$NH$_2$)$_3$](Ph$_4$B)$_2$·3Me$_2$SO	1.30	25.4°	[278]
K$_3$[Fe(O$_2$C$_6$H$_4$)$_3$]1$\frac{1}{2}$H$_2$O	1.30	22.4°	[922]
[Co(HOCH$_2$CH$_2$OH)$_3$](SO$_4$)	1.29	24.1°	[48]
K[Ni(phen)$_3$][Co(C$_2$O$_4$)$_3$]2H$_2$O (cation)	1.29	22.9°	[197]
[Ni(phen)$_3$][Mn(CO)$_5$]$_2$	1.28	24.1°	[454]
[Fe(O$_2$C$_{14}$H$_8$)$_3$]C$_{14}$H$_8$O$_2$	1.28	21.8°	[176]
K[Cd(MeCOCHCOMe)$_3$]H$_2$O	1.28	0.4°	[500]
[Os(phen)$_3$](ClO$_4$)$_2$	1.27	26.1°	[1086]
[Ru(bipy)$_3$](PF$_6$)$_2$	1.27	25.9°	[935]
[Ni(bipy)$_3$]$_2$Cl$_2$(C$_4$H$_4$O$_6$)nH$_2$O	1.27	24.1°	[1077]
[Ni(bipy)$_3$](SO$_4$)7.5H$_2$O	1.27	24.0°	[1078]
[Zn(HOCH$_2$CH$_2$OH)$_3$](SO$_4$)	1.27	23.3°	[49]
[W(OCH$_2$CH$_2$O)$_3$]	1.27	18.4°	[969]
[Mo(O$_2$C$_{14}$H$_8$)$_3$]	1.27	0.0°	[875]
[Fe(O$_2$C$_7$H$_5$)$_3$]	1.26	19.3°	[524]
Rb$_2$[Na(CF$_3$COCHCOCF$_3$)$_3$]	1.24	2°	[433]
[Ni(S$_2$CNBu$_2$)$_3$]Br	1.24	22.7°	[422]
[Co(S$_2$CNC$_4$H$_8$O)$_3$]2C$_6$H$_6$	1.24	22.6°	[194]
[Co(S$_2$CNH$_2$)$_3$]	1.24	22.6°	[914]
[Co(S$_2$COC$_6$H$_2$Me$_3$)$_3$]	1.24	22.5°	[227]
[Fe(S$_2$COEt)$_3$]	1.24	19.9°	[1080]
	1.22	20.5°	[577]
[Co(S$_2$CC$_5$H$_8$N)$_3$]Me$_2$CO	1.23	22.6°	[749]
[Co(S$_2$CNEt$_2$)$_3$]	1.23	21.8°	[148, 782]
[Co(S$_2$COEt)$_3$]	1.23	21.6°	[783]
[Co(S$_2$CNC$_4$H$_8$O)$_3$]CH$_2$Cl$_2$	1.23	21.5°	[547]
[Fe(S$_2$CNC$_4$H$_4$)$_3$]$\frac{1}{2}$CH$_2$Cl$_2$	1.23	22.3°	[98]
[Fe(S$_2$CNEt$_2$)$_3$] (79 K)	1.23	20.3°	[709]
(297 K)	1.21	18.8°	[709]
[Fe{S$_2$CN(CH$_2$Ph)$_2$}$_3$] (150 K)	1.23	21.2°	[20]
(295 K)	1.21	20.5°	[20]
[Fe(S$_2$CNC$_4$H$_8$O)$_3$]2C$_6$H$_6$	1.22	21.3°	[194]
[Fe(S$_2$CSBu)$_3$]	1.22	21.1°	[719]
[Fe(S$_2$CNMePh)$_3$]	1.22	20.4°	[548]
[Fe(S$_2$CNC$_4$H$_8$)$_3$](ClO$_4$)	1.22	19.0°	[765]
[Fe(S$_2$CNEt$_2$)$_3$](I$_5$)	1.22	16.5°	[1086]
[Mn(S$_2$CNC$_5$H$_{10}$)$_3$](ClO$_4$)CHCl$_3$	1.22	19.5°	[164]

Complex		b	θ	Ref
[Fe(S$_2$CNMe$_2$)$_3$]	(150 K)	1.22	20.0°	[21]
	(295 K)	1.20	19.1°	[21]
[Fe(S$_2$CNC$_2$H$_4$OH)$_3$]	(150 K)	1.22	17.7°	[22]
	(295 K)	1.20	16.9°	[22]
[Fe(S$_2$CNBu$_2$)$_3$]C$_6$H$_6$		1.21	20.1°	[798]
[Cr(S$_2$CNC$_4$H$_8$O)$_3$]2 C$_6$H$_6$		1.21	21.3°	[194]
[Cr(S$_2$CNC$_4$H$_8$)$_3$]$\frac{1}{2}$C$_6$H$_6$		1.21	20.7°	[992]
[Cr(S$_2$CSEt)$_3$]		1.21	21.1°	[230, 721]
[Cr(S$_2$COEt)$_3$]		1.21	20.3°	[784]
[Fe(S$_2$CNC$_4$H$_8$)$_3$]	(150 K)	1.20	19.5°	[20]
	(295 K)	1.20	18.7°	[20, 548]
[Rh(S$_2$CNC$_4$H$_8$O)$_3$]2 C$_6$H$_6$		1.20	22.8°	[194]
[Rh(S$_2$CNEt$_2$)$_3$]		1.20	22.1°	[911]
[Ir(S$_2$CNC$_4$H$_8$)$_3$]$\frac{1}{2}$C$_6$H$_6$		1.20	22.7°	[922]
[Cr(S$_2$CNEt$_2$)$_3$]		1.20	20.0°	[912]
[Ga(S$_2$CNEt$_2$)$_3$]		1.20	19.4°	[376]
(BzPh$_3$P)$_2$[Fe{S$_2$CC(CO$_2$Et)$_2$}$_3$]		1.20	18.1°	[270, 568]
[Sc(O$_2$C$_7$H$_5$)$_3$]		1.20	17.0°	[46]
[Er(BuCOCHCOBu)$_3$]		1.20	0.0°	[1069]
[Sm{S$_2$P(C$_6$H$_{11}$)$_2$}$_3$]		1.20	13.2°	[785]
[Ir(S$_2$CNC$_4$H$_8$O)$_3$]2 C$_6$H$_6$		1.19	23.0°	[194]
[Ir(S$_2$CNEt$_2$)$_3$]		1.19	22.3°	[909]
[Ru(S$_2$CNC$_4$H$_8$O)$_3$]2$\frac{1}{2}$CHCl$_3$		1.19	20.4°	[906]
[Ru(S$_2$CNEt$_2$)$_3$]		1.19	19.0°	[880]
(Me$_4$N)[Ni(S$_2$COBu)$_3$]		1.19	20.7°	[235]
[Fe(S$_2$CNBu$_2$)$_3$]		1.19	16.8°	[576]
(Bu$_4$N)[Ni(S$_2$COBu)$_3$]		1.19	21.8°	[324]
(Me$_4$N)[Ni(S$_2$COC$_6$H$_{11}$)$_3$]Me$_2$CO		1.19	20.7°	[1015]
[Fe(S$_2$CNC$_4$H$_8$)$_3$]$\frac{1}{2}$C$_6$H$_6$		1.19	18.5°	[992]
[Fe(S$_2$CNC$_4$H$_8$O)$_3$]CHCl$_3$		1.19	16.9°	[195]
[Fe(S$_2$CNC$_4$H$_8$O)$_3$]PhMe		1.19	15.8°	[195]
[Fe(S$_2$CNC$_4$H$_8$O)$_3$]CH$_2$Cl$_2$		1.18	16.8°	[547]
[Pr{S$_2$P(C$_6$H$_{11}$)$_2$}$_3$]		1.18	12.6°	[785]
[Hg(phen)$_3$](CF$_3$SO$_3$)$_2$		1.15	14.9°	[308]
[In(S$_2$CNEt$_2$)$_3$]		1.14	16.9°	[376]
[In(S$_2$CNC$_5$H$_{10}$)$_3$]		1.14	16.5°	[542]
[Tl(S$_2$CNMe$_2$)$_3$]H$_2$O		1.12	16.6°	[4]
[Tl(S$_2$CNEt$_2$)$_3$]		1.12	14.8°	[641]
[Co(NO$_3$)$_3$]		1.12	20.9°	[560]
[Co(PhN$_3$Ph)$_3$]PhMe		1.08	20.7°	[259]
[Co(PhN$_3$Ph)$_3$]		1.07	20.8°	[676]

The four structures in Fig. 8.4 near b = 1.2 and θ_{min} = 0 that deviate markedly from the theoretical curve are discussed in more detail in Sect. 8.F. Complexes of manganese(III) and copper(II), and dithiolate complexes, have been deleted from Table 8.1 and Fig. 8.4, and are discussed separately in sections D and E respectively.

C. Distortions to Lower Symmetry

Figure 8.2 shows that molecules are expected to be more easily distorted from the most stable structure as the normalized bite is decreased. This is confirmed by the experimental data shown in Fig. 8.4, which shows a greater scattering of structures as b is decreased.

Extension of the repulsion energy calculations shows that distortions from D$_3$ symmetry are also increasingly possible as b is decreased.

The general stereochemistry is shown in Fig. 8.5. A twofold axis passes through the midpoint of the AD ligand, and the angles of twist θ_1 and θ_2 are obtained by projecting onto the plane passing through the midpoint of each bidentate ligand.

Typical potential energy surfaces are shown in Fig. 8.6 calculated for b = 2$^{1/2}$, as in the regular octahedron, and in Fig. 8.7 calculated for b = 1.12, as in the tris(dithiocarbamato)thallium complexes. It can be seen that as the normalized bite is decreased the surface becomes much flatter, allowing significant departure from the condition $\theta_1 = \theta_2$.

These surfaces may be compared with the three individual twist angles θ_1, θ_2, and θ_3, for complexes with b $\sim$ 1.41 and b $\sim$ 1.12 respectively (Table 8.2). As expected, there are much larger distortions from D$_3$ symmetry in the latter case due to crystal packing, but results are consistent with the general observation that structures usually lie within the first one or two contour lines of such surfaces.

D. Manganese(III) and Copper(II) Complexes

Manganese(III) and copper(II) complexes have been excluded from Table 8.1, since experience with complexes containing unidentate ligands leads to the expectation that the octahedral stereochemistry will be tetragonally distorted (Sect. 6.B.3). It may be noted that the trigonal twisting of an octahedron due to three bidentate ligands does not remove the Jahn-Teller degeneracy.

If, as a rough approximation, any Jahn-Teller distortions are ignored and the bond angles are averaged assuming D$_3$ symmetry, then the values of b and θ

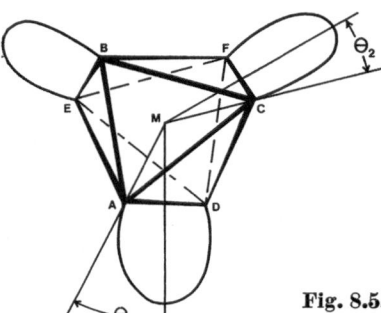

Fig. 8.5. General stereochemistry for [M(bidentate)$_3$] containing only a twofold axis

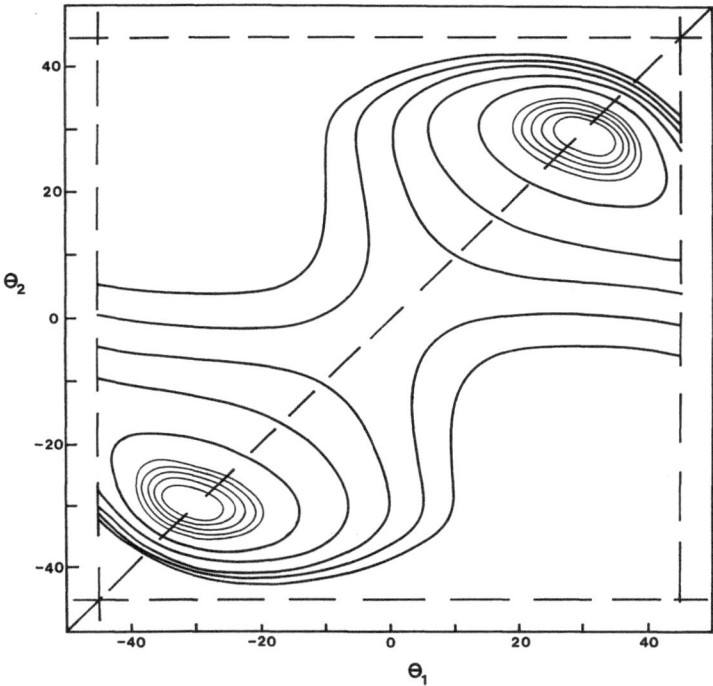

Fig. 8.6. Projection of the potential energy surface for [M(bidentate)$_3$] onto the $\theta_1 - \theta_2$ plane (in degrees). The five *faint contour lines* are for successive 0.02 increments above the minima, and the five *heavy contour lines* are for 0.2 increments above the minima, at $\theta_1 = \theta_2 = \pm 30°$. b = $2^{1/2}$, n = 6

(Table 8.3) are again in reasonable accord with the repulsion energy calculations (Fig. 8.8).

The difference between the shortest and longest bond length is much less than that observed in complexes containing unidentate ligands, but the type of distortion is very variable. In $(C_{14}H_{19}N_2)[Cu(CF_3COCHCOCF_3)_3]$, $[Cu(HOCH_2CH_2OH)_3](SO_4)$, $[Cu(NH_2CH_2CH_2NH_2)_3]Cl_2 \cdot {}^3/_4 C_2H_8N_2$, and $[Cu(phen)_3](ClO_4)_2$ there is a fairly

Table 8.2. Individual twist angles for [M(bidentate)$_3$]$^{x\pm}$

Complex	b	θ_1	θ_2	θ_3
$(Et_3NH)[P(O_2C_6H_4)_3]$	1.43	29.4°	29.4°	29.4°
$[Al(PhCOPCOPh)_3]$	1.42	29.2°	29.2°	29.7°
$[Cr(NH_2CH_2CH_2CH_2NH_2)_3][Ni(CN)_5]2H_2O$	1.41	28.5°	28.9°	29.6°
$[Co(NH_2CH_2CH_2CH_2NH_2)_3]Br_3$	1.40	27.9°	27.9°	27.9°
$[In(S_2CNEt_2)_3]$	1.14	14.1°	14.1°	21.5°
$[In(S_2CNC_5H_{10})_3]$	1.14	14.1°	15.9°	19.2°
$[Tl(S_2CNMe_2)_3]$	1.12	13.7°	14.7°	21.4°
$[Tl(S_2CNEt_2)_3]$	1.12	11.6°	11.6°	21.3°

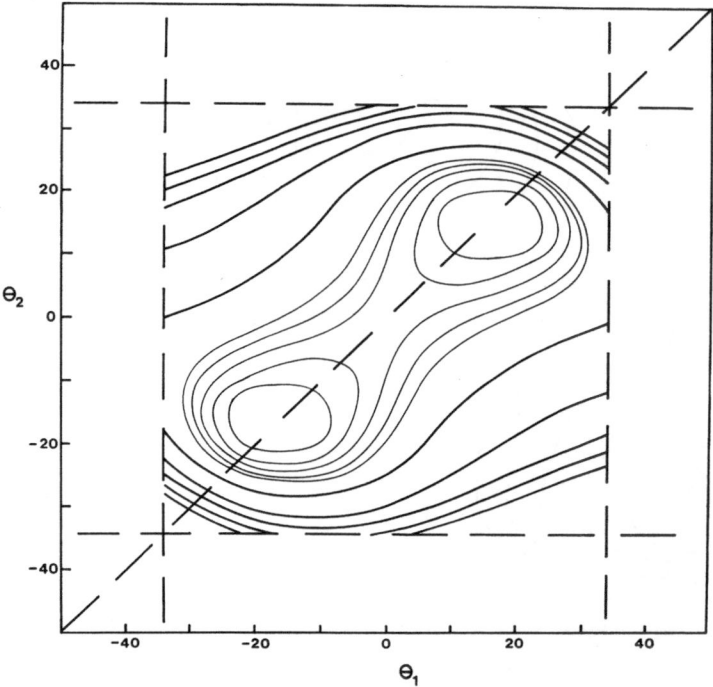

Fig. 8.7. Projection of the potential energy surface for [M(bidentate)$_3$] onto the $\theta_1 - \theta_2$ plane (in degrees). The five *faint contour lines* are for successive 0.02 increments above the minima, and the five *heavy contour lines* are for 0.2 increments above the minima, at $\theta_1 = \theta_2$ $= \pm 16.4°$. b = 1.12, n = 6

Table 8.3. Structural parameters for [M(bidentate)$_3$]$^{x\pm}$ (M = Mn^{III}, Cu^{II})

Complex	b	θ	Ref.
β-[Mn(MeCOCHCOMe)$_3$]	1.41	30.1°	[421]
γ-[Mn(MeCOCHCOMe)$_3$]	1.40	29.1°	[1023]
[Mn(O$_2$C$_7$H$_5$)$_3$]¼ PhMe	1.27	24.5°	[62]
[Mn(S$_2$CNEt$_2$)$_3$]	1.19	20.1°	[549]
[Mn{S$_2$CH(NC$_4$H$_8$O)}$_3$]CH$_2$Cl$_2$	1.18	17.6°	[193]
[Mn(S$_2$CNC$_4$H$_8$O)$_3$]CHCl$_3$	1.17	17.7°	[195]
[Cu{OP(NMe$_2$)$_2$CH$_2$P(NMe$_2$)$_2$O}$_3$](ClO$_4$)$_2$	1.41	30.6°	[793]
[Cu{OP(NMe$_2$)$_2$OP(NMe$_2$)$_2$O}$_3$](ClO$_4$)$_2$	1.39	29.0°	[616]
(C$_{14}$H$_{19}$N$_2$)[Cu(CF$_3$COCHCOCF$_3$)$_3$]	1.38	28.7°	[1057]
[Cu(HOCH$_2$CH$_2$OH)$_3$](SO$_4$)	1.30	—	[50]
[Cu(NH$_2$CH$_2$CH$_2$NH$_2$)$_3$](SO$_4$)	1.30	24.5°	[283]
[Cu(NH$_2$CH$_2$CH$_2$NH$_2$)$_3$](SO$_4$) (120 K)	1.30	25.1°	[106]
[Cu(NH$_2$CH$_2$CH$_2$NH$_2$)$_3$]Cl$_2 \cdot \frac{3}{4}$ C$_2$H$_8$N$_2$	1.29	24.7°	[106]
[Cu(phen)$_3$](ClO$_4$)$_2$	1.26	25.9°	[43]
[Cu(bipy)$_3$](ClO$_4$)$_2$	1.25	25.5°	[42]

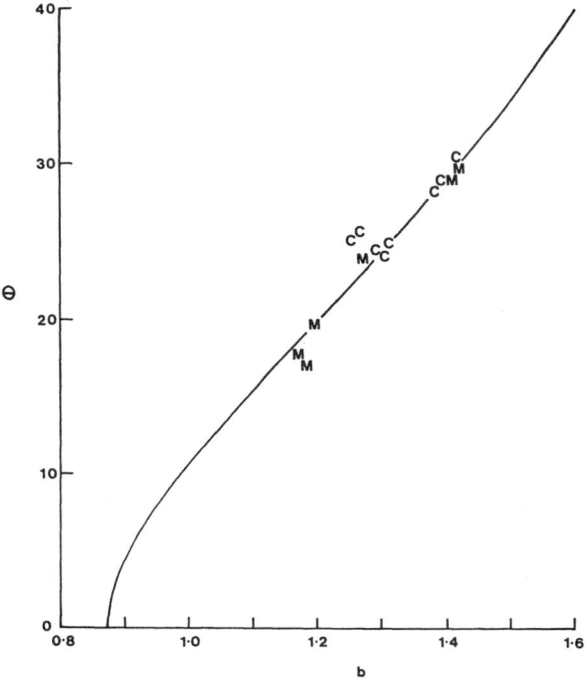

Fig. 8.8. Angle of twist θ (degrees) and normalized bite b for complexes of the type [Mn^{III}(bidentate)$_3$]$^{x\pm}$ (M) and [Cu^{II}(bidentate)$_3$]$^{x\pm}$ (C); theoretical curve for n = 6 also shown

obvious tetragonal elongation of the octahedron, two *trans* bonds being 8 to 16% longer than the other four. On the other hand, [Cu{OP(NMe$_2$)$_2$OP(NMe$_2$)$_2$O}$_3$](ClO$_4$)$_2$ and [Cu(NH$_2$CH$_2$CH$_2$NH$_2$)$_3$](SO$_4$) (at room temperature) have six equal copper-ligand bond lengths, and it is again necessary to propose a rapid oscillation between the three mutually perpendicular tetragonal distortions at each lattice site

Table 8.4. Structural parameters for tris(dithiolate) complexes

Complex	b	θ	Ref.
(Ph$_4$As)$_2$[Fe{S$_2$C$_2$(CN)$_2$}$_3$]	1.39	~ 24.5°	[979]
(Me$_4$N)$_2$[V{S$_2$C$_2$(CN)$_2$}$_3$]	1.33	17.0°	[1016]
[Mo(S$_2$C$_2$H$_2$)$_3$]	1.33	0°	[993]
[Mo{Se$_2$C$_2$(CF$_3$)$_2$}$_3$]	1.33	0°	[877]
(Ph$_4$As)$_2$[Mo{S$_2$C$_2$(CN)$_2$}$_3$]	1.32	14.0°	[161]
(Ph$_4$As)$_2$[W{S$_2$C$_2$(CN)$_2$}$_3$]	1.32	14.0°	[161]
[Mo(S$_2$C$_6$H$_4$)$_3$]	1.31	0°	[272]
[V(S$_2$C$_2$Ph$_2$)$_3$]	1.31	0°	[389]
[Re(S$_2$C$_2$Ph$_2$)$_3$]	1.30	0°	[390]
(Ph$_4$As)[Ta(S$_2$C$_6$H$_4$)$_3$]	1.30	~ 16.0°	[764]
(Ph$_4$As)[Nb(S$_2$C$_6$H$_4$)$_3$]	1.29	0.6°	[273]
(Me$_4$N)$_2$[Zr(S$_2$C$_6$H$_4$)$_3$]	1.28	19.4°	[274]

(see Sect. 6.B.3). In [Cu(bipy)$_3$](ClO$_4$)$_2$ the two *trans*-ligands are not pushed away equally, one bond being 10% and the other 21% longer than the other four bonds. Even more complex distortions are found for many of the complexes in Table 8.3.

E. Dithiolate Complexes

Tris(bidentate) complexes [M(S$_2$C$_2$R$_2$)$_3$]$^{x-}$ (R=H, CN, Ph, CF$_3$) form complexes that are much further distorted away from the octahedron toward the trigonal prism than would be expected from simple repulsion energy calculations. A second problem is the considerable difficulty in formulating these complexes, which may be described as either dithiolate or dithiolene:

Dithiolene   Dithiolate

The latter is obviously unrealistic for complexes such as [V(S$_2$C$_2$Ph$_2$)$_3$], which would require the formation of vanadium(VI). Evidence from C—C and C—S bond lengths show that the bonding is intermediate between these extremes, with extensive electron delocalization between the metal and ligands. Similar complexes are formed with benzenedithiolate, o-C$_6$H$_4$S$_2^{x-}$.

Values of the normalized bite b and the twist angle θ for these complexes are given in Table 8.4. In sharp contrast to other tris(chelate) complexes, it is very clear that θ is not a simple function of b. In all cases θ is lower than predicted, and in many cases the trigonal prism with $\theta = 0°$ is observed. The stereochemistry closest to that expected from repulsion energy calculations is that of [Fe{S$_2$C$_2$(CN)$_2$}$_3$]$^{2-}$, for which $\theta = 24.5°$, compared with $\theta \sim 28°$ expected for b = 1.39. These dithiolate ligands are clearly different from other bidentate ligands, even other sulfur donors such as MeSCH=CHS$^-$, dithioacetylacetonate, dithiooxalate, dithiophosphates, dithiocarbamates, xanthates, and thioxanthates. (Table 8.1).

It is not clear why dithiolate ligands stabilize the trigonal prism. As the octahedron is twisted toward the trigonal prism the S··S distance between the three ligands necessarily decreases. Bonding within these S$_3$ triangles, together with the utilization of metal d-orbitals, so that the ligands approach a sexidentate, may well be important. Complexes closer to the trigonal prism appear to have shorter C—S bonds, suggesting an increased dithiolene contribution relative to the dithiolate contribution [637].

F. Some Exceptional Structures

Of the 158 compounds in Table 8.1, four have angles of twist grossly at variance with predictions (Fig. 8.4), and it is necessary to seek other explanations for their structures:

	b	θ
$[Mo(O_2C_{14}H_8)_3]$	1.27	0.0°
$K[Cd(MeCOCHCOMe)_3]H_2O$	1.28	0.4°
$Rb_2[Na(CF_3COCHCOCF_3)_3]$	1.24	2°
$[Er(Bu^tCOCHCOBu^t)_3]$	1.20	0.0°

The ligands in $[Mo(O_2C_{14}H_8)_3]$ can be considered as the oxidized o-quinone, as the anion of the reduced dihydroxyphenanthrene, or as the intermediate semi-quinone, with simultaneous change in the formal oxidation state of the molybdenum atom (compare with the dithiolene and dithiolate complexes in the previous section):

o-quinone   dianion

In contrast, the structure of $[Fe(O_2C_{14}H_8)_3]C_{14}H_8O_2$ is normal (b = 1.28, θ = 21.8°), as are a number of catecholate $[M(O_2C_6H_4)_3]^{x-}$ and substituted catecholate complexes (b = 1.30—1.43, θ = 22.4—29.4°) (Table 8.1).

In the trigonal prismatic molybdenum-phenanthrenequinone complex, two ligands behave normally with a planar $MoO_2C_{14}H_8$ structure. However, the dihedral angle between the third ligand and its MoO_2 plane is only 120°. This third chelate ring has shorter C—O bonds (1.31 Å compared with 1.34 Å), and longer Mo—O bonds (1.98 Å compared with 1.95 Å), suggesting a greater quinone character. This bending of one chelate ring allows the molecules to stack together as pairs with strong charge transfer interactions (Fig. 8.9).

The remaining three exceptional structures contain very large metal atoms, and are very uncrowded six-coordinate molecules (see Sect. 13.C.4). Two observations may be noted [637]. Firstly, K+ and Rb+ cations lie outside the triangular faces of the prisms, which will contract these faces and unscrew the octahedron into a trigonal prism. Secondly, the structures may be associated with the bulk of the alkyl groups on the β-diketonates.

Clearly, additional work on large metal atoms of low oxidation state, with bulky β-diketonate ligands, would be of considerable interest.

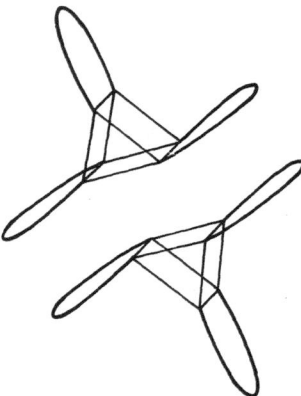

Fig. 8.9. A pair of trigonal prismatic [Mo(O$_2$C$_{14}$H$_8$)$_3$] molecules, showing dimer formation

G. Some Comments on Ligand Design

If the only way of obtaining the normalized bite of a bidentate ligand were from the crystal structure, the repulsion energy approach to stereochemistry would be limited. No matter how useful the theory may be in rationalizing the relation between different stereochemical parameters, it would have little predictive use since these other parameters are precisely determined at the same time as the normalized bite. Therefore, in order to utilize fully this repulsion approach to stereochemistry, it is essential to be able to make some estimate of the expected normalized bite of any coordinated ligand.

Figure 8.10 again displays some of the data in Tables 8.1 and 8.3, and Figs. 8.4 and 8.8. The data are now restricted to complexes of the first row transition metals from chromium to copper to avoid large variations in central atom size, the 12-coordinate metallic radii being in the range 1.2—1.3 Å.

To a first approximation the ligands divide into three groups. The first group is contained within the limits b = 1.37$_5$ to b = 1.50 and consists of those ligands that form six-membered chelate rings such as acetylacetonate and trimethylenediamine. The second group is contained within the limits b = 1.25 to b = 1.37$_5$ and contains the complexes with five-membered chelate rings such as o-phenanthroline and ethylenediamine. The third group is contained within b = 1.05 to b = 1.25 and contains the complexes with four-membered rings such as nitrate and dithiocarbamates. This strikingly simple and important correlation is further illustrated in Fig. 8.11 with three representative examples: [Co(MeCOCHCOMe)$_3$], [Co(NH$_2$CH$_2$CH$_2$NH$_2$)$_3$]$^{3+}$, and [Co(NO$_3$)$_3$].

Extension to flexible seven-membered rings does not necessarily increase the normalized bite, as indicated by [Co(NH$_2$CH$_2$CH$_2$CH$_2$NH$_2$)$_3$]Br$_3$, for which the parameters are b = 1.40 and θ = 27.9°. This bite is achieved by substantial buckling of the seven-membered ring.

Further subdivisions can be made according to the size of the nonmetal atoms in the chelate ring. For example, the four-membered chelate ring group may be divided into four subgroups. The subgroup containing ligands of lowest normalized bite (b $= 1.05-1.12_5$) consists of $(PhN_3Ph)^-$ and NO_3^-, where three small second-row elements complete the chelate ring. There are no known examples of the second subgroup. The third subgroup of normalized bite $1.17-1.25$ contains ligands with one second-row element and two larger elements in the chelate ring, such as the dithiocarbamates and xanthates.

The fourth subgroup contains the dithiophosphates with three large ring atoms, and this subgroup intrudes into the group containing the five-membered chelate rings. A comparison of the ring geometries of $[Co(NO_3)_3]$, $[Co(S_2COEt)_3]$, and $[Co\{S_2P(OMe)_2\}_3]$ is shown in Fig. 8.12. The central cobalt atom in $[Co\{(OH)_2Co(NH_2CH_2CH_2NH_2)_2\}_3](SCN)_4(NO_3)_2 \cdot 2H_2O$ [329] and $[Co\{(OH)_2Co(NH_2CH_2CH_2NH_2)_2\}_3](S_2O_6)_3 \cdot 8H_2O$ [1044] can be considered to be linked to three large four-membered chelate rings, each of which contains a different cobalt atom and two oxygen atoms (Fig. 8.12). The structural parameters, $b = 1.32$, $\theta = 27.5°$, and $b = 1.33$, $\theta = 27.8°$ respectively, are in good agreement with prediction (compare with Fig. 8.4).

The smallest normalized bite for a four-membered chelate ring that can be envisaged is probably the borohydride ion. Although crystal structure data are not available, the electron-diffraction data of gaseous $[Al(BH_4)_3]$ have been

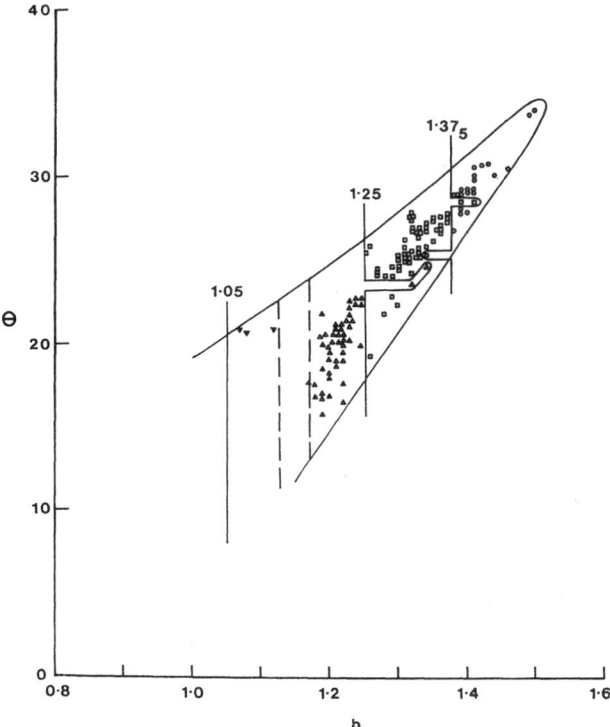

Fig. 8.10. Angle of twist θ (degrees) and normalized bite b for complexes of the type $[M(bidentate)_3]^{x\pm}$ (M=Cr, Mn, Fe, Co, Ni, Cu)

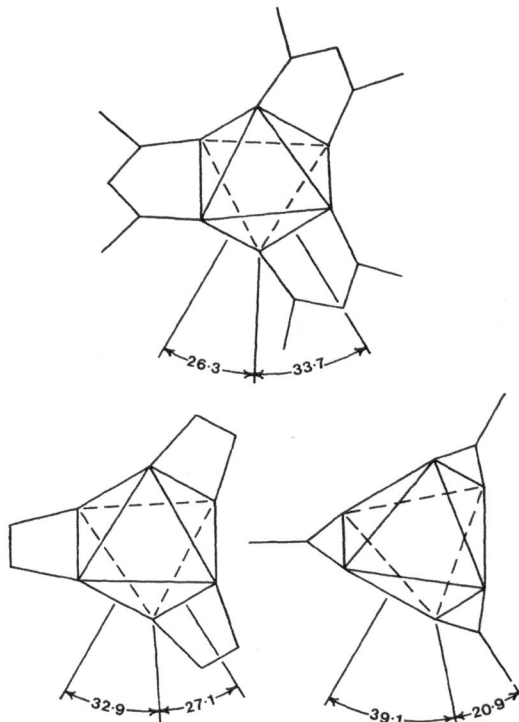

Fig. 8.11. [Co(acac)$_3$], [Co(en)$_3$]$^{3+}$, and [Co(NO$_3$)$_3$]; angles in degrees

interpreted as showing approximately trigonal prismatic coordination of three bidentate borohydride groups, b = 1.20, θ ∼ 0° [38].

Unusually large normalized bites may be achieved with five-membered chelate rings by again incorporating large atoms into the ring. The intrusion of the five-membered ring region into the six-membered ring region in Fig. 8.10 is due to a dithiooxalate complex containing two sulfur atoms in the chelate ring, all other five-membered rings in Fig. 8.10 containing only carbon, nitrogen, and oxygen atoms. Comparative ring geometries in [Co(C$_2$O$_4$)$_3$]$^{3-}$ and [Co(S$_2$C$_2$O$_2$)$_3$]$^{3-}$ are also shown in Fig. 8.12.

Six-membered chelate rings have greater flexibility to accommodate larger atoms without large increases in normalized bite, as shown by (NH$_4$)$_2$[Pt(S$_5$)$_3$]2 H$_2$O, b = 1.44, θ = 30.3°. A comparison of the ring geometries in [Al(MeCOCHCOMe)$_3$] and [Al(PhCO·P·COPh)$_3$] is shown in Fig. 8.12.

The size of the metal atom is also important in determining the normalized bite, the smaller normalized bites being obtained with the larger metal atoms. Thus, the lowest normalized bite for a six-membered ring containing a transition metal ion (Table 8.1) is that observed for [Sc(MeCOCHCOMe)$_3$] in which the normalized bite of 1.31 is similar to complexes of the other transition metals containing five-membered chelate rings. Similarly, the lowest normalized bite for a transition metal complex containing a five-membered ring (Table 8.1) is b = 1.20, observed for the tropolonate complex [Sc(O$_2$C$_7$H$_5$)$_3$]. The metal-oxygen distance and the

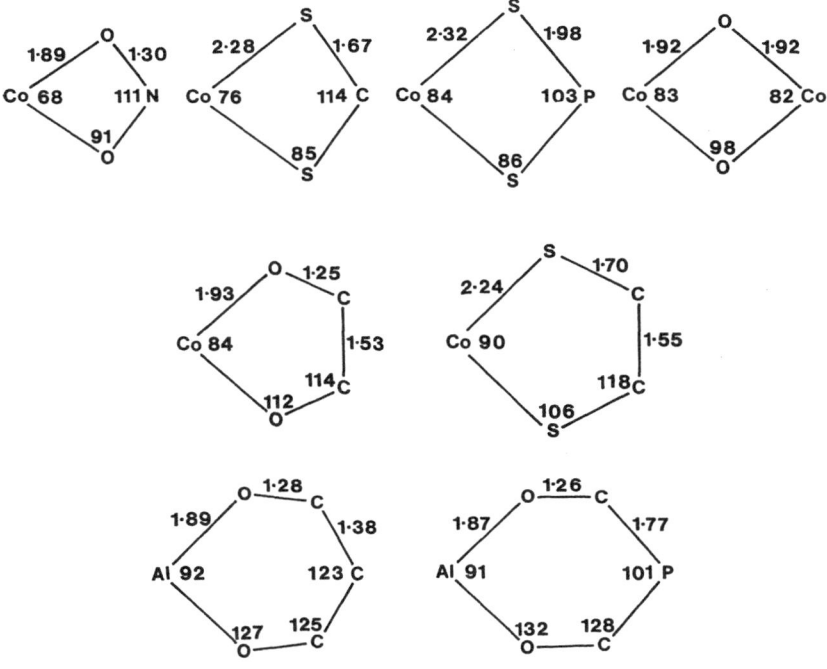

Fig. 8.12. Ring geometries, degrees and Å, in: $[Co(NO_3)_3]$, $[Co(S_2COEt)_3]$, $[Co\{S_2P(OMe)_2\}_3]$, $[Co\{(OH)_2Co(en)_2\}_3]^{6+}$, $[Co(C_2O_4)_3]^{3-}$, $[Co(S_2C_2O_2)_3]^{3-}$, $[Al(MeCOCHCOMe)_3]$, and $[Al(PhCOPCOPh)_3]$

O——O intraligand distance are not completely independent of each other. For example in β-diketonate complexes, an increase in the metal ion radius of about 10% causes a general loosening of the coordination sphere, with a small increase of about 2 to 5% in the O——O distance.

Conversely very small central atoms will confer unusually large normalized bites on the chelated ligand. For example, transition metal catecholate complexes have M—O ∼ 2.0 Å and b ∼ 1.3, whereas in the complexes of the smaller phosphorus(V) and silicon (IV), the M—O distance is reduced to 1.7_5 Å, and b increased to 1.4.

H. Intramolecular Rearrangements

The two most plausible modes for the intramolecular racemization of the optically active tris-chelate complexes involve twisting of opposite triangular faces of the octahedron in opposite directions. The first such reaction (Fig. 8.13a) involves a trigonal twist about the C_3-axis, with the formation of a trigonal prismatic transition state where the bidentate ligands span the three parallel edges. The potential energy curves in Fig. 8.2 correspond to the reaction coordinates for this type of reaction. The second mechanism (Fig. 8.13b) involves a trigonal

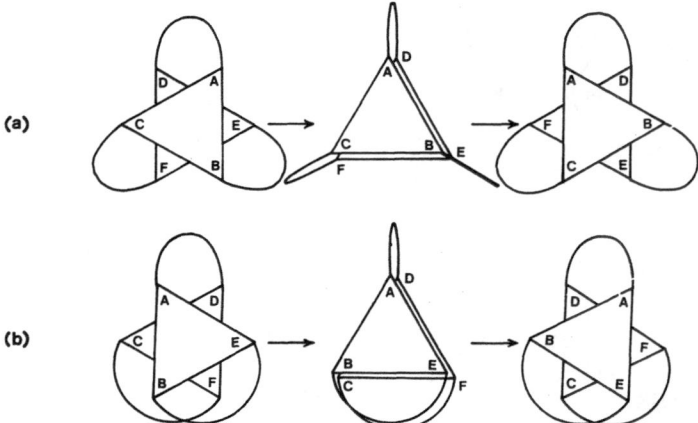

Fig. 8.13a, b. Racemization of tris(bidentate) complexes: **a** twist about the C_3-axis; **b** twist about a pseudo-C_3-axis

twist about one of the pseudo-C_3-axes, in which the stereochemistry of the transition state is a trigonal prism with two of the bidentate ligands spanning triangular edges.

The relative energy of these two transition-state trigonal prisms depends on the value of the normalized bite. In both transition states the upper bidentate ligand AD in Fig. 8.13 lies above the BEFC rectangle formed by the other two bidentate ligands. The favorable transition state will be when AD is projected above the long edges of this rectangle. For $b < 1.3$, the long edges are BC and EF which are not spanned by the bidentate ligands, and the first mechanism will be favored, the twist about the C_3 axis. For $b > 1.3$ the long edges are BE and CF which are spanned by the bidentate ligands, and it is the second mechanism which is favored.

The activation energy for these intramolecular twists are shown as a function of b in Fig. 8.14. Much higher energies of activation are found for other transition states, such as a planar hexagon, or a skew-trapezoidal bipyramid, in which one bidentate ligand is normal to a trapezoid formed by the other two bidentate ligands, the metal atom being in the plane of the trapezoid (Fig. 8.14). These intramolecular reactions will inevitably be associated with some degree of bond stretching, and oversimplified conclusions based on the above geometric guidelines should be avoided. Electronic effects will also be important in transition metal complexes.

Regardless of the proposed mechanism, it is clear from Fig. 8.14 that the activation energy will decrease as the normalized bite of the bidentate ligand decreases. Experimentally it is observed that complexes containing four-membered chelate rings racemize very rapidly, for example, many tris(dithiocarbamates) are fluctional at room temperature. Lower reaction rates are observed for five-membered ring systems. For example, the activation energy for racemization of substituted tropolonate complexes decreases along the series [Al(trop)₃] (~ 50

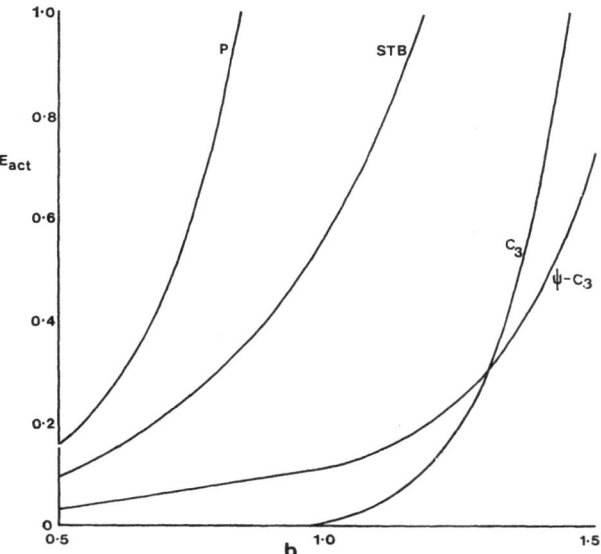

Fig. 8.14. Activation energies for racemization of tris(bidentate) complexes as a function of normalized bite b. C_3, trigonal twist about the C_3-axis; ψ-C_3, trigonal twist about a pseudo-C_3 axis, STB skew-trapezoidal bipyramidal intermediate; P, planar intermediate. n = 6

kjoule mol^{-1}) > [Ga(trop)$_3$] ($\sim$ 40 kjoule mol^{-1}) > [In(trop)$_3$] ($\sim$ 30 kjoule mol^{-1}) due to increasing metal-oxygen distances and decreasing normalized bites. Trigonal twisting of six-membered ring systems is much slower, and the β-diketonate complexes generally isomerize through a bond-breaking mechanism with five-coordinate intermediates.

I. Unsymmetrical Bidentate Ligands

The difference in repulsion energy between the *fac*- and *mer*-isomers of [M(unsymmetric bidentate)$_3$]$^{x\pm}$, as a function of effective bond length ratio R(A/D) and normalized bite b, is shown in Fig. 8.15. The stabilization of the *mer*-isomer is small, although greater than that found for complexes of the type [M(unidentate A)$_3$(unidentate B)$_3$] (Sect. 6.C).

Many examples of both isomers are known.

J. [M(bidentate A)(bidentate B)$_2$]

The general stereochemistry for a complex having one bidentate ligand different from the other two is shown in Fig. 8.16. The normalized bite of the AD ligand is b_1, and the BE and CF ligands are b_2. The angular coordinates are defined relative to a pseudo-threefold axis normal to the plane containing the metal atom and the

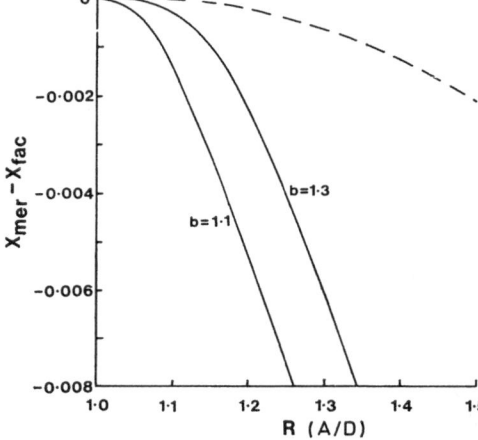

Fig. 8.15. Stabilization of *mer*-[M(asymmetric bidentate)₃] as a function of bond-length ratio R(A/D) and normalized bite b; *broken line* corresponds to [M(unidentate A)₃(unidentate B)₃]. n = 6

midpoints of the three bidentate ligands. A twofold axis passes through the midpoint of AD. The individual angles of twist for the bidentate ligands, θ_1 and θ_2, are defined as before. The introduction of an additional angle θ_X (Fig. 8.16) is required to completely define the stereochemistry.

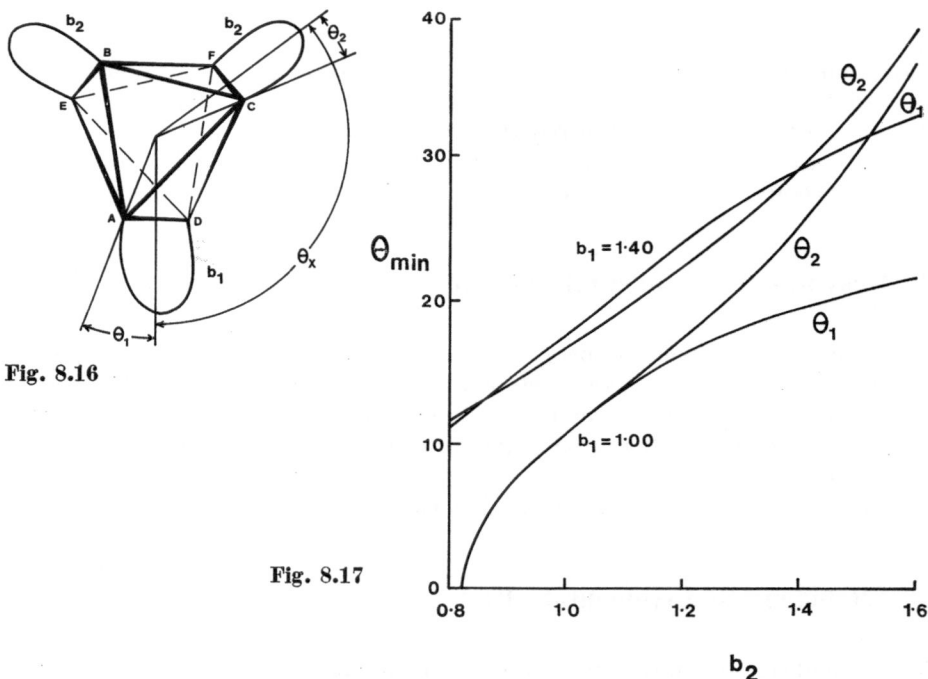

Fig. 8.16

Fig. 8.17

Fig. 8.16. General stereochemistry for [M(bidentate A)(bidentate B)₂]

Fig. 8.17. Individual twist angles (degrees) for [M(bidentate A)(bidentate B)₂] as a function of normalized bites b_1 and b_2, calculated for n = 6

The decrease in the individual angles of twist on decreasing the individual normalized bites is shown in Fig. 8.17, for selected values of b_1. Provided b_1 and b_2 are not too dissimilar, the dependence of the *average* angle of twist upon the *average* normalized bite agrees very well with that calculated in Sect. A. For example, at $b_1 = 1.40$, $b_2 = 1.00$ (average b = 1.13), Fig. 8.17 shows $\theta_1 = 17.6$, $\theta_2 = 16.8$ (average $\theta = 17.1°$), compared with $\theta = 17.0°$ at b = 1.13 calculated for [M(bidentate)$_3$] (Fig. 8.3). Similarly at $b_1 = 1.00$, $b_2 = 1.40$ (average b = 1.27), Fig. 8.17 shows $\theta_1 = 19.6$, $\theta_2 = 25.1$ (average $\theta = 23.3°$), compared with $\theta = 23.0°$ at b = 1.27 found for [M(bidentate)$_3$]. That is, each bidentate ligand does not adopt an individual twist angle appropriate to its normalized bite (for example $\theta = 10.8$ and $29.3°$ at b = 1.00 and 1.40 respectively, as found for [M(bidentate)$_3$] (Fig. 8.3)), but the total twist of the molecule corresponds to the total repulsion interactions in the molecule. This distortion is achieved by tilting the ABC triangular face relative to the DEF face.

CHAPTER 9

Six-Coordinate Compounds Containing Tridentate Ligands

A. [M(tridentate)(unidentate)$_3$]

For six-coordinate compounds containing one tridentate ligand, there are two minima on the potential energy surface, corresponding to the *meridional* and *facial* octahedral isomers (Fig. 9.1) [430, 431].

For flexible tridentate ligands, and for effective bond length ratios R(unidentate/tridentate) ≤ 1.0 and normalized bites b < 1.3, it is predicted that the *fac*-isomer will be more stable than the *mer*-isomer (Fig. 9.2). However the tridentate angle ABC is also an important factor, and is shown in Fig. 9.3. Flexible ligands such as $NH_2CH_2CH_2NHCH_2CH_2NH_2$ can clearly be *mer* or *fac*. On the other hand ligands such as terpyridyl can only be *mer*, while ligands such as $[S(CH_2COO)_2]^{2-}$ can only be *fac*.

In the *mer*-isomer, the bond angle between the unidentate ligands, DME=EMF, decreases as b decreases, and increases as R(unidentate/tridentate) decreases. Molecules of known structure can be fitted against calculations and values of R(unidentate/tridentate) obtained in the usual way. For example the bond angles in [Ga(terpy)Cl$_3$] [97] and [Ti{MeN(CH$_2$C$_5$H$_3$NMe)$_2$}Cl$_3$] [256] yield R(Cl$^-$/tridentate) $= 0.8-0.9$ (n $= 6$), as expected for a charged unidentate ligand and an uncharged tridentate ligand. It is also predicted that the M-unidentate bond *trans* to the hinging atom will be shorter than the other two M-unidentate bonds [431] and this is observed: [Ga(terpy)Cl$_3$], ME/MD $= 0.94$; [Ti{MeN(CH$_2$C$_5$H$_3$NMe)$_2$}Cl$_3$], ME/MD $= 0.96$.

A similar fitting procedure can be used for the *fac*-isomer. For example the O=Mo=O bond angles of $\sim 105°$ found in [Mo{NH(CH$_2$CH$_2$NH$_2$)$_2$}O$_3$] [265] and K$_3$[Mo{N(CH$_2$COO)$_2$}O$_3$] [191] yield R(O^{2-}/tridentate) ~ 0.7 (n $= 6$), which is again as expected.

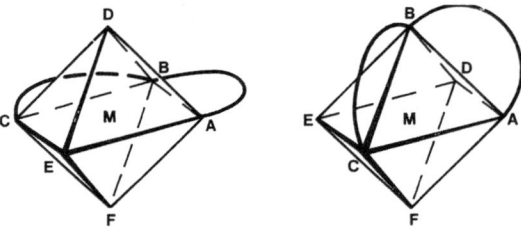

Fig. 9.1. *mer*- and *fac*-isomers of [M(tridentate)(unidentate)$_3$].

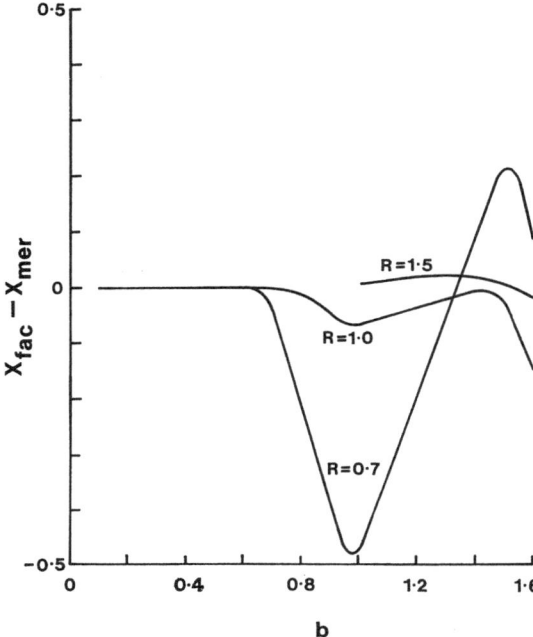

Fig. 9.2. Difference in stability between *mer-* and *fac-*[M(tridentate)(unidentate)$_3$] as a function of normalized bite, b, and effective bond length ratio, R(unidentate/tridentate). n = 6

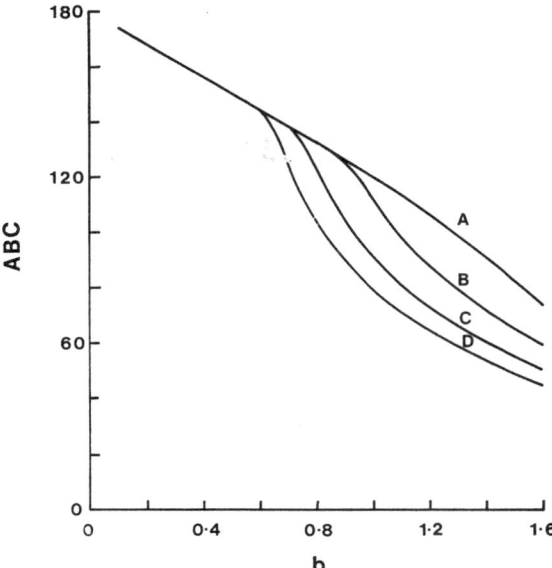

Fig. 9.3. Tridentate angle ABC (degrees) in [M(tridentate)(unidentate)$_3$] complexes as a function of normalized bite, b. A, *mer*-complexes; B, *fac*-complexes, R(unidentate/tridentate) = 1.5; C, *fac*-complexes, R(unidentate/tridentate) = 1.0; D, *fac*-complexes. R(unidentate/tridentate) = 0.7. n = 6

B. [M(tridentate)$_2$]

For six-coordinate compounds containing two tridentate ligands, there are
three minima on the potential energy surface corresponding to the *meridional*,
symmetrical-facial, and *unsymmetrical-facial* octahedral isomers (Fig. 9.4) [425].
The *mer*-isomer exists as a minimum all at values of the normalized bite, whereas
the *sym-fac* and *unsym-fac* exist only above b = 1.1.

For flexible tridentate ligands, it is predicted that the *mer*-isomer is significantly
stabilized below b ~ 1.3, and the *unsym-fac* isomer above b ~ 1.5 (Fig. 9.5).
These predictions may again be modified by the design of the tridentate ligand,
the tridentate angle ABC being shown as a function of b in Fig. 9.6.

These predictions are in agreement with compounds of known crystal structure
(Table 9.1). The *mer*-isomer is seen to be stabilized with ligands of small normalized
bite, b < ~ 1.3, and/or with relatively rigid tridentate ligands such as terpyridyl
in which the tridentate angle ABC cannot fall below ~ 100°. On the other hand,
flexible tridentate ligands such as $NH(CH_2CH_2NH_2)_2$ form all three isomers,
with ABC = 62−100°. The predicted order of isomer stability is in agreement with the observed equilibrium mixture of $[Co\{NH(CH_2CH_2NH_2)_2\}_2]^{3+}$

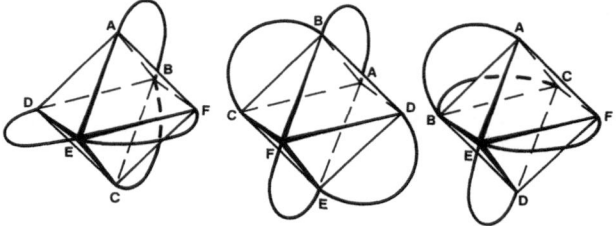

Fig. 9.4. *mer-*, *sym-fac-*
and *unsym-fac*-isomers
of [M(tridentate)$_2$]

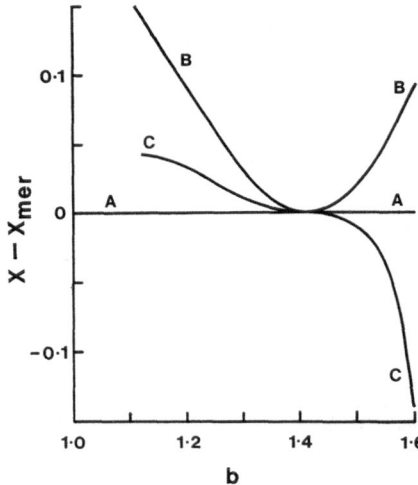

Fig. 9.5. Relative repulsion energy coefficients
for [M(tridentate)$_2$] as a function of normalized
bite, b. A, *mer*-complexes; B, *sym-fac*-com-
plexes; C, *unsym-fac*-complexes. n = 6

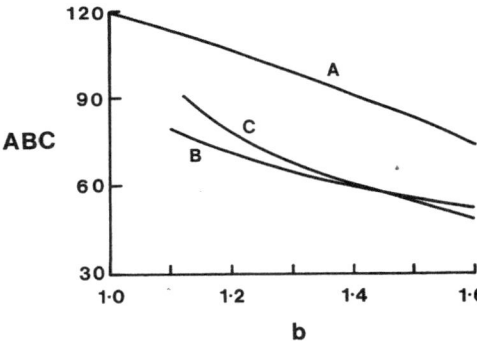

Fig. 9.6. Tridentate angle ABC (degrees) in [M(tridentate)$_2$] complexes as a function of normalized bite, b. A, *mer*-complexes; B, *sym-fac*-complexes; C, *unsym-fac*-complexes. n = 6

isomers in solution, of ~ 10 (*mer*): ~ 3 (*unsym-fac*): ~ 1 (*sym-fac*) [635]. Tridentate ligands such as S(CH$_2$CH$_2$NH$_2$)$_2$, with large donor atoms at the hinging B site and which form five-membered chelate rings, must have small ABC angles and can only exist as the *fac*-isomers (Table 9.1) [976].

One point of interest is the stereochemistry of the *unsym-fac*-isomer, as at low values of the normalized bite (b $\sim 1.1-1.2$), ABDE approaches a rectangle. The resulting stereochemistry, viewed normal to this ABDE plane, is shown in Fig. 9.7 b. The stereochemistry is not closely related to the regular octahedron, but could be considered as being intermediate between a trigonal prism (Fig. 9.7 a) and

Table 9.1. Structures of [M(tridentate)$_2$]$^{x\pm}$

Complex	b	ABC	Structure	Ref.
[Zn(purpurate)$_2$]4 H$_2$O	1.25	103	*mer*	[1087]
[Co{C$_5$H$_3$(CMe :NNH$_2$)$_2$}$_2$]I$_2$·H$_2$O	1.26	108	*mer*	[436]
Rb[Cr{C$_5$H$_3$N(COO)$_2$}$_2$]	1.27	102	*mer*	[457]
[Cr(terpy)$_2$](ClO$_4$)$_3$·H$_2$O	1.27	—	*mer*	[1097]
[Co(terpy)$_2$]Br$_2$·3 H$_2$O	1.27	101	*mer*	[767]
[Ni{NH(CH$_2$CNH$_2$NOH)$_2$}$_2$]Cl·2 H$_2$O	1.27	66	*sym-fac*	[284]
[Co(terpy)$_2$](SCN)$_2$	1.28	101	*mer*	[908]
[Zn{NH(CH$_2$CH$_2$NH$_2$)$_2$}$_2$](NO$_3$)$_2$	1.28	100	*mer*	[1124]
[Mo{S(CH$_2$CH$_2$S)$_2$}$_2$]	1.29	90	*unsym-fac*	[602]
[Ni{NH(CH$_2$CH$_2$NH$_2$)$_2$}$_2$]Cl$_2$·H$_2$O	1.31	98	*mer*	[111]
[Ni{NH(CH$_2$CONH$_2$)$_2$}$_2$](ClO$_4$)$_2$	1.31	68	*unsym-fac*	[978]
[Ni{(C$_2$H$_5$O)N(CH$_2$CH$_2$OH)$_2$}$_2$](NO$_3$)$_2$	1.31	67	*sym-fac*	[837]
Cs$_2$[Ni{NH(CH$_2$COO)$_2$}$_2$]4 H$_2$O	1.31	65	*sym-fac*	[756]
Li$_2$[Ni{NH(CH$_2$COO)$_2$}$_2$]4 H$_2$O	1.32	64	*sym-fac*	[756]
[Ni{C$_5$H$_3$N(CMeNO)$_2$}$_2$]	1.32	98	*mer*	[1004]
K$_2$[Ni{S(CH$_2$COO)$_2$}$_2$]3 H$_2$O	1.35	63	*unsym-fac*	[311]
[Co{NH(CH$_2$CH$_2$NH$_2$)$_2$}$_2$]Br$_3$	1.35	96	*mer*	[843]
[Co{NH(CH$_2$CH$_2$NH$_2$)$_2$}$_2$](NO$_3$)$_3$·H$_2$O	1.36	95	*mer*	[955]
[Co{NH(CH$_2$CH$_2$NH$_2$)$_2$}$_2$][Co(CN)$_6$]2 H$_2$O	1.36	64	*unsym-fac*	[672]
K[Co{NH(CH$_2$COO)$_2$}$_2$]2.5 H$_2$O	1.37	64	*unsym-fac*	[262]
[Co{NH(CH$_2$CH$_2$NH$_2$)$_2$}$_2$]Br$_3$	1.38	62	*sym-fac*	[662]
[Co{S(CH$_2$CH$_2$NH$_2$)$_2$}$_2$]Cl$_3$·2 H$_2$O	1.38	57	*unsym-fac*	[523]
[Ni{NH(CH$_2$CH$_2$CH$_2$NH$_2$)$_2$}$_2$](ClO$_4$)$_2$	1.43	89	*mer*	[111]

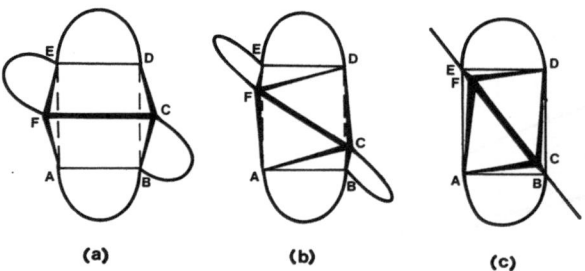

Fig. 9.7a—c. *unsym-fac*-[M(tridentate)₂]: **a** trigonal prism; **b** intermediate; **c** skew trapezoidal bipyramid

a skew-trapezoidal bipyramid with A and D as the apical sites (Fig. 9.7c). The structure of *unsym-fac*-[Mo{S(CH₂CH₂S)₂}₂] (b = 1.29) has been described as a trigonal prism, although significant distortions arise from differences in Mo—S bond lengths, and in the normalized bite between different chelate rings. The use of unsymmetric tridentate ligands produce three *unsym-fac* isomers, one of which will be close to a trigonal prism. For example, in the acetylacetonebenzoylhydrazone complex [V(PhCONNCMeCHCOMe)₂] [323], the two five-membered chelate rings (b = 1.22) form a rectangular face (ABDE in Fig. 9.7), with the six-membered chelate rings BC and EF (b = 1.33) completing the trigonal prism in Fig. 9.7a, rather than the structure in 9.7b.

CHAPTER 10

Seven-Coordinate Compounds Containing only Unidentate Ligands

A. Six Symmetrical Structures

There are no regular or semiregular polyhedra available for seven-coordination, and the starting point therefore is a consideration of the planar heptagon, three non-uniform polyhedra, the pentagonal bipyramid, capped trigonal prism, and end-capped trigonal prism (Sect. 2.A.3), and two chemical coordination polyhedra, the capped octahedron and hexagonal pyramid (Sect. 2.B.3). The repulsion energy coefficient X, calculated for n = 6, are listed:

	X		X
Capped octahedron	3.230	End-capped trigonal prism	4.484
Capped trigonal prism	3.231	Hexagonal pyramid	6.964
Pentagonal bipyramid	3.266	Planar heptagon	17.000

Attention will be focussed on the first three structures which have closely comparable energies, and are shown in Fig. 10.1. The remaining three structures have prohibitively high repulsion energy, and are neither expected nor experimentally observed.

It is important to note that the structures shown in Fig. 10.1 contain two or three different types of ligand site, and different sized faces and edge lengths.

The edges between the axial site of a pentagonal bipyramid and an equatorial site are 20% longer than the edges connecting equatorial sites, and the repulsion

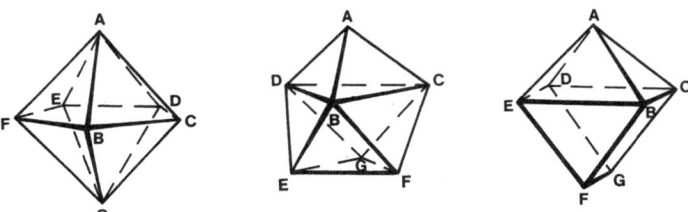

Fig. 10.1. The pentagonal bipyramid, capped octahedron, and capped trigonal prism

experienced by the two axial atoms is considerably less than that experienced by the five equatorial atoms:

$$AB = 1.4142r, \qquad BC = 1.1756r$$

$$Y_A = 0.320, \qquad Y_B = 0.525 \qquad (n = 6)$$

The stereochemistry of the capped octahedron is defined by polyhedral edge lengths, or by the angular parameters ϕ_B and ϕ_E, the angles the M—B and M—E bonds make with the threefold axis. The capping atom is associated with the greatest repulsion energy, although the difference between the different sites is not nearly as great as for the pentagonal bipyramid. Parameters calculated for $n = 6$ are shown:

$$\phi_B = 74.6°, \qquad \phi_E = 130.3°$$

$$AB = 1.2124r, \qquad BE = 1.2676r$$

$$EF = 1.3218r, \qquad BC = 1.6701r$$

$$Y_A = 0.514, \qquad Y_B = 0.454, \qquad Y_E = 0.452$$

Similarly the capped trigonal prism is defined by the angles the M—B and M—F bonds make to the twofold axis, ϕ_B and ϕ_F, and θ_B (defining $\theta_F = 0$):

$$\phi_B = 79.4°, \qquad \phi_F = 143.3°, \qquad \theta_B = 48.7°,$$

$$AB = 1.277r, \qquad BC = 1.297r, \qquad BF = 1.233r,$$

$$FG = 1.195r, \qquad BE = 1.478r,$$

$$Y_A = 0.482, \qquad Y_B = 0.437, \qquad Y_F = 0.501$$

B. Relations Between the Pentagonal Bipyramid, Capped Octahedron, and Capped Trigonal Prism

The three polyhedra in Fig. 10.1 are of comparable stability, and it is shown in this section that there are no potential energy barriers between them. Two types of interconversion are illustrated, the first in which a twofold axis is maintained, and the second in which a mirror plane is maintained throughout the transformations.

The transformation of a pentagonal bipyramid to a capped trigonal prism with retention of a twofold axis is shown in Fig. 10.2. The twofold axis passes through atom A, while the θ_B and θ_C coordinates are defined relative to $\theta_F = 0$ and $\theta_G = 180°$. The potential energy surface for this transformation is shown projected onto the $\theta_B - \theta_C$ plane in Fig. 10.3. Movement along the flat valley corresponds to free rotation of the four BCDE atoms relative to the two FG atoms.

The general stereochemistry containing a mirror plane is shown in Fig. 10.4. Atoms A, B, and C lie on the mirror plane, and the axes are defined by $\phi_A = 0$, $\theta_B = 0$, $\theta_C = 180°$. The potential energy surface projected onto the $\phi_B - \phi_C$

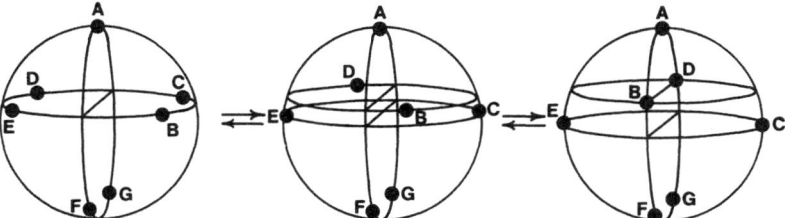

Fig. 10.2. Transformation of the capped trigonal prism (*left*) to the pentagonal bipyramid (*right*) with retention of a twofold axis

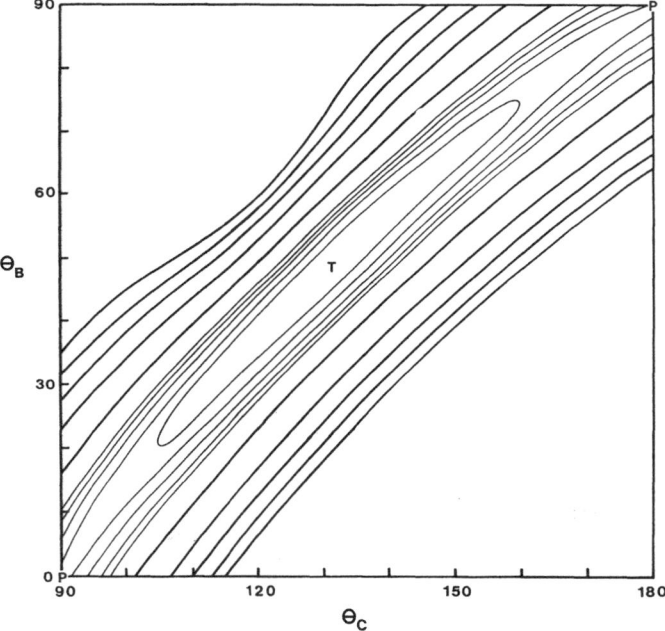

Fig. 10.3. Projection of the potential energy surface for [M(unidentate)$_7$] onto the $\theta_B - \theta_C$ plane (in degrees), with retention of a twofold axis. The five *faint contour lines* are for successive 0.02 increments above the minimum, and the five *heavy contour lines* are for successive 0.2 increments above the minimum, at T. n = 6. The locations of the pentagonal bipyramid (P) and capped trigonal prism (T) are indicated

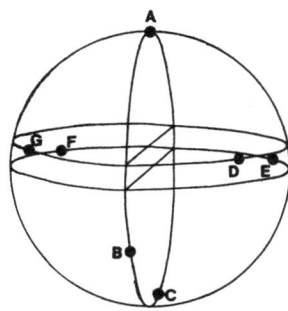

Fig. 10.4. General stereochemistry for [M(unidentate)$_7$] with retention of a mirror plane

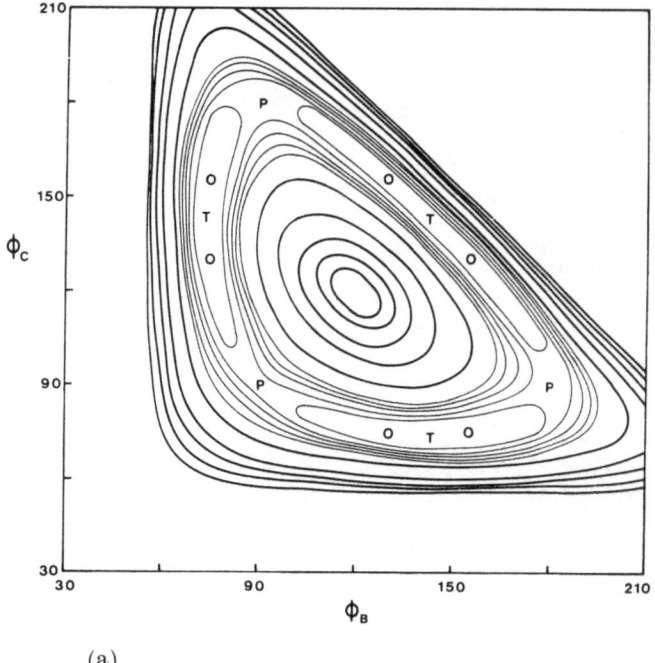

(a)

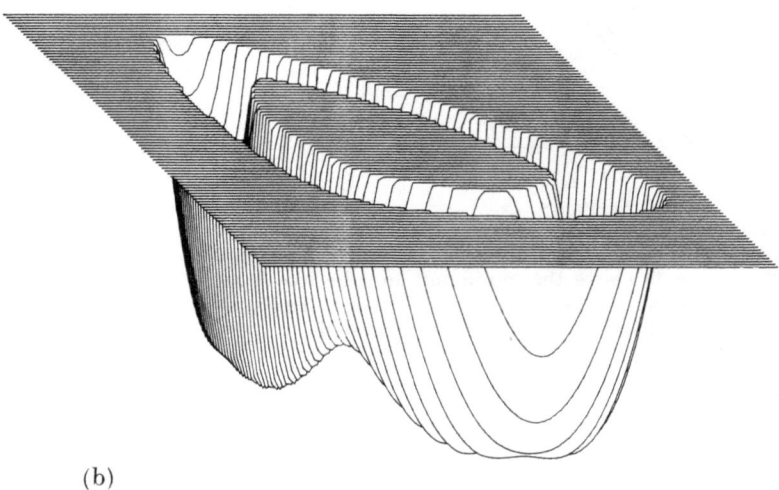

(b)

Fig. 10.5. a Projection of potential energy surface for [M(unidentate)$_7$] onto the $\theta_B - \theta_C$ plane (in degrees), with retention of a mirror plane. The five *faint contour lines* are for successive 0.02 increments above the minima, and the five *heavy contour lines* for successive 0.2 increments above the minima, at T. n = 6. The locations of the pentagonal bipyramids (P), capped octahedra (O), and capped trigonal prisms (T) are indicated. **b** As in **a**, but with truncation at X = 0.1

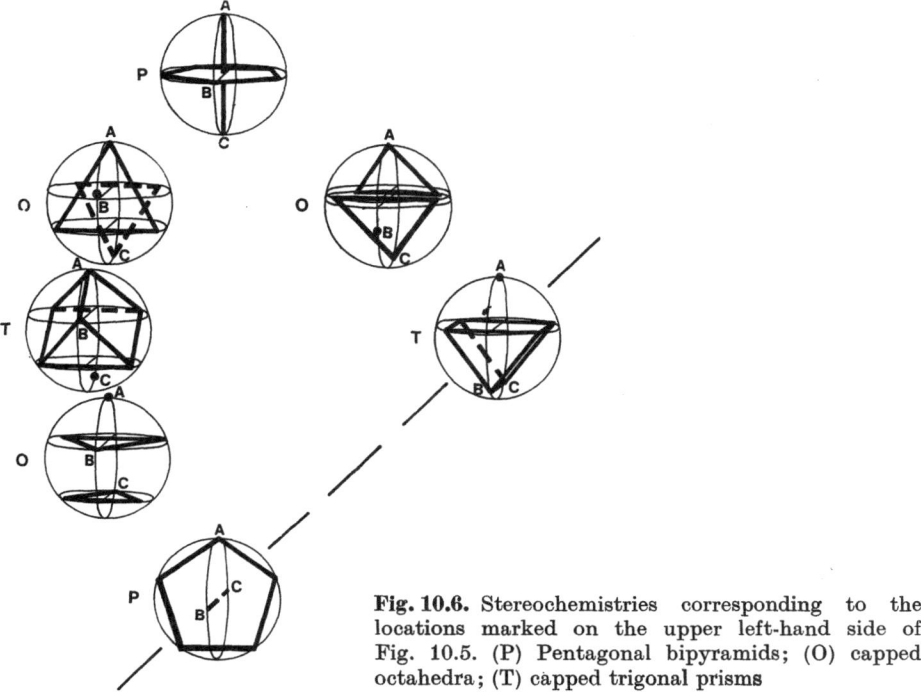

Fig. 10.6. Stereochemistries corresponding to the locations marked on the upper left-hand side of Fig. 10.5. (P) Pentagonal bipyramids; (O) capped octahedra; (T) capped trigonal prisms

plane is shown in Fig. 10.5. The surface is symmetrical across the line $\phi_B = \phi_C$, and the continuous "reaction coordinate" lies in the bottom of the "moat" shown. The locations of the pentagonal bipyramid (P), capped octahedron (O), and capped trigonal prism (T) are marked. Eeach stereochemistry occurs more

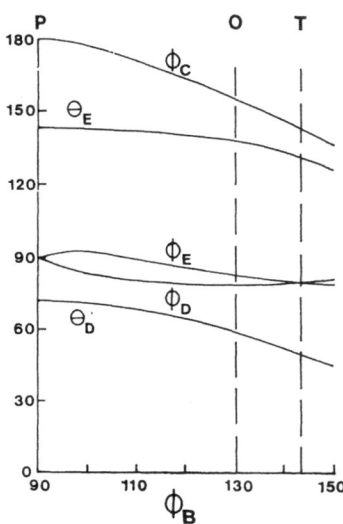

Fig. 10.7. Angular coordinates (degrees) corresponding to movement along the bottom of the potential energy trough in the upper righthand portion of Fig. 10.5. n = 6. The locations of the pentagonal bipyramid (P), capped octahedron (O), and capped trigonal prism (T) are indicated

than once on the potential energy surface, but this apparent redundancy is useful when comparison is made with related surfaces (see later).

The stereochemistry at each of these locations is shown in Fig. 10.6. It can be seen that the capped octahedron can be considered as an intermediate between the pentagonal bipyramid and the capped trigonal prism, but it is much closer in energy and stereochemistry to the latter.

The angular parameters corresponding to the change from the pentagonal bipyramid through the capped octahedron to the capped trigonal prism, depicted on the righthand side of Fig. 10.6, is shown in Fig. 10.7. It can be seen that the change in stereochemistry, which involves very large atom movements, is smooth and continuous. At $\phi_B = 121.7°$, AMB = DMG and ADBG becomes a square, with the formation of the so-called "4 : 3 piano stool" stereochemistry.

C. Comparison with Experiment

Structurally characterized molecules with seven identical ligands are restricted mainly to some fluorocomplexes $[MF_7]^{x-}$, aquo complexes $[M(H_2O)_7]^{2+}$, isonitrile complexes $[M(CNR)_7]^{2+}$, and cyanide complexes $[M(CN)_7]^{x-}$. The stereochemistries are distributed along the long valley connecting the capped trigonal prism and the pentagonal bipyramid. The relative metal-ligand bond lengths of the fluoro complexes are in accord with the predictions from simple repulsion theory, but not those of the isonitrile or cyanide complexes.

The X-ray data on IF_7 are not sufficient to obtain the precise stereochemistry [179, 334]. The ^{19}F NMR shows equivalence of all seven fluorine atoms, consistent with rapid intramolecular rearrangement [76, 476, 818]. The cubic unit cell observed in $(NH_4)_3[ZrF_7]$ cannot contain an ordered seven-coordinate anion, and the structure has been described as a distorted pentagonal bipyramid [590].

The anion in $K_2[NbF_7]$ has approximately capped trigonal prismatic stereochemistry [162]. The averaged structural parameters assuming full C_{2v} symmetry are $\phi_B = 78.6°$, $\phi_F = 143.1°$, and $\theta_B = 48.2°$ (atom labels according to Fig. 10.1) and are in excellent agreement with those calculated, $\phi_B = 79.4°$, $\phi_F = 143.3°$, and $\theta_B = 48.7°$ (Sect. 10.A). If the bond angles are not averaged assuming full C_{2v} symmetry, the structure is distorted towards a pentagonal bipyramid along the reaction coordinate shown in Fig. 10.5. The two fluorine atoms *trans* to the capping atom move around the mirror plane so that AMB = $\phi_B = 140.5°$ and AMC = $\phi_C = 145.6°$ (atom labels according to Figs. 10.4 and 10.6), which corresponds to the molecule moving along the potential energy surface (Fig. 10.5) about 20% of the way to the capped octahedron, or about 5% of the way to the pentagonal bipyramid.

In the calcium polyiodide compound $[Ca(H_2O)_7](I_{10})$ the stereochemistry appears to be intermediate between a capped octahedron and a pentagonal bipyramid, whereas $[Sr(H_2O)_7](I_{12})$ contains a capped trigonal prismatic cation [1046].

A close approximation to a capped trigonal prism is also observed for $[Mo(CNBu)_7](PF_6)_2$ [720]. There is a crystallographic mirror plane normal to the AFG plane (Fig. 10.1), the relevant angular parameters averaged assuming full

C_{2v} symmetry being $\phi_B = 82.0°$, $\phi_F = 144.2°$, and $\theta_B = 50.2°$. However, the corresponding methyl compound $[Mo(CNMe)_7](BF_4)_2$, lies considerably further along the potential energy valley, and is closer to a capped octahedron than a capped trigonal prism, $\phi_B = 134.4°$, $\phi_C = 154.1°$ [146]. The tungsten compound $[W(CNBu)_7](W_6O_{19})$ is similar [697]. Solutions of $[Mo(CNBu)_7]^{2+}$ yield only a single ^{13}C NMR signal even down to $-135°C$, showing the expected stereochemical nonrigidity [730].

The structures of $K_4[V(CN)_7]2H_2O$ [715], $K_5[Mo(CN)_7]H_2O$ and $Na_5[Mo(CN)_7]$ $\cdot 10H_2O$ [350], and $K_4[Re(CN)_7]2H_2O$ [961] are close to pentagonal bipyramids.

The structure of $[Er(O:C_7H_{10}O)_7](ClO_4)_3$ has been described as a distorted pentagonal bipyramid [220]. Structural details are not yet available, but the distortion appears considerable ($O_{apical}-Er-O_{apical} = 171°$, $O_{apical}-Er-O_{equatorial} = 82-107°$).

D. [M(unidentate A)₆(unidentate B)]

Figure 10.8 shows a typical potential energy surface calculated for one effective bond length shorter than the other six:

$$MA = R = 0.8, MB = MC = MD = ME = MF = MG = 1.0.$$

A mirror plane has been assumed, and Fig. 10.8 should be compared with Fig. 10.5, which was calculated for seven equal bond lengths. It can be seen that continuous "moat" in Fig. 10.5 contracts to two "ponds" at $\phi_B = 92.6°$ and $\phi_C = 180°$ (and vice versa), corresponding to a pentagonal bipyramid with the short bond in one of the axial sites. Also shown in Fig. 10.8 are the locations of the capped octahedron with the short bond in the capping site at $\phi_B = 78.5°$, $\phi_C = 131.4°$ (O), the capped trigonal prism with the short bond in the capping site at $\phi_B = \phi_C = 144.1°$ (T), and the pentagonal bipyramid with the short bond occupying one of the equatorial sites at $\phi_B = \phi_C = 91.4°$ (P').

A typical potential energy surface calculated for a single long bond, $MA = R = 1.2$, is shown in Fig. 10.9. Destablization of the pentagonal bipyramid with the single long bond in an axial site (at $\phi_B = 87.9°$, $\phi_C = 180.0°$ and vice versa) results in a bridging of the "moat" at this stereochemistry with the creation of two "lagoons", one deeper than the other. The deepest minimum at $\phi_B = 70.4°$, $\phi_C = 129.1°$ corresponds to the capped octahedron with the long bond occupying the capping site. This minimum is fairly flat, and the capped trigonal prism with the long bond occupying one of the sites *trans* to the capping atom is of comparable energy. The saddle at $\phi_B = \phi_C = 87.9°$ shows that the pentagonal bipyramid with the long bond in an equatorial site is also readily accessible. The second, less deep minimum at $\phi_B = \phi_C = 142.5°$ is the capped trigonal prism with the long bond in the capping site.

Although little structural information concerning compounds of the type [M(unidentate A)₆(unidentate B)] is available, there are known examples of all the predicted structures.

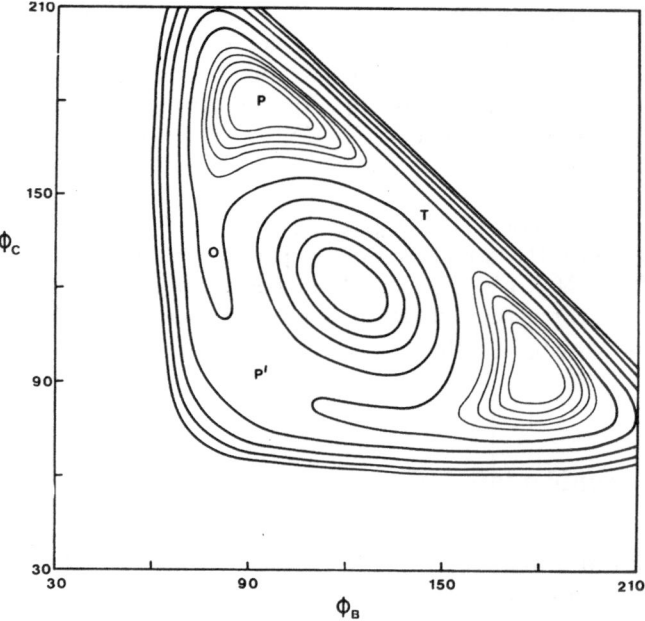

Fig. 10.8. Projection of the potential energy surface for [M(unidentate A)$_6$(unidentate B)] onto the $\phi_B - \phi_C$ plane (in degrees) with retention of a mirror plane. The five *faint contour lines* are for successive 0.02 increments above the minima, and the five *heavy contour lines* are for successive 0.2 increments above the minima, at P. R = 0.8, n = 6. The locations of the pentagonal bipyramids (P and P'), capped octahedron (O), and capped trigonal prism (T) are indicated

The anion in K$_3$[NbOF$_6$] is seven-coordinate, although the data are insufficient to establish the stereochemistry [1101].

The molybdenum isonitrile complexes [Mo(CNBu)$_6$I]I [718] and [Mo(CNBu)$_6$Br]Br [688] have capped trigonal prismatic stereochemistries, with the halogen atoms in the unique capping sites, $\phi_B = 145.1°$, $\phi_C = 145.2°$, and $\phi_B = 145.1°$, $\phi_C = 145.7°$, respectively. On the other hand, [Mo(CNBu)$_6$(SnCl$_3$)](ϕ_3BCNBϕ_3) lies in the alternative potential energy hollow in Fig. 10.9 at $\phi_B = 72.4°$, $\phi_C = 111.8°$, with a stereochemistry intermediate between a pentagonal bipyramid with the tin atom in an equatorial site, and a capped octahedron with the tin atom in the axial site.

The anions in K$_4$[V(CN)$_6$(NO)]$^1/_2$KOH$\cdot{}^1/_2$H$_2$O [351] and K$_3$Na[V(CN)$_6$(NO)]2 H$_2$O [610] are pentagonal bipyramidal with the nitrosyl ligand in one of the axial sites, as would be expected for this type of molecule.

Capped octahedral stereochemistry, with crystallographic threefold symmetry, is observed for [U(Me$_3$PO)$_6$Cl]Cl$_3$ [132].

The structure of [Er{OC(NHMe)$_2$}$_6$(H$_2$O)](ClO$_4$)$_3$ has been described as a pentagonal bipyramid with the water molecule in one of the equatorial sites [744].

NMR studies on [WF$_6$(Me$_3$P)] in acetonitrile or sulfur dioxide show fluxional behavior down to $-85°$ [1040].

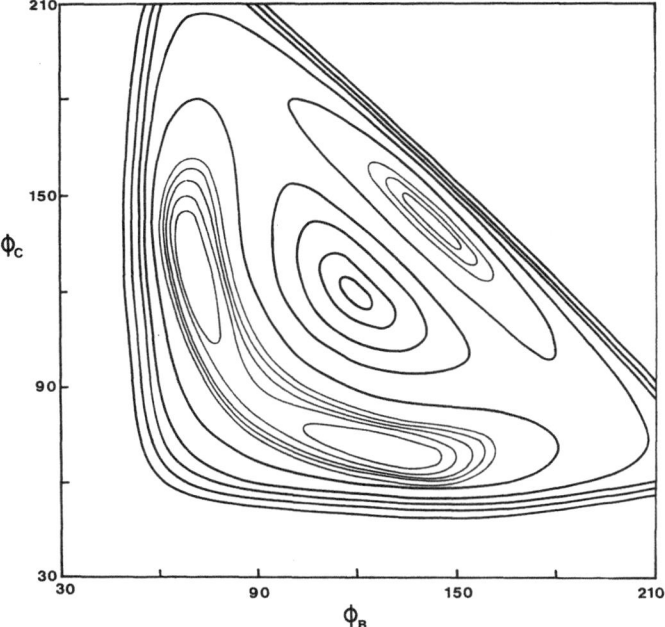

Fig. 10.9. Projection of the potential energy surface for [M(unidentate A)₆(unidentate B)] onto the $\phi_B - \phi_C$ plane (in degrees) with retention of a mirror plane. The five *faint contour lines* are for successive 0.02 increments above the minima, and the five *heavy contour lines* are for succesive 0.2 increments above the minima. R = 1.2, n = 6

E. [M(unidentate)₆(lone pair)]

The following compounds of the p-block elements contain seven electron pairs in the valence shell, but the six unidentate ligands are octahedrally arranged, with only small distortions [639]:

Group IV	Group V	Group VI
	$Sb^{III}Cl_6^{3-}$	$Te^{IV}Cl_6^{2-}$
	$Sb^{III}Br_6^{3-}$	$Te^{IV}Br_6^{2-}$
	$Bi^{III}Cl_6^{3-}$	
	$Bi^{III}Br_6^{3-}$	
	$Bi^{III}I_6^{3-}$	
$[Pb^{II}(ligand)_6]^{2+}$	$[Bi^{III}(ligand)_6]^{3+}$	

In these compounds the nonbonding pair of electrons does not have any marked influence on the stereochemistry, which is in sharp contrast to the complexes of the p-block elements with lower coordination numbers $[MX_3(\text{lone pair})]^{x\pm}$ (Sect. 3.A.3), $[MX_4(\text{lone pair})]^{x\pm}$ (Sect. 4.B.4), and $[MX_5(\text{lone pair})]^{x\pm}$ (Sect. 6.A). At first sight, it might be expected from these repulsion energy calculations, and from Fig. 10.8 in particular, that there would be a pentagonal bipyramidal arrangement of electron pairs with the lone pair occupying one of the axial sites. It is clear that in these sterically crowded molecules the nonbonding pair of electrons plays no significant stereochemical role. However the lone pair is observed to be stereochemically active in compounds in which the steric crowding is reduced by the introduction of chelate groups, as in [Sb^{III}(S$_2$COEt)$_3$(lone pair)] (Chap. 11). Even the presence of chelate groups does not allow stereochemical activity of the lone pair as the coordination number further increases, as shown by the undistorted structures observed for compounds such as [Te^{IV}(S$_2$CNEt$_2$)$_4$] (Chap. 13).

Xenon hexafluoride, on the other hand, is less sterically crowded than the above MX$_6^{x\pm}$ compounds, as shown by the polymeric structure in the solid state (five Xe—F bonds of about 1.85 Å, with two bridging atoms, Xe—F = 2.2—2.6 Å) [183], and by a nonoctahedral structure in the vapor phase in which the lone pair of electrons exhibits some stereochemical influence [75, 180]. The model proposed for the vapor state, however, is not the pentagonal bipyramid with the electron pair in an axial site, but a fluxional structure in which an octahedron is distorted by the lone pair projecting towards one face (that is, a capped octahedron with the lone pair in a capping site) and/or by the lone pair projecting towards one edge (that is, a pentagonal bipyramidal structure with the lone pair in an equatorial site).

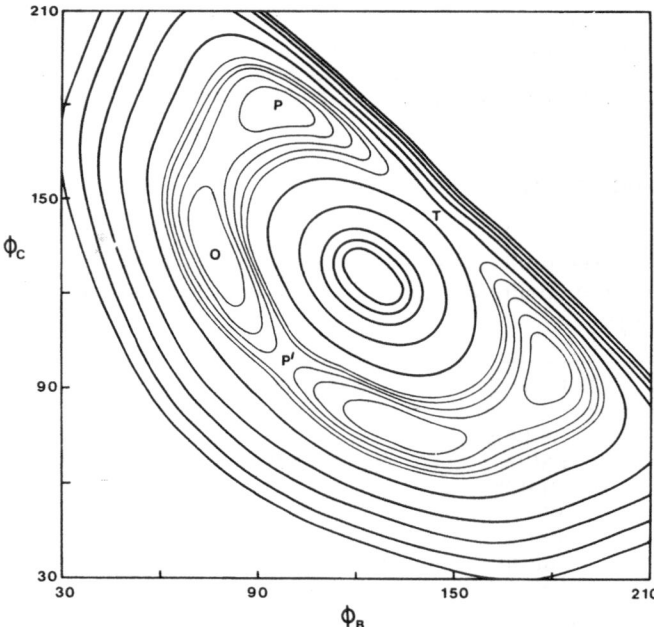

Fig. 10.10. Projection of the potential energy surface for [M(unidentate A)$_6$(unidentate B)] onto the $\phi_B - \phi_C$ plane (in degrees) with retention of a mirror plane. The five *faint contour lines* are for successive 0.02 increments above the minima, and the five *heavy contour lines* are for successive 0.2 increments above the minima, at O. R = 0.2, n = 6. The locations of the pentagonal bipyramids (P and P′), capped octahedron (O), and capped trigonal prism (T) are incicated

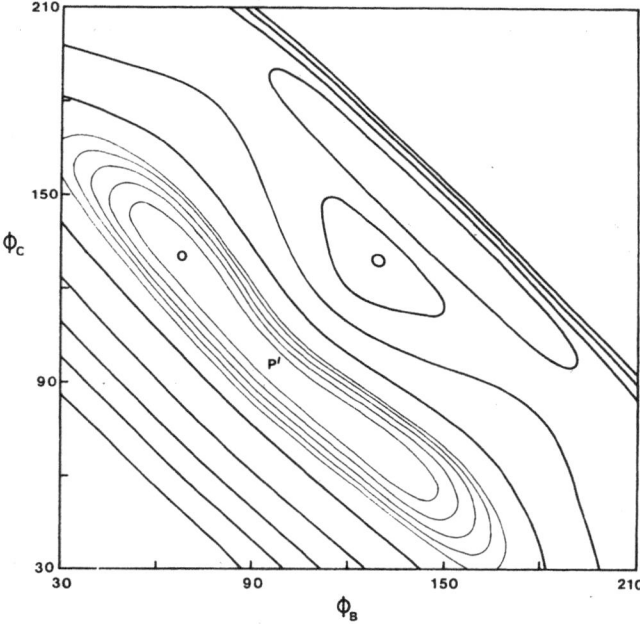

Fig. 10.11. Projection of the potential energy surface for [M(unidentate A)$_6$(unidentate B)] onto the $\phi_B - \phi_C$ plane (in degrees) with retention of a mirror plane. The five *faint contour lines* are for successive 0.02 increments above the minima, and the five *heavy contour lines* are for successive 0.2 increments above the minima, at O. R = 0.1, n = 6. The locations of the pentagonal bipyramid (P'), and capped octahedra (O) are indicated

This behavior is reproduced from the repulsion energy calculations if a small effective bond length ratio R(lone pair/F$^-$) is used. Typical surfaces, calculated for R(lone pair/F$^-$) = 0.2 and 0.1, are shown in Figs. 10.10 and 10.11, respectively. They may be compared with Figs. 10.9, 10.5, and 10.8, for R(A/B) = 1.2, 1.0, and 0.8, respectively. At R(lone pair/F$^-$) = 0.2, the capped octahedron at $\phi_B = 77.5°$, $\phi_C = 132.4°$ with the lone pair in the capping site (marked O in Fig. 10.10) is of comparable stability to the pentagonal bipyramid at $\phi_B = 96.3°$, $\phi_C = 180°$ with the lone pair in an axial site (marked P in Fig. 10.10). Less stable are the capped trigonal prism with the lone pair in the capping site at $\phi_B = \phi_C = 144.8°$ (T) and the pentagonal bipyramid with the lone pair in an equatorial site at $\phi_B = \phi_C = 98.5°$ (P'). At R(lone pair/F$^-$) = 0.1 (Fig. 10.11), the potential energy surface

Fig. 10.12. Transformation of the capped octahedron (*left*) to the pentagonal bipyramid (*center*) and an alternative capped octahedron (*right*) for [M(unidentate)$_6$(lone pair)], corresponding to Fig. 10.11

shows continued stabilization of the capped octahedron(O). Two such capped octahedra are connected by a low saddle (P′), and the movement across this saddle corresponds to the stereochemical change shown in Fig. 10.12. This is the same as that deduced for the gas-phase structure of XeF_6. Even with R(lone pair/F^-) = 0.1, the structure is considerably distorted from a regular octahedron, the F—Xe—F bond angles varying from 79° to 106°.

F. [M(unidentate A)₅(unidentate B)₂]

Potential energy surfaces in which the effective bond lengths to two ligands is 0.8, the other five being defined as unity, are shown in Fig. 10.13 where a twofold axis is retained (compare with Fig. 10.3 calculated for [M(unidentate)₇]), and in Fig. 10.14 in which a mirror plane is retained (compare with Fig. 10.5). In both surfaces the single deep minimum corresponds to a pentagonal bipyramid with the two short effective bond lengths in the two axial sites.

This predicted structure has been observed in $K_3[UO_2F_5]$ [1112], $[UO_2(H_2O)_5]$ $\cdot (ClO_4)_2 \cdot 2H_2O$ [26], $[UO_2\{OC(NH_2)_2\}_5](NO_3)_2$ [1116], and $Cs_3[UO_2(NCS)_5]$ [53], although the pentagonal plane in the last compound was described as highly puckered. However the hydrogen maleate complex [Ca(OOCCH:CHCOOH)₂(H₂O)₅] has distorted capped trigonal prismatic stereochemistry, with the charged ligands occupying two of the less hindered capped face sites [583].

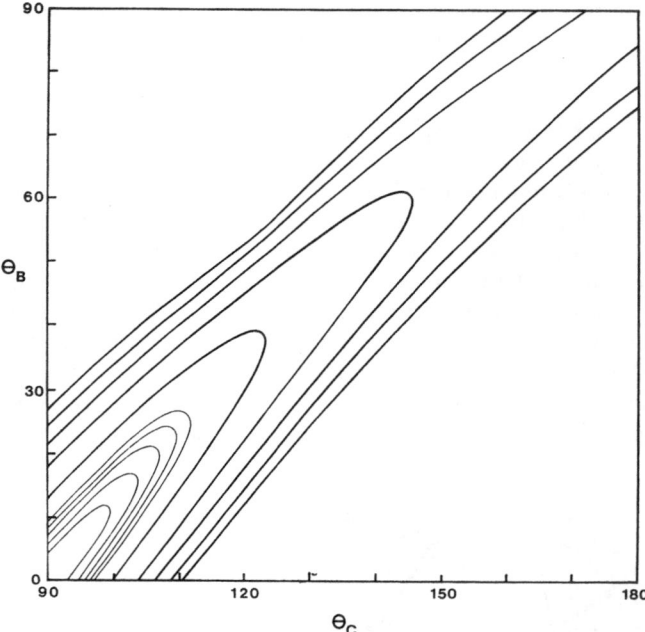

Fig. 10.13. Projection of the potential energy surface for [M(unidentate A)₅(unidentate B)₂] onto the $\theta_B - \theta_C$ plane (in degrees) with retention of a twofold axis. The five *faint contour lines* are for successive 0.02 increments above the minimum, and the five *heavy contour lines* are for successive 0.2 increments above the minimum. R = 0.8, n = 6

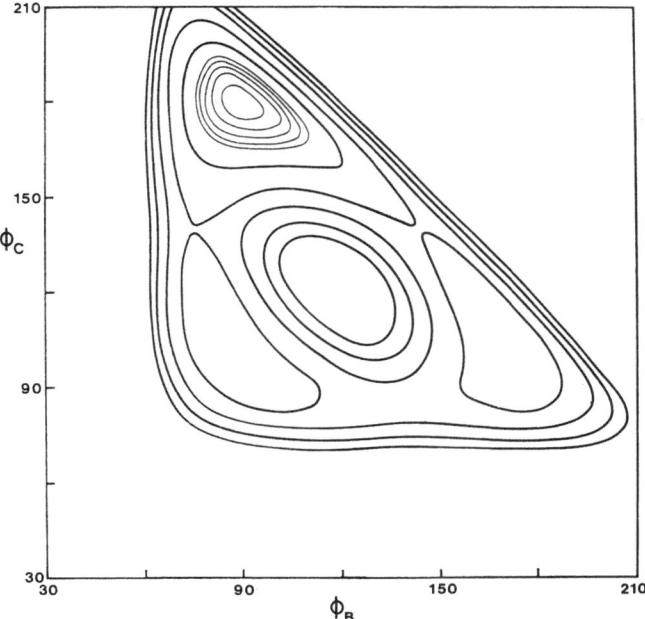

Fig. 10.14. Projection of the potential energy surface for [M(unidentate A)$_5$(unidentate B)$_2$] onto the $\phi_A - \phi_B$ plane (in degrees) with retention of a mirror plane. The five *faint contour lines* are for successive 0.02 increments above the minimum, and the five *heavy contour lines* are for successive 0.2 increments above the minimum. R = 0.8, n = 6

G. [M(unidentate A)$_4$(unidentate B)$_3$]

The four known molecules of this stoichiometry are the closely related [MoBr$_4$(Me$_2$PhP)$_3$] [362], [MoCl$_4$(Me$_2$PhP)$_3$]EtOH [760], [MoCl$_4$(MeCl$_2$P)$_3$]$^1/_3$CS$_2$ [178], and (Et$_4$N)[WBr$_3$(CO)$_4$] [363]. All have exact or reasonable threefold symmetry and capped octahedral stereochemistry (Fig. 10.15).

In each compound three halogen atoms occupy the three equivalent sites *trans* to the capping atom, but the angle these bonds make with the threefold axis $\phi_E \sim 126°$, is significantly greater than that predicted for seven equal ligands, $\phi_E = 130.3°$ (Sect. 10.A), because of greater repulsion from these three halide ions. The experimental bond angles can be accurately fitted to calculated bond angles using R(X$^-$/ligand) $\sim$ 0.9 (n = 6).

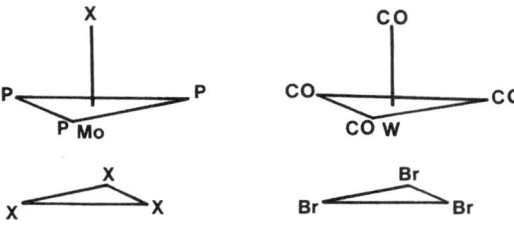

Fig. 10.15. Capped octahedral isomers of [MoBr$_4$(Me$_2$PhP)$_3$], [MoCl$_4$(Me$_2$PhP)$_3$]EtOH, [MoCl$_4$(MeCl$_2$P)$_3$]$^1/_3$CS$_2$ and (Et$_4$N)[WBr$_3$(CO)$_4$]. X=Cl, Br

CHAPTER 11

Seven-Coordinate Compounds Containing Chelate Groups

A. [M(bidentate)(unidentate)$_5$]

1. Introduction

The potential energy surfaces for [M(bidentate)(unidentate)$_5$] are complex, and contain a number of minima. A summary of the more important stereochemistries, corresponding to calculated potential energy minima, is given in Fig. 11.1.

Below a normalized bite of 1.1 there is a shallow potential energy trough that incorporates the capped trigonal prism with the bidentate ligand spanning the

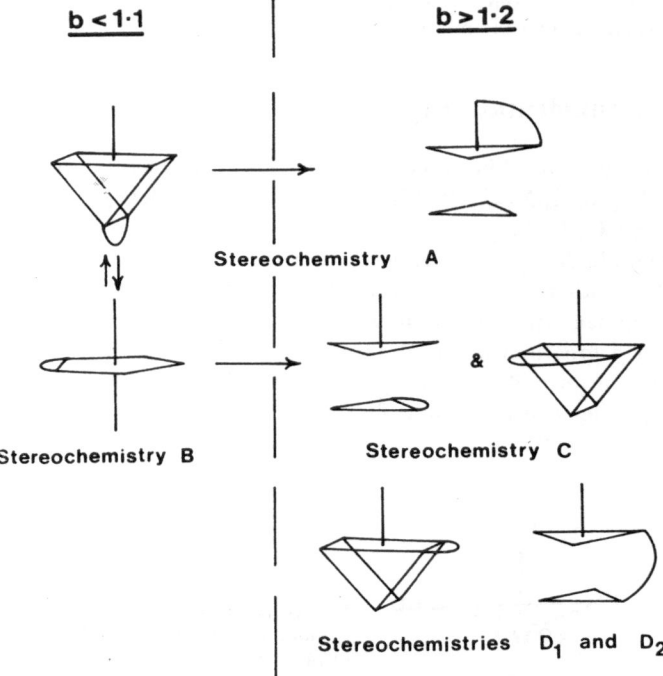

Fig. 11.1. Stereochemistries for [M(bidentate)(unidentate)$_5$]

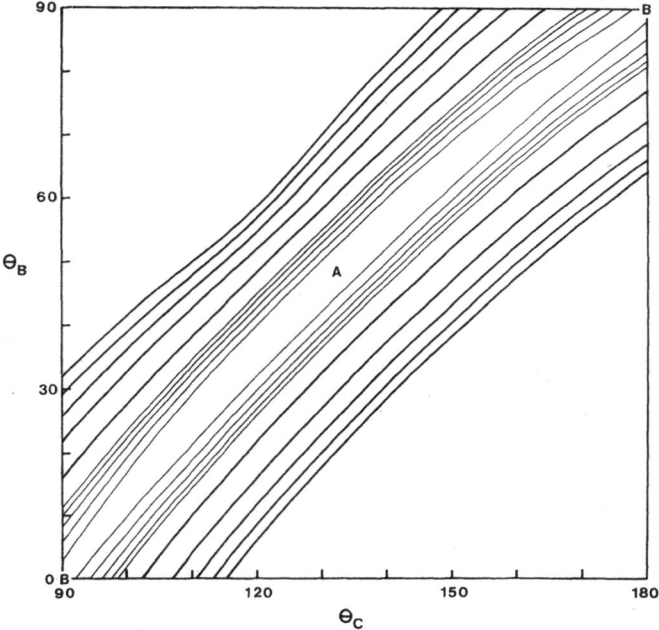

Fig. 11.2. Projection of the potential energy surface for [M(bidentate)(unidentate)$_5$] onto the $\theta_B - \theta_C$ plane (in degrees) with retention of a twofold axis. The five *faint contour lines* are for successive 0.02 increments above the minimum, and the five *heavy contour lines* are for 0.2 increments above the minimum, at A. b = 1.1, n = 6. The locations of stereochemistries A and B are indicated

prism edge *trans* to the capping atom (stereochemistry A) and the pentagonal bipyramid with the bidentate ligand spanning one of the pentagonal edges (stereochemistry B) (Fig. 11.1). A typical potential energy surface, calculated assuming a twofold axis, is shown in Fig. 11.2, and should be compared with Fig. 10.3 which was calculated for [M(unidentate)$_7$]. The transformation is similar to that shown in Fig. 10.2, but with the bidentate ligand spanning the F and G sites, and can be envisaged as free rotation of the bidentate ligand about the twofold axis.

As the value of the normalized bite increases, both these stereochemistries progressively distort and momentarily traverse various isomers of the capped trigonal prism, capped octahedron, or pentagonal bipyramid. Both stereochemistry A and stereochemistry B/C contain a mirror plane; the mirror plane contains the bidentate ligand and one unidentate ligand in stereochemistry A and three unidentate ligands in stereochemistry C.

A third important potential energy minimum appears at high values of the normalized bite, corresponding to stereochemistry D in Fig. 11.1. Stereochemistry D$_1$ also contains a mirror plane, which in this case contains only one unidentate ligand. It is very closely related to the unsymmetrical capped octahedral stereo-

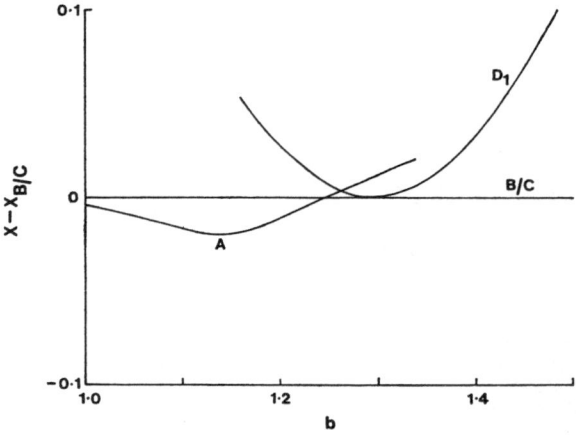

Fig. 11.3. Repulsion energy coefficient, above that corresponding to stereochemistries B/C, for [M(bidentate)(unidentate)$_5$], as a function of b. n = 6

chemistry D_2, with the bidentate ligand spanning one of the edges between the capped face and uncapped face (Fig. 11.1).

The relative energies of these three stereochemistries are shown in Fig. 11.3. The differences in energy are not large, although it can be seen that stereochemistries A and B/C are favored for b < 1.2, and stereochemistry C is favored for b > 1.4.

Other possible stereochemistries contain no elements of symmetry, have not been experimentally observed, and need not be discussed further [639].

2. Stereochemistry A

Stereochemistry A, which is the capped trigonal prism at low values of b, distorts as the normalized bite is increased above b = 1.19 (n = 6). This is shown in Fig. 11.4 by $\phi_C > \phi_B$, $\phi_D \neq \phi_E$ and sharp increases in θ_D and θ_E, the atom labels and coordinates being defined in Fig. 11.5. These changes correspond to the bidentate ligand on the mirror plane moving towards the unidentate ligand on the mirror plane, towards the formation of a pentagonal bipyramid with the bidentate ligand spanning an axial and equatorial site (Fig. 11.5). However the pentagonal bipyramid at b = $2^{1/2}$ is not reached, stereochemistry A ceasing to exist as a potential energy minimum above b = 1.34 (n = 6). This nonappearance of the pentagonal bipyramid as a stable minimum corresponds to the slightly higher repulsion energy associated with this stereochemistry for [M(unidentate)$_7$].

The capped trigonal prism — pentagonal bipyramid interconversion also traverses the capped octahedron at b = 1.21. Atom B is the capping atom, and CDG and AEF are the two triangular planes normal to the threefold axis. The relevant angular parameters are marked in Fig. 11.4, and are the same as those calculated for [M(unidentate)$_7$] (Sect. 10.A). It can be seen that this is expected to be a reasonable description of the structure only over a small range of b. It should also be noted that the potential energy surface is fairly flat near b ∼ 1.2, with changes in ϕ_B and ϕ_C of up to ∼ 30° possible.

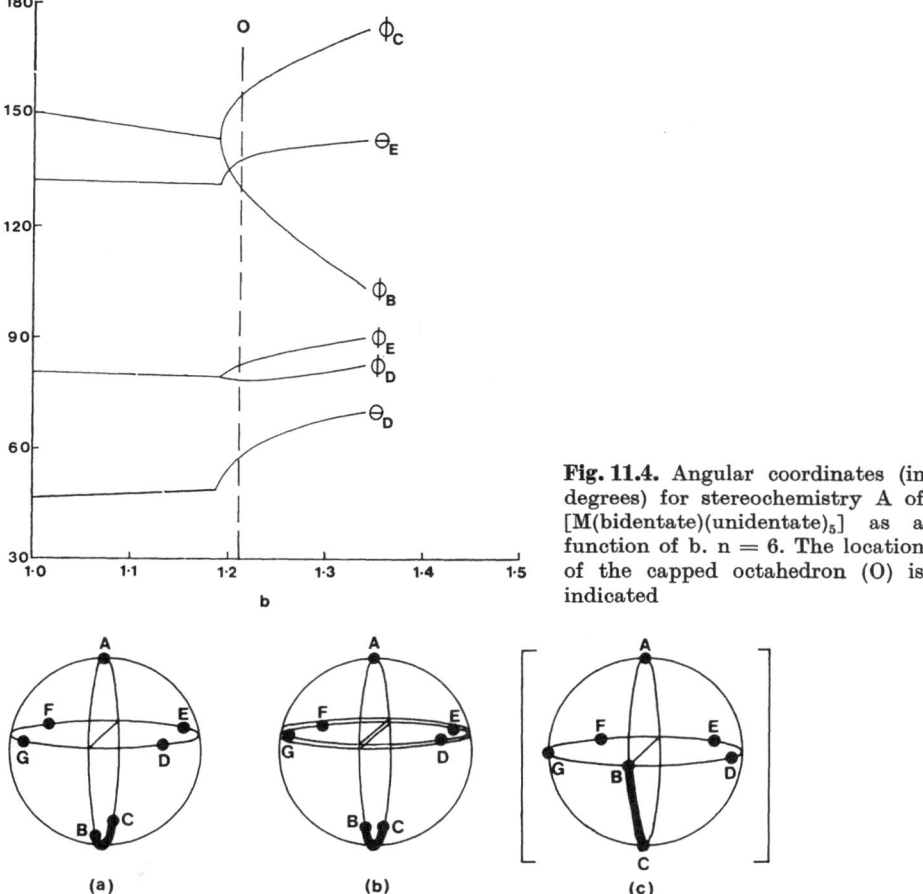

Fig. 11.4. Angular coordinates (in degrees) for stereochemistry A of [M(bidentate)(unidentate)$_5$] as a function of b. n = 6. The location of the capped octahedron (O) is indicated

Fig. 11.5a—c. Stereochemistry A of [M(bidentate)(unidentate)$_5$]. **a** b = 1.1; **b** b = 1.2; **c** b = $2^{1/2}$ (not a discrete minimum)

3. Stereochemistries B and C

At small normalized bites, stereochemistry B is a pentagonal bipyramid with the bidentate ligand spanning one of the pentagonal edges. The effect of increasing the normalized bite on the potential energy surface is shown in Figs. 11.6 and 11.7. A mirror plane is retained, and these surfaces are directly comparable to the much flatter surface calculated for [M(unidentate)$_7$], shown in Fig. 10.5.

Above b = 1.2, the minimum at ϕ_B = 90°, ϕ_C = 180° corresponding to the pentagonal bipyramid (stereochemistry B), splits into two minima, corresponding to two different orientations of stereochemistry C. These minima move further apart on further increasing b to 1.2 and 1.4 (Figs. 11.6 and 11.7).

Angular parameters are given as a function of b in Fig. 11.8, with stereochemistries and atom labels according to Fig. 11.9. As before, this continuous change in stereochemistry from the pentagonal bipyramid traverses the capped

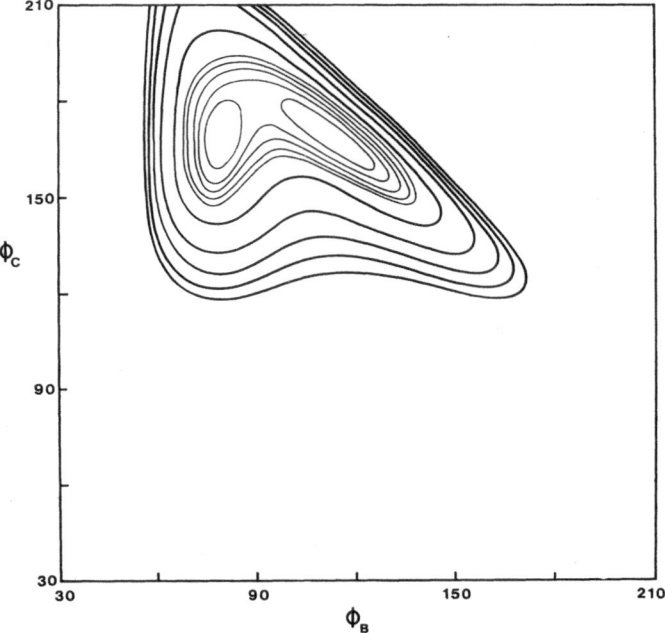

Fig. 11.6. Projection of the potential energy surface for [M(bidentate)(unidentate)$_5$] onto the $\phi_B - \phi_C$ plane (in degrees) with retention of a mirror plane. The five *faint contour lines* are for successive 0.02 increments above the minima, and the five *heavy contour lines* are for successive 0.2 increments above the minima. b = 1.2, n = 6

octahedron and capped trigonal prism. The capped octahedron is formed at b = 1.32. where B is the capping atom and CDG and AEF are the equilateral triangles normal to the threefold axis. The capped trigonal prism occurs at b = 1.48. The relevant parameters are marked in Fig. 11.8.

4. Stereochemistry D

Stereochemistry D exists as a potential energy minimum only above b = 1.16 (n = 6), and is reasonably well represented as a capped trigonal prism (Fig. 11.1). There is negligible difference in energy between this capped trigonal prism (stereochemistry D$_1$) and the capped octahedron shown in Fig. 11.1 (stereochemistry D$_2$). The difference in stereochemistry is also small [639]. The capped octahedral description rather than the more symmetrical capped trigonal prismatic description is expected to be favored for complexes with a greater diversity of ligands.

5. Comparison with Experiment

Compounds of the type [M(bidentate)(unidentate)$_5$] that have been structurally characterized are given in Table 11.1. It can be seen that there are very few compounds containing five identical unidentate ligands, and it is necessary to

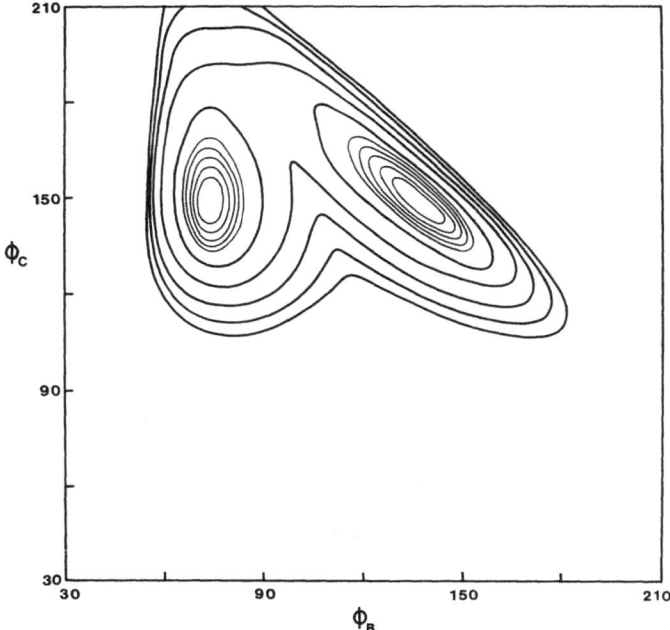

Fig. 11.7. Projection of the potential energy surface for [M(bidentate)(unidentate)$_5$] onto the $\phi_B - \phi_C$ plane (in degrees) with retention of a mirror plane. The five *faint contour lines* are for successive 0.02 increments above the minima, and the five *heavy contour lines* are for successive 0.2 increments above the minima. b = 1.4, n = 6

include examples of more complex stoichiometry, where distortions from the idealized structure can be relatively large [639].

As predicted from Fig. 11.3, bidentate ligands with very small normalized bites such as peroxide and nitrate form isomer A or B.

At larger normalized bites, that is, for ligands forming five- or six-membered chelate rings and/or having large atoms in the chelate rings, stereochemistries A, C and D are observed, as predicted.

By far the most common type of structurally characterized complex is the substituted carbonyl halide [M(bidentate)X$_2$(CO)$_3$]. The majority contain five- or six-membered chelate rings and can be adequately represented by stereochemistry D$_1$ with distortions depending on the particular mix of ligands present. However, it does not follow that all such substituted carbonyl halides must be of stereochemistry D, as this potential energy minimum disappears at low b. It is therefore important to note that [W(Ph$_2$PCH$_2$PPh$_2$)I$_2$(CO)$_3$], which has the lowest value of b for these compounds, has the pentagonal bipyramidal stereochemistry B. This point is further emphasized by Fig. 11.10, which compares the stereochemistries of [W(Ph$_2$PCH$_2$PPh$_2$)I$_2$(CO)$_3$], [Mo(Ph$_2$PCH$_2$CH$_2$PPh$_2$)I$_2$(CO)$_3$]CH$_2$Cl$_2$, and [Mo(Ph$_2$PCH$_2$CH$_2$CH$_2$PPh$_2$)I$_2$(CO)$_3$]; in all cases the metal-ligand bonds have been made equal, and the view is such that the chelate ring is on the same projection to the page.

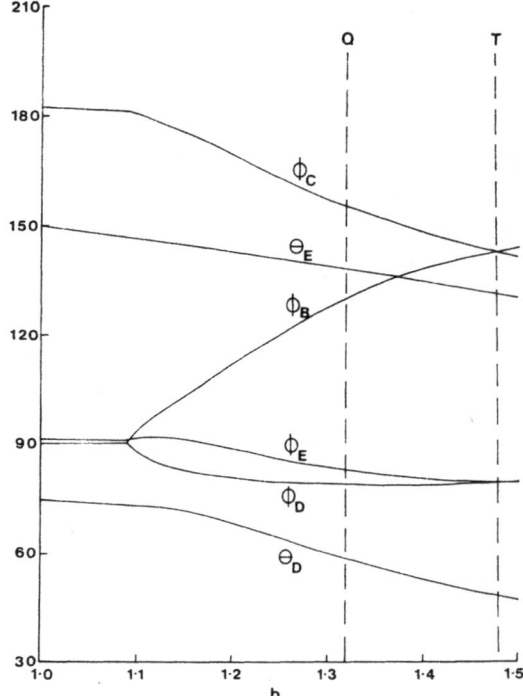

Fig. 11.8. Angular coordinates (in degrees) for stereochemistries B and C of [M(bidentate)(unidentate)$_5$] as a function of b. n = 6. The locations of the capped octahedra (O) and capped trigonal prism (T) are indicated

Whether or not these [M(bidentate)(unidentate)$_5$] complexes show fluxional behavior in solution depends both on the normalized bite of the bidentate ligand and on the particular mix of donor atoms present. For example, the ^{1}H NMR spectra down to $-70\,°C$, the ^{19}F NMR spectra down to $-70\,°C$, and the room-temperature ^{13}C NMR spectra of [Mo{Me$_2$AsC(CF$_3$):C(CF$_3$)AsMe$_2$}X$_2$(CO)$_3$], where

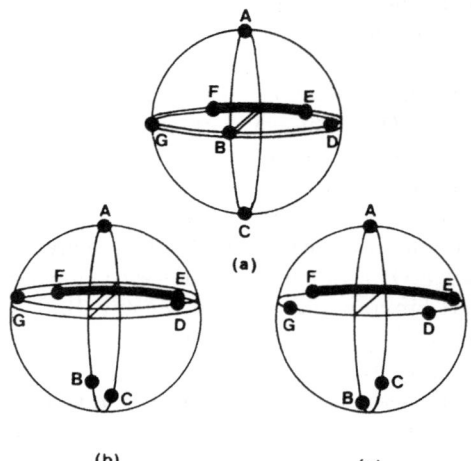

Fig. 11.9 a—c. Stereochemistries B and C of [M(bidentate)(unidentate)$_5$]. **a** b = 1.0; **b** b = 1.3; **c** b = 1.5

Table 11.1 Stereochemistries of M(bidentate)(unidentate)$_5$]

Complex	b	Stereo-chemistry	Ref.
$(C_9H_8NO)_2[Nb(O_2)F_5]3 H_2O$	0.61	B	[950]
$(C_9H_8NO)_2[W(O_2)OF_4]3 H_2O$	0.61	B	[950]
$[Mo\{C(NHBu)(BuNH)C\}(CNBu)_4I]I$	0.66	A	[687]
$(NH_4)_3[Mo(O_2)OF_4]F$	0.71	B	[693]
$K_6[Ta(O_2)F_5][Ta_2O(O_2)_2F_8]$	0.73	B	[768]
$K_2[Mo(O_2)OF_4]H_2O$	0.74	B	[493]
$K_3[Ta(O_2)F_5](HF_2)$	0.81	B	[951]
$(Et_4N)[Ta(O_2)F_4(ONC_5H_4Me)]$	0.81	B	[316]
$[Sn(NO_3)Ph_2(Me_2SO)_3](NO_3)$	0.87	B	[251]
$[Mo(H_2BH_2)H(PMe_3)_4]$	0.89	B	[60]
$[Mo(Ph_2PCH_2PPh_2)Cl_2(CO)_2(Ph_2PCH_2PPh_2)]C_6H_6$	1.05	D	[366]
$[W(Ph_2PCH_2PPh_2)I_2(CO)_3]$	1.09	B	[450]
$[U\{ON(Ph)C(Ph)O\}O_2Cl(C_4H_8O_2)]$	1.10	B	[997]
$[Mo(Ph_2AsCH_2AsPh_2)Cl_2(CO)_2(Ph_2AsCH_2AsPh_2)]$	1.10	D	[366]
$[W(Ph_2AsCH_2AsPh_2)I_2(CO)_3]$	1.11	D	[364]
$[Ta(bipy)Cl_2Me_3]$	1.12	C	[353]
$[Mo(Ph_2AsCH_2AsPh_2)Br_2(CO)_2(Ph_2AsCH_2AsPh_2)]$	1.18	D	[341]
$[W\{C_6H_4(AsMe_2)_2\}OCl_4]$	1.18	B	[348]
$[Mo(bipy)(HgCl)Cl(CO)_3]$	1.18	D	[153]
$[W(bipy)(GeBr_3)Br(CO)_3]$	1.2	D	[276]
$[Mo\{rac-C_6H_4(AsMePh)_2\}I_2(CO)_3]CHCl_3$	1.21	D	[317]
$[W(Ph_2PCH_2CH_2PPh_2)I_2(CO)_3]$	1.22	D	[352]
$[W\{Me_2AsC(CF_3):C(CF_3)AsMe_2\}I_2(CO)_3]$	1.23	D	[781]
$[Mo\{meso-C_6H_4(AsMePh)_2\}I_2(CO)_3]$	1.24	D	[317]
$[Mo(Ph_2PCH_2CH_2PPh_2)I_2(CO)_3]CH_2Cl_2$	1.24	D	[450]
$[Mo(Ph_2PCH_2CH_2PPh_2)Br_2(CO)_3]Me_2CO$	1.24	D	[342]
$[Re(bipy)Br_3(CO)_2]$	1.24	D	[343]
$[W\{C_6H_4(AsMe_2)_2\}I(CO)_4]I_3$	1.26	D	[358]
$[Mo(Ph_2PCH_2CH_2PPh_2)(SnCl_3)(CO)_4][SnCl_5(H_2O)]C_6H_6$	1.28	C	[385]
$[Mo(Ph_2PCH_2CH_2CH_2PPh_2)I_2(CO)_3]$	1.33	D	[450]

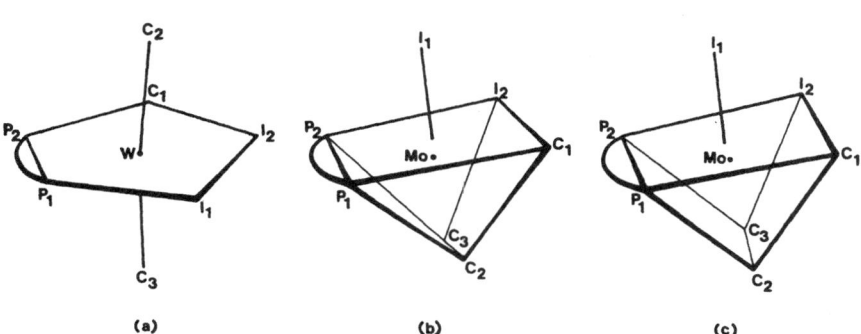

Fig. 11.10. The stereochemistries of **a** $[W(Ph_2PCH_2PPh_2)I_2(CO)_3]$; **b** $[Mo(Ph_2PCH_2CH_2PPh_2)I_2(CO)_3]CH_2Cl_2$; **c** $[Mo(Ph_2PCH_2CH_2CH_2PPh_2)I_2(CO)_3]$

X is Br or I, show rapid scrambling of all carbonyl and halide ligands [286]. Substitution of one carbonyl group by a tertiary phosphine again yielded nonrigid compounds. However, further substitution of a second carbonyl yielded [Mo{Me$_2$AsC(CF$_3$):C(CF$_3$)AsMe$_2$}Br$_2$(CO)(PR$_3$)$_2$] which was stereochemically rigid at room temperature, but became nonrigid on heating.

B. [M(bidentate)$_2$(unidentate)$_3$]

1. The Theoretical Stereochemistries

Minimization of the calculated repulsion energy again shows the existence of several isomers, and that as the normalized bite is varied, there is again a smooth transformation between structures based on the symmetrical polyhedra.

The predicted stereochemistries are summarized in Fig. 11.11.

Below a normalized bite of 1.1, a deep minimum on the potential energy surface corresponds to the pentagonal bipyramid, with the two bidentate ligands spanning equatorial edges (stereochemistry I, Fig. 11.11). As the normalized bite is increased to b = 1.14, the pentagonal plane commences to pucker with loss of both morror planes but retention of the twofold axis through one unidentate ligand.

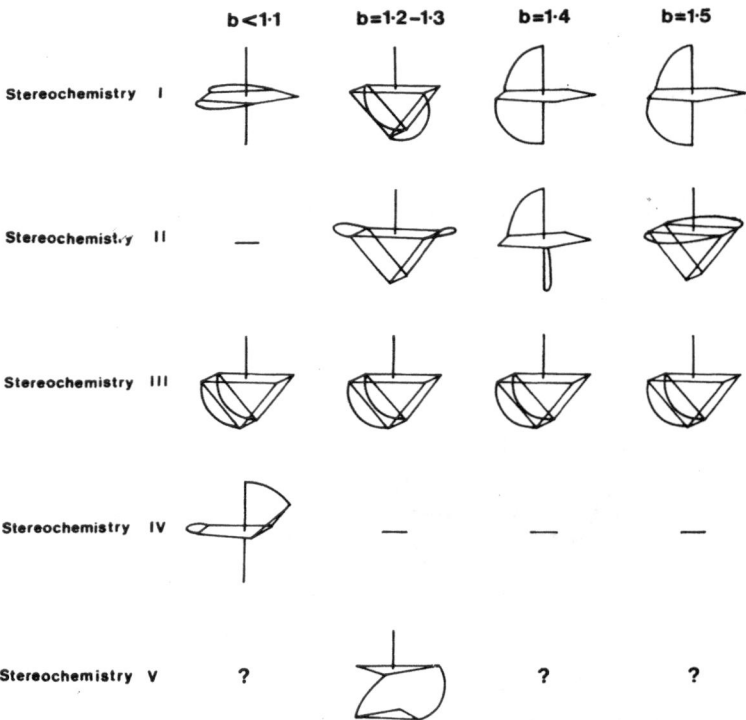

Fig. 11.11. Stereochemistries for [M(bidentate)$_2$(unidentate)$_3$]

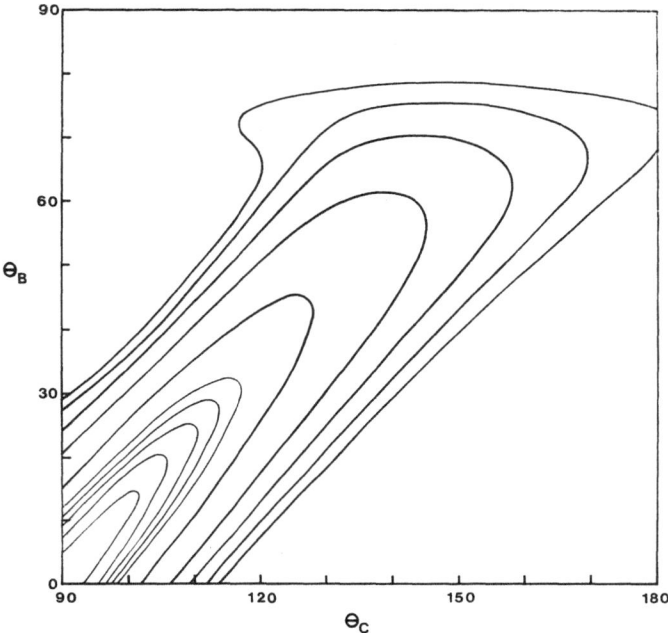

Fig. 11.12. Projection of the potential energy surface for stereochemistry I of [M(bidentate)$_2$-(unidentate)$_3$] onto the $\theta_B - \theta_C$ plane (in degrees) with retention of a twofold axis. The five *faint contour lines* are for successive 0.02 increments above the minimum, and the five *heavy contour lines* are for successive 0.2 increments above the minimum. b = 1.0, n = 6

As the normalized bite is further increased, the stereochemistry continuously changes. A trigonal prism is formed first, followed by a pentagonal bipyramid with the bidentate ligands spanning adjacent equatorial-axial edges (Fig. 11.11).

These progressive changes are illustrated by the potential energy surfaces in Figs. 11.12–11.14, the stereochemistries at the minima being shown in Fig. 11.15. These surfaces may be compared with the much more elongated minima in the potential energy surfaces calculated for [M(bidentate)(unidentate)$_5$] (Fig. 11.2), and [M(unidentate)$_7$] (Fig. 10.3). A potential energy surface in which a mirror plane is enforced is shown in Fig. 11.16 as a function of ϕ_B and ϕ_C, which are the two bond angles between the unidentate ligands, AMB and BMC. The three deep minima correspond to the three possible orientations of the pentagonal bipyramid, and this surface should be compared with Fig. 10.5 calculated for [M(unidentate)$_7$], or Figs. 11.6 and 11.7 calculated for [M(bidentate)(unidentate)$_5$].

Stereochemistry II, like stereochemistry I, has the three unidentate ligands coplanar with the central metal atom with a twofold axis passing through the central unidentate ligand (Fig. 11.11). Stereochemistry II only occurs as a potential energy minimum for relatively high values of the normalized bite. Stereochemistry II exists as a very long narrow trough, which contains the two trigonal prisms with the bidentate ligands forming the capped face (Fig. 11.11), and the intermediate pentagonal bipyramid with the bidentate ligands linking axial-equatorial sites

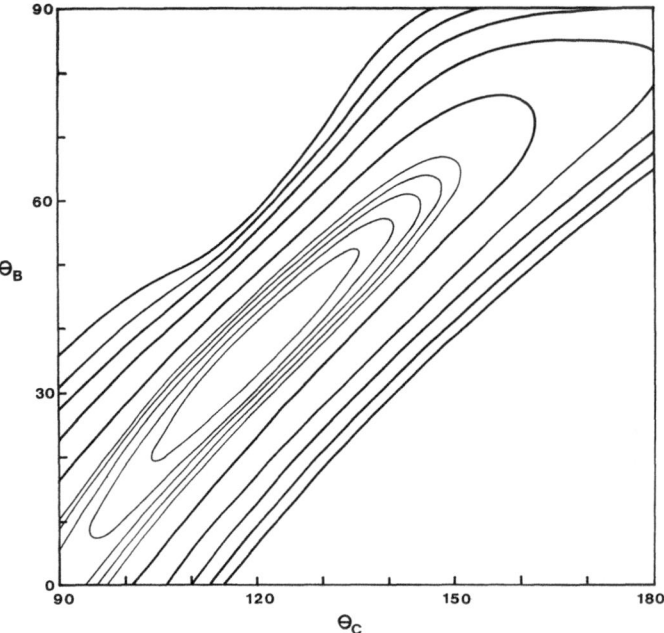

Fig. 11.13. Projection of the potential energy surface for stereochemistry I of [M(bidentate)$_2$-(unidentate)$_3$] onto the $\theta_B - \theta_C$ plane (in degrees) with retention of a twofold axis. The five *faint contour lines* are for successive 0.02 increments above the minimum, and the five *heavy contour lines* are for successive 0.2 increments above the minimum. b = 1.2; n = 6

(Fig. 11.11). The trough is very shallow, and corresponds to virtually free rotation of the two bidentate ligands around the twofold axis. One end of the trough exists as a slight minimum over range b = 1.28—1.41, the other above b = 1.34 (n = 6). The instability of stereochemistry II at low b is shown by the potential energy surface in Fig. 11.16. The capped trigonal prism exists as a high-energy saddle between each pair of pentagonal bipyramids, stereochemistry I.

Stereochemistry III contains a mirror plane that incorporates only a single unidentate ligand (Fig. 11.11), in contrast to the mirror planes in stereochemistries I and II which contain all three unidentate ligands, or one unidentate ligand and both bidentate ligands.

The final type of mirror plane possible is one which contains one unidentate ligand and one bidentate ligand, as in stereochemistry IV (Fig. 11.11). This stereochemistry is not closely related to the pentagonal bipyramid, capped octahedron, or capped trigonal prism, although for convenience the relation to the pentagonal bipyramid has been emphasized in Fig. 11.11. An important distinguishing feature of stereochemistry IV is that the dihedral angle between the two chelate rings is 90°, compared with much smaller values, or zero, observed in stereochemistries I to III. Stereochemistry IV is only present as a potential energy minimum at low values of the normalized bite.

A large dihedral angle between the bidentate ligands is also found in stereo-

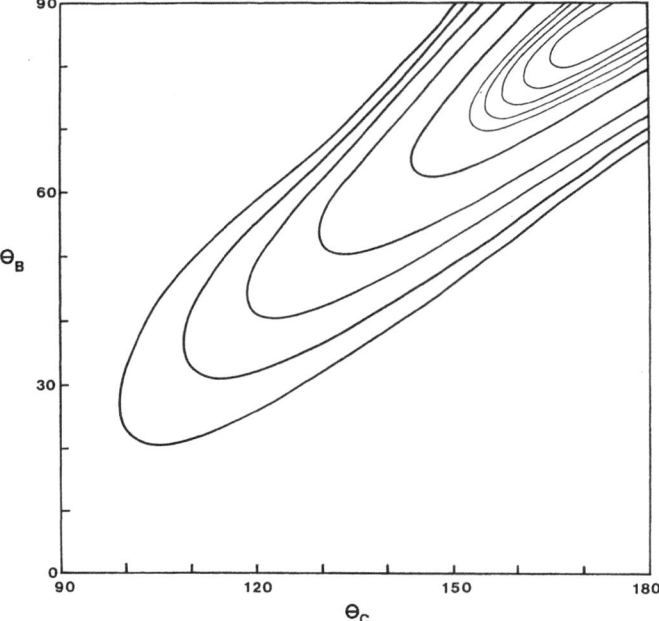

Fig. 11.14. Projection of the potential energy surface for stereochemistry I of [M(bidentate)$_2$ (unidentate)$_3$] onto the $\theta_B - \theta_C$ plane (in degrees), with retention of a twofold axis. The five *faint contour lines* are for successive 0.02 increments above the minimum, and the five *heavy contour lines* are for successive 0.2 increments above the minimum. b = 1.4; n = 6

chemistry V (Fig. 11.11). This stereochemistry contains no symmetry elements, and detailed structural parameters and the range of existence as a potential energy minimum have not yet been determined.

The relative energies of the first four stereochemistries, and their range of

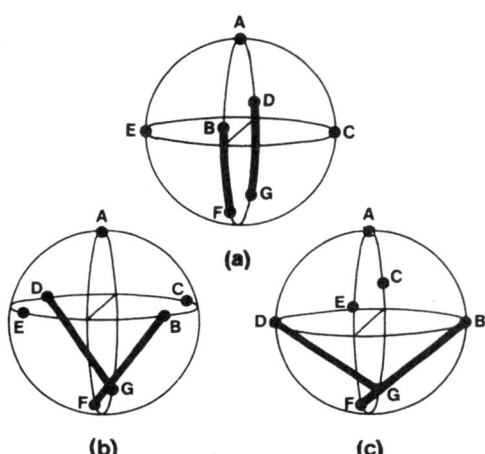

Fig. 11.15 a—c. Stereochemistry I of [M(bidentate)$_2$(unidentate)$_3$]. **a** b = 1.0; **b** b = 1.2; **c** b = 1.4

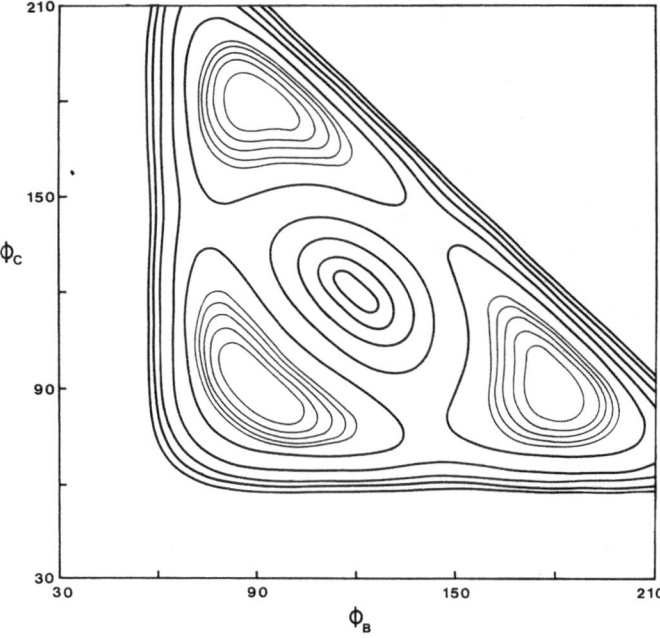

Fig. 11.16. Projection of the potential energy surface for stereochemistry I of $[M(\text{bidentate})_2$-(unidentate)$_3]$ onto the $\phi_B - \phi_C$ plane (in degrees) with retention of a mirror plane. The five *faint contour lines* are for successive 0.02 increments above the minima, and the five *heavy contour lines* for successive 0.2 increments above the minima. $b = 1.1$. $n = 6$

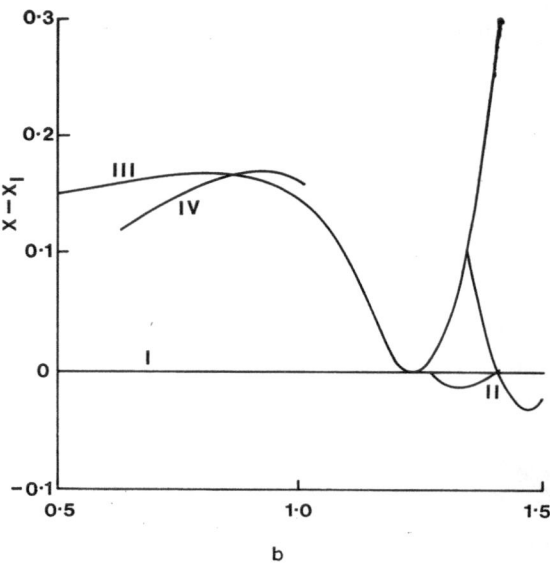

Fig. 11.17. Repulsion energy coefficients, above that corresponding to stereochemistry I, for $[M(\text{bidentate})_2(\text{unidentate})_3]$ as a function of b. $n = 6$

existence as discrete minima, are shown in Fig. 11.17. It can be seen that below a normalized bite of 1.1, stereochemistry I is predicted to be the most stable structure, while for b > 1.2 stereochemistry II is slightly more stable than stereochemistry I. Isomer III is possible only over the small range b ∼ 1.2−1.3. Isomer IV exists only at low at low b, and has high repulsion energies. Stereochemistry V is expected to be stable at least in the vicinity of b = 1.27, corresponding to a regular capped octahedron. Stereochemistry II separates the unidentate ligands more effectively from each other does stereochemistry I, and stereochemistry II becomes more stable than stereochemistry I as R(unidentate/bidentate) decreases [639].

2. Comparison with Experiment

All structurally characterized compounds of stoichiometry [M(bidentate)$_2$-(unidentate)$_3$] are given in Table 11.2 in order of increasing normalized bite. In excellent agreement with the predictions from the repulsion energy calculations, Fig. 11.17, it can be seen that stereochemistry I is the dominant isomer at low normalized bites, and stereochemistry II is the dominant isomer at high normalized bites. Isomers III and IV are unknown.

Table 11.2. Stereochemistries of [M(bidentate)$_2$ (unidentate)$_3$]

Complex	b	Stereo-chemistry	Ref.
[Cr(O$_2$)$_2$(NH$_3$)$_3$]	0.76	I	[1019]
K$_3$[Cr(O$_2$)$_2$(CN)$_3$]	0.77	I	[1020]
[Mo(O$_2$)$_2$O{OP(NMe$_2$)$_3$}(C$_5$H$_5$N)]	0.75	I (distorted)	[702]
[Mo(O$_2$)$_2${OP(NMe$_2$)$_3$}(H$_2$O)]	0.77	I (distorted)	[702]
[Hg(O$_2$CCF$_3$)$_2$(C$_5$H$_5$N)$_3$]	0.85	I	[515]
K$_3$[Cu(NO$_2$)$_5$]	0.85	I	[653]
[Cd(NO$_3$)$_2$(C$_9$H$_7$N)$_2$(H$_2$O)]	0.87	I (distorted)	[208]
[Cd(NO$_3$)$_2$(C$_5$H$_5$N)$_3$]	0.87	I	[207]
[Zn(NO$_3$)$_2$(C$_5$H$_5$N)$_3$]	0.90	I	[206]
[Co(NO$_3$)$_2$(C$_5$H$_5$N)$_3$]	0.94	I	[206]
(Et$_2$NH$_2$)[U(OSCNEt$_2$)$_2$O$_2$(OEt)]	0.97	I	[864]
(Pr$_2$NH$_2$)[U(OSCNPr$_2$)$_2$O$_2$(OEt)]	0.97	I	[865]
(Me$_4$N)[Sn(O$_2$CMe)$_5$]	0.98	I (distorted)	[30]
[Ta(O$_2$CNMe$_2$)$_2$(NMe$_2$)$_2$(OOCNMe$_2$)]	1.01	I	[231]
[Ta(PrNCMeNPr)$_2$Cl$_3$] (ortho)	1.00	I	[359]
[Ta(PrNCMeNPr)$_2$Cl$_3$] (mono)	1.02	I	[356]
[Ta(PrNCMeNPr)$_2$Cl$_2$Me]C$_6$H$_6$	1.01	I (distorted)	[360]
[Ta(C$_6$H$_{11}$NCMeNC$_6$H$_{11}$)$_2$Cl$_2$Me]	1.01	V	[357, 639]
[Ta(C$_6$H$_{11}$NCMeNC$_6$H$_{11}$)$_2$Cl$_3$]	1.02	I	[361]
(Et$_4$N)[Te(S$_2$COEt)$_3$(lone pair)$_2$]	1.03	I	[580]
[V(NO$_3$)$_2$O(ONO$_2$)(NCMe)]	1.02	I (distorted)	[384]
[U(S$_2$CMe)$_2$O$_2$(OPPh$_3$)]	1.02	I	[130]
[U(S$_2$CNEt$_2$)$_2$O$_2$(ONMe$_3$)]	1.03	I	[447]
[U(S$_2$CNEt$_2$)$_2$O$_2$(OPPh$_3$)]	1.03	I	[499]
[U(S$_2$CNEt$_2$)$_2$O$_2$(OAsPh$_3$)]	1.04	I	[499]

(Continued)

Table 11.2 (continued)

Complex	b	Stereo-chemistry	Ref.
[Nb(MeNCMeS)$_2$Cl$_3$]	1.05	I	[355]
[U(O$_2$C$_7$H$_5$)$_2$O$_2$(C$_5$H$_5$N)]	1.06	I	[310]
[U(Se$_2$CNEt$_2$)$_2$O$_2$(OAsPh$_3$)]	1.07	I	[281, 1119]
[U(oxine)$_2$O$_2$(Hoxine)]CHCl$_3$	1.07	I	[517]
[Nb(S$_2$CNEt$_2$)$_2$(OMe)$_2$Cl]	1.11	I	[801]
[Sm{S$_2$P(OEt)$_2$}$_2$(OPPh$_3$)$_3$][S$_2$P(OEt)$_2$]	1.12	I	[882]
[Mo(S$_2$CNEt$_2$)$_2$(NPh)Cl$_2$]CHCl$_3$	1.12	I	[742]
[Mo(S$_2$CNEt$_2$)$_2$OCl$_2$]	1.13	I	[325]
[Mo(S$_2$CNEt$_2$)$_2$OBr$_2$]	1.13	I	[325]
[U(MeCOCClCOMe)$_2$O$_2$(OPPh$_3$)]	1.14	I	[1036]
β-[U(CF$_3$COCHCOCF$_3$)$_2$O$_2${OP(OMe)$_3$}]	1.14	I	[1038]
α-[U(CF$_3$COCHCOCF$_3$)$_2$O$_2${OP(OMe)$_3$}]	1.15	I	[1037]
(Et$_4$N)[U(PhCOCHCOPh)$_2$O$_2$(ONO$_2$)]	1.15	I	[498]
[U(MeCOCHCOMe)$_2$O$_2${OC(Me)(C$_3$H$_6$N)}]	1.15	I	[941]
[U(MeCOCHCOMe)$_2$O$_2${OC(Me)(C$_4$H$_8$N)}]	1.16	I	[514]
[U(MeCOCHCOMe)$_2$O$_2${OC(Me)(C$_5$H$_{10}$N)}]	1.16	I	[829]
[U(MeCOCHCOMe)$_2$O$_2${OC(Me)(C$_6$H$_{12}$N)}]	1.16	I	[940]
[U(MeCOCHCOMe)$_2$O$_2${OC(Me)(CH$_2$COMe)}]	1.16	I	[713]
[U(MeCOCHCOMe)$_2$O$_2$(OPPh$_3$)]C$_6$H$_6$	1.16	I	[1036]
[U(CF$_3$COCHCOCF$_3$)$_2$O$_2$(NH$_3$)]	1.18	I	[620]
Cs[Nb(C$_2$O$_4$)$_2$O(H$_2$O)$_2$]2 H$_2$O	1.19	I	[664]
(NH$_4$)[Nb(C$_2$O$_4$)$_2$O(H$_2$O)$_2$]3 H$_2$O	1.19	I	[460]
Cs[Ti(C$_2$O$_4$)$_2$(H$_2$O)$_3$]2 H$_2$O	1.20	I	[345]
[W(Ph$_2$PCH$_2$PPh$_2$)$_2$I(CO)$_2$]I	1.21	II	[365]
[Mo{C$_6$H$_4$(AsMe$_2$)$_2$}$_2$Cl(CO)$_2$]I$_3$·2CHCl$_3$	1.23	II	[354]
[Ta(Me$_2$PCH$_2$CH$_2$PMe$_2$)$_2$H(CO)$_2$]	1.23	II^a	[778]
[Re(Ph$_2$PCH$_2$CH$_2$PPh$_2$)$_2$H$_3$]	1.33	II^a	[10]

^{a} Hydrogen atoms not located. Probably stereochemistry II.

The most convenient way of classifying structures as stereochemistry I or stereochemistry II is from the values of the three coplanar unidentate-metal-unidentate bond angles:

Stereochemistry I

$BMC = AMB \sim 90°$

$AMC \sim 180°$

Stereochemistry II

$BMC \sim 80°$

$AMB = AMC \sim 140°$

Distortions occur when all three unidentate ligands are not identical. In the extreme case when one is an O^{2-} ligand, it is always located in site A of stereo-

Fig. 11.18. (Et₄N)[Te(S₂COEt)₃]

chemistry I, and AMB increases to ∼ 95° and BMC decreases to ∼ 85°. In dioxa complexes there is an O^{2-} ligand in each of the A and C sites, and the structures are much more regular as a result of the opposing action of these axial oxygen atoms on the five equatorial atoms.

A similar structure is found for the tellurium(II) complex (Et₄N)[Te(S₂COEt)₃], if it is assumed that two nonbonding pairs of electrons occupy the two axial site (Fig. 11.18). Each bidentate ligand is unsymmetrically bonded, one tellurium-sulfur bond being 15% longer than the other. The iodine(III) complex [I(O₂CMe)₂Ph(lone pair')₂] similarly has a planar structure, but now one iodine-oxygen bond of each chelate ring is 30% longer than the other, and the ligand can be considered as a unidentate [25].

Many of these seven-coordinate molecules show rapid intramolecular rearrangements at room temperature, but become rigid at low temperatures [231, 257, 478, 778].

C. [M(bidentate)₃(unidentate)]

1. The Theoretical Stereochemistries

The repulsion energy calculations show the existence of three major stereochemistries. If the unidentate ligand A is placed at $\phi = 0°$ and the bidentate ligands are labelled BC (with $\theta_B = 0°$), DE, and FG, the relation among the three stereochemistries is most readily depicted by projecting the potential energy surface onto the $\phi_D - \theta_D$ plane (Fig. 11.19). The stereochemistries associated with the points indicated in Fig. 11.19 are shown in Fig. 11.20, and as polyhedra in Fig. 11.21.

Stereochemistry A contains a threefold axis and is a capped octahedron with the unidentate ligand in the unique capping site. This is the only way three bidentate ligands can be arranged around a seven-coordinate metal atom so that all three ligands span equivalent polyhedral edges.

Stereochemistry B₁ corresponds to a fairly shallow minimum on the potential energy surface and contains no elements of symmetry. The polyhedron chosen to describe best the stereochemistry is somewhat arbitrary, two common examples being shown in Fig. 11.21. However the closely related stereochemistry B₂ contains a mirror plane which contains only the unidentate ligand (Fig. 11.21).

Stereochemistry C contains a mirror plane through ADE, and is intermediate between a pentagonal bipyramid (BCDFG as the pentagonal plane) and a capped trigonal prism (A the capping atom, BCFG as the capped face). As has been mentioned several times previously, this stereochemistry is also related to the

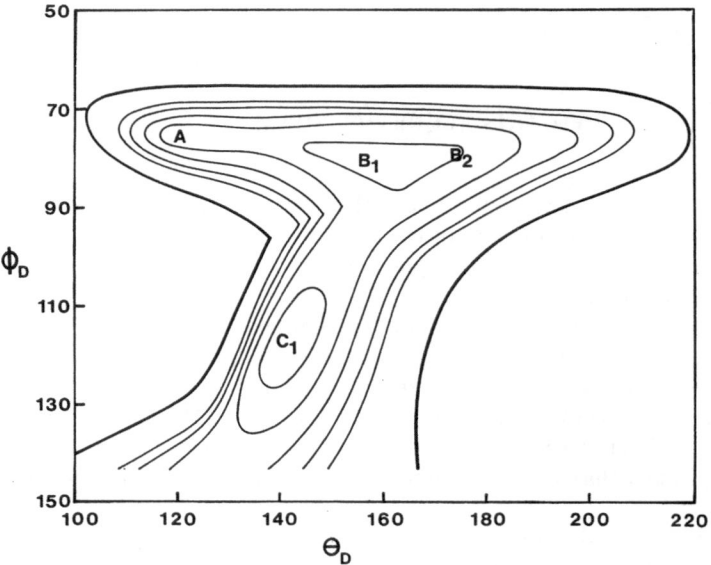

Fig. 11.19. Projection of the potential energy surface for [M(bidentate)$_3$(unidentate)] onto the $\phi_D - \theta_D$ plane (in degrees). The five *faint contour lines* are for successive 0.02 increments above the minimum, and the *heavy contour line* is 0.2 above the minimum, at C$_1$. b = 1.2, n = 6. The locations of stereochemistries, A, B$_1$, B$_2$, and C$_1$ are indicated

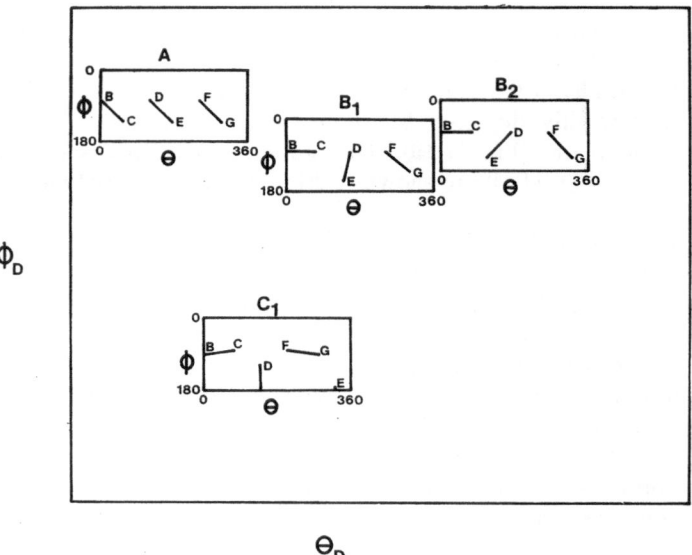

Fig. 11.20. Projections of the four stereochemistries for [M(bidentate)$_3$(unidentate)] that were located in Fig. 11.19. In all cases the unidentate ligand at $\phi = 0°$ is not shown

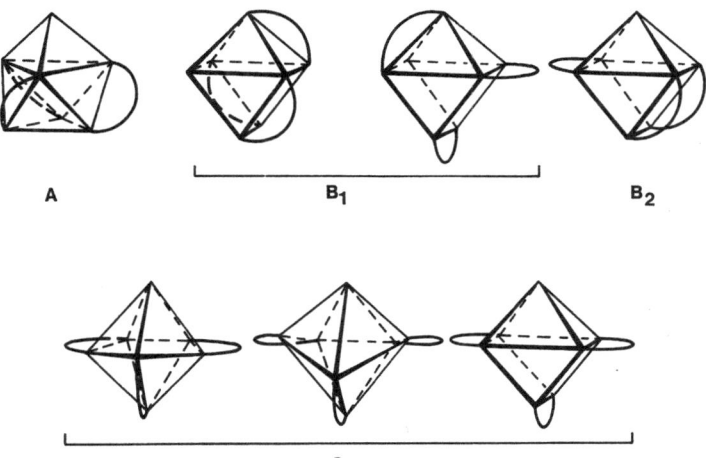

A   B₁   B₂

C

Fig. 11.21. Stereochemistries for [M(bidentate)₃(unidentate)]

capped octahedron, with D as the capping atom and CEF and ABG as the two triangular planes normal to the threefold axis.

It is clear from the potential energy surface (Fig. 11.19) that structures between these ideal limits are also expected.

The form of the potential energy surface is critically dependent on the effective bond length ratio R(unidentate/bidentate). As R(unidentate/bidentate) becomes less than unity, stereochemistry C becomes very stable, a typical potential energy surface being shown in Fig. 11.22. Conversely, as R(unidentate/bidentate) increases, it is the other end of the potential energy trough that deepens, corresponding to the capped octahedron (Fig. 11.23). The relative energies of these two limits, stereochemistry A and stereochemistry C, are shown in Fig. 11.24. It can be seen that they are of comparable stabilities at very low values of R(unidentate/bidentate) as well as at R (unidentate/bidentate) ~ 1.0, which is of particular relevance when discussing [M(bidentate)₃(lone pair)].

2. Comparison with Experiment

The results of all crystal structure determinations are given in Table 11.3. In agreement with the flat potential energy surface of Fig. 11.19, all three stereochemistries have been observed. It is also apparent that uncharged ligands such as H_2O, R_2CO, and R_3N with R(unidentate/bidentate) > 1.0 favor stereochemistries A and B in agreement with Fig. 11.23, whereas charged unidentate ligands such as Cl^-, Me^-, and O^{2-} with R(unidentate/bidentate) < 1.0 favor stereochemistry C in agreement with Fig. 11.22.

The experimental bond angles for any compound can be fitted against those calculated as a function of b and R(unidentate/bidentate), and hence a value for the effective bond length ratio obtained [639]. For example, for the complexes [M(bidentate)₃(uncharged ligand)] in Table 11.3, the value of R(unidentate/bidentate) is close to unity, indicating no gross differences in bond type. However

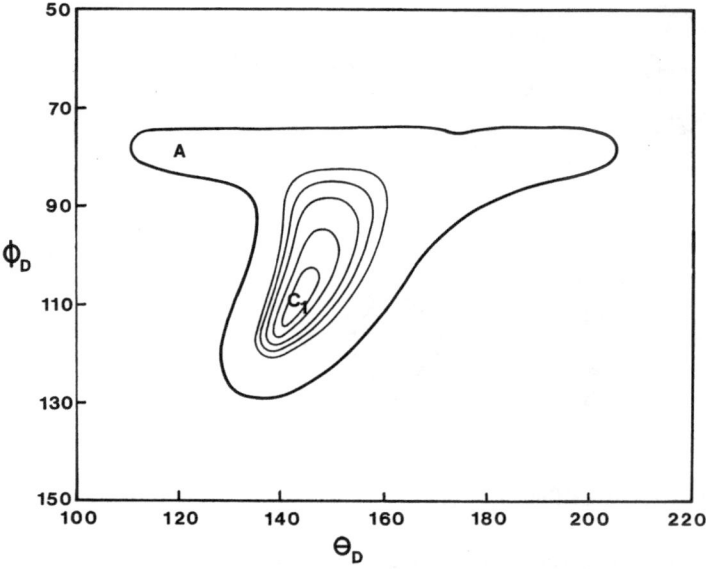

Fig. 11.22. Projection of the potential energy surface for [M(bidentate)₃(unidentate)] onto the $\phi_D - \theta_D$ plane (in degrees). The five *faint contour lines* are for successive 0.02 increments above the minimum, and the *heavy contour line* is 0.2 above the minimum, at C_1. b = 1.2, R = 0.8, n = 6. The locations of stereochemistries A and C_1 are indicated

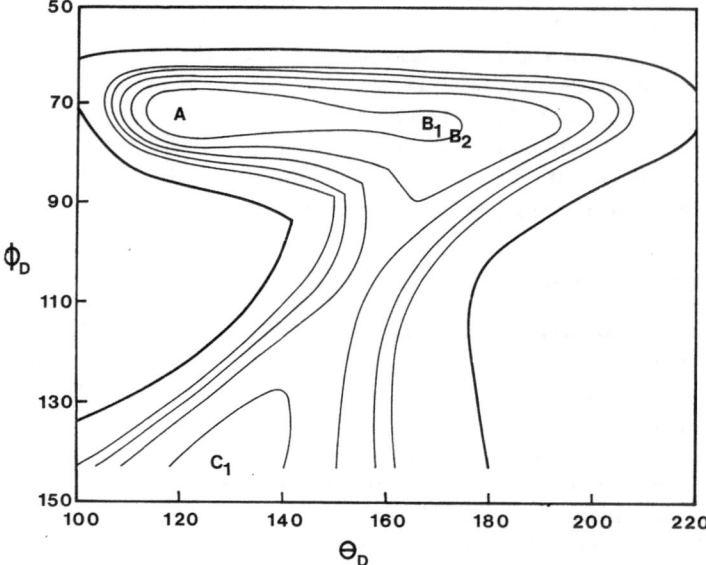

Fig. 11.23. Projection of the potential energy surface for [M(bidentate)₃(unidentate)] onto the $\phi_D - \theta_D$ plane (in degrees). The five *faint contour lines* are for successive 0.02 increments above the minimum, and the *heavy contour line* is 0.2 above the minimum, at A. b = 1.2, R = 1.2, n = 6. The locations of stereochemistries A, B_1, B_2, and C_1 are indicated

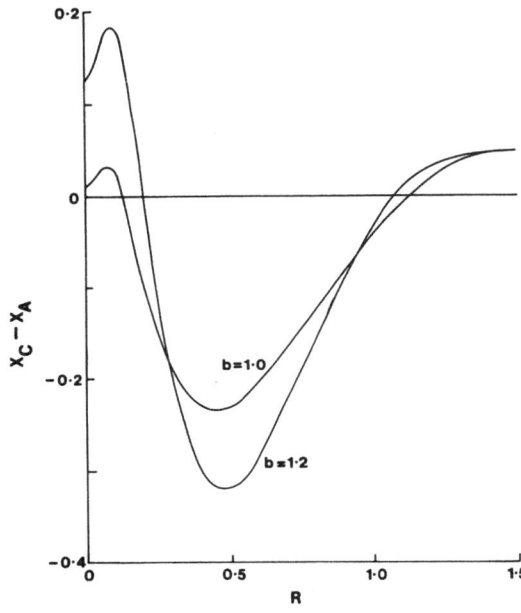

Fig. 11.24. Repulsion energy coefficient of stereochemistry C relative to stereochemistry A of [M(bidentate)$_3$(unidentate)] as a function of R. n = 6

much lower values are obtained for charged ligands, as expected:

$$R(X^-/\text{bidentate}) \sim 0.7 \qquad (n = 6)$$
$$R(O^{2-}/\text{bidentate}) \sim 0.4 \qquad (n = 6)$$
$$R(\text{lone pair}/\text{bidentate}) \sim 0.1 \qquad (n = 6)$$

The expected asymmetries of the bidentate ligands in compounds of stereochemistries A and C are fairly independent of b, but increase as R(unidentate/

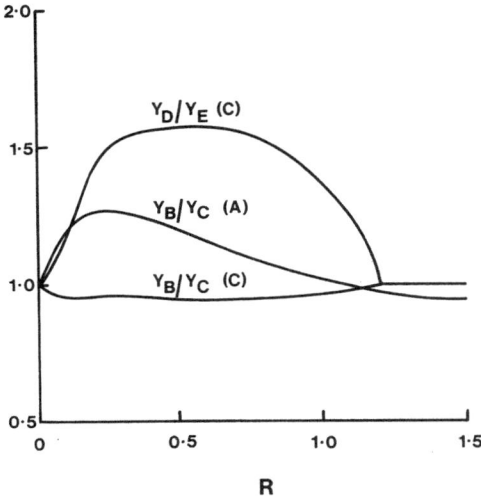

Fig. 11.25. Ratio of the individual atom repulsion coefficients for the two ends of each bidentate ligand in stereochemistries (A) and (C) of [M(bidentate)$_3$-(unidentate)] as a function of R(unidentate/bidentate). b = 1.2, n = 6

Table 11.3. Stereochemistries of [M(bidentate)$_3$(unidentate)]

Complex	b	Ref.
Stereochemistry A		
[Sb(S$_2$COEt)$_3$(lone pair)]		[490]
[As(S$_2$CNEt$_2$)$_3$(lone pair)]	1.13	[911]
[Bi{S$_2$P(OPr)$_2$}$_3$(lone pair)]	1.18	[699]
[Eu(BuCOCHCOBu)$_3$(NC$_7$H$_{13}$)]	1.18	[202]
[Ho(PhCOCHCOPh)$_3$(H$_2$O)]	1.21	[1117]
[Y(PhCOCHCOMe)$_3$(H$_2$O)]	1.21	[267]
Stereochemistry B		
[Dy(BuCOCHCOBu)$_3$(H$_2$O)]	1.20	[415]
[Eu(BuCOCHCOBu)$_3${OS(CH$_2$)$_2$CMe$_2$}]		[1066]
[Yb(MeCOCHCOMe)$_3$(H$_2$O)]$\frac{1}{2}$C$_6$H$_6$	1.22	[1081]
[Lu(BuCOCHCOBu)$_3$(NC$_5$H$_4$Me)]	1.22	[1079]
[Yb(MeCOCHCOMe)$_3${MeCOCH:C(NH$_2$)Me}]	1.23	[929]
[Lu(C$_3$F$_7$COCHCOBu)$_3$(H$_2$O)]	1.23	[122]
[Yb(MeCOCHCOMe)$_3$(H$_2$O)]	1.24	[288]
Stereochemistry C		
[Sn(NO$_3$)$_3$Me]		[168]
[Co{P(OCH$_2$)$_3$CMe}$_5$][Co(NO$_3$)$_3$(NCMe)]	0.98	[23]
(Et$_4$N)[Pb(S$_2$COEt)$_3$(lone pair)]	0.98	[822]
[Te(S$_2$CNEt$_2$)$_3$Ph]	1.05	[416]
[Sb(S$_2$COEt)$_3$(lone pair)]$\frac{1}{2}$C$_{10}$H$_8$N$_2$	1.07	[643]
[Te{S$_2$CN(Me)C$_2$H$_4$OH}$_3$Br]	1.10	[597]
[Ti(SOCNEt$_2$)$_3$Cl]	1.10	[545]
[Ta(S$_2$CNEt$_2$)$_3$S]	1.11	[868]
[Nb(S$_2$CNEt$_2$)$_3$O]	1.11	[318]
[Mo(S$_2$CNEt$_2$)$_3$N]	1.11	[592]
[Mo(S$_2$CNMe$_2$)$_3$(NS)]	1.12	[592]
[Mo(S$_2$CNMe$_2$)$_3$(N$_2$Ph)]CH$_2$Cl$_2$	1.12	[1100]
[Mo(S$_2$CNMe$_2$)$_3$(N$_2$C$_6$H$_4$NO$_2$)]$\frac{1}{2}$CH$_2$Cl$_2 \cdot \frac{1}{2}$H$_2$O	1.12	[1100]
[V(S$_2$CNEt$_2$)$_3$O]	1.13	[318]
[Mo(S$_2$CNBu$_2$)$_3$(NO)]	1.13	[149]
[Mo{S$_2$CN(CH$_2$)$_5$}$_3$(N$_2$EtPh)](Ph$_4$B)		[762]
[{Mo(S$_2$CNEt$_2$)$_3$N}$_2$Mo(S$_2$CNEt$_2$)$_3$]		[117]
[Re(S$_2$CNEt$_2$)$_3$(CO)]	1.13	[443]
[Ti(S$_2$CNMe$_2$)$_3$Cl]	1.14	[716]
A$_3$[Sb(C$_2$O$_4$)$_3$(lone pair)]4 H$_2$O (A=NH$_4$, K)	1.14	[888]
[Ru(S$_2$CNMe$_2$)$_3$I](I$_2$)	1.15	[775]
[Al(H$_2$BH$_2$)$_3$(NMe$_3$)]	1.15	[64]
[Ru(S$_2$CNEt$_2$)$_3$Cl]	1.16	[479]
[Sb(S$_2$PPh$_2$)$_3$(lone pair)]	1.18	[90]
(NH$_4$)$_3$[Nb(C$_2$O$_4$)$_3$O]H$_2$O	1.19	[770]
[Sn(O$_2$C$_7$H$_5$)$_3$(OH)]2 H$_2$O$\cdot\frac{1}{2}$MeOH	1.19	[854]
[Sn(O$_2$C$_7$H$_5$)$_3$Cl]CHCl$_3$	1.20	[854]
[Zr(MeCOCHCOMe)$_3$Cl]	1.25	[1076]

bidentate) decreases (Fig. 11.25). The predicted behavior is experimentally observed. For compounds containing uncharged unidentate ligands, the bonding of the bidentate ligands is fairly symmetric, but markedly unsymmetric in compounds containing a lone pair of electrons, with low R(lone pair/bidentate) values. For example, in compounds with stereochemistry A, the interaction with the lone pair lengthens the three bonds *cis* to it, as expected from $Y_B/Y_C > 1.0$:

	MB/MC
[Sb(S₂COEt)₃ (lone pair)]	1.19
[As(S₂CNEt)₃ (lone pair)]	1.21
[Bi{S₂P(OPr)₂}₃ (lone pair)]	1.06

In these compounds the large differences in the M—S bonds introduces differences into the S—C bonds, which can be associated with partial localization of the double bond (Fig. 11.26).

Figure 11.25 shows that for compounds with stereochemistry C, the BC (and FG) bidentate ligand will be more symmetrically bonded than the DE bidentate ligand, $Y_B/Y_C < 1.0$ and $Y_D/Y_E \gg 1.0$. These predicted distortions are again observed, for example:

	MB/MC	MD/ME
(Et₄N)[Pb(S₂COEt)₃(lone pair)]	0.94	1.10
[Sb(S₂COEt)₃(lone pair)] ½ C₁₀H₈N₂	0.91	1.25
A₃[Sb(C₂O₄)₃(lone pair)] 4 H₂O (A = NH₄, K)	0.97	1.11
[Sb(S₂PPh₂)₃(lone pair)]	0.88	1.30

Stereochemistry C is intermediate between a pentagonal bipyramid and a capped trigonal prism, and the repulsion energy calculations show that as the normalized bite increases, the bidentate ligand DE progressively swings further away from the pentagonal bipyramid position towards the capped trigonal prism position. This rotation as a function of normalized bite is shown in the isomorphous pair [Nb(S₂CNEt₂)₃O] and [V(S₂CNEt₂)₃O] [318], where the decrease in metal-sulfur distance from 2.60 to 2.50 Å results in an increase in normalized bite from 1.11 to 1.13. The decrease in ϕ_D and ϕ_E, as measured by the O=M—S angles, is from 95° to 98° and from 161° to 166°, respectively (Fig. 11.27).

As would be expected from the calculated potential energy surfaces, the rate of intramolecular rearrangement of seven-coordinate compounds such as [M(S₂CNR₂)₃X] is very dependent upon the nature of X[116, 545].

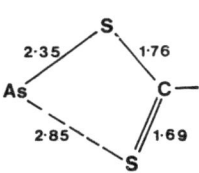

Fig. 11.26. Chelate ring geometry (Å) in [As(S₂CNEt₂)₃ (lone pair)]

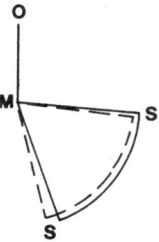

Fig. 11.27. MADE geometry in [Nb(S₂·CNEt₂)₃O] (——) and [V(S₂CNEt₂)₃O] (- - - -)

CHAPTER 12

Eight-Coordinate Compounds Containing only Unidentate Ligands

A. Seven Symmetrical Structures

Seven polyhedra with eight vertices were described in Chap. 2. With the exception of the cube which is a regular polyhedron, all polyhedra will distort by decreasing the size of the larger faces, and in the case of the non-uniform and chemical coordination polyhedra, by placing all vertices on the surface of a sphere. The resulting bicapped trigonal prism then becomes identical to the square antiprism. The repulsion energy coefficients X, calculated for n = 6, are listed for the remaining structures:

	X
Square antiprism	5.185
Triangular dodecahedron	5.245
Cube	5.758
Double triangular prism	6.504
Hexagonal bipyramid	7.785
Bi-(end-capped) trigonal prism	9.938

The two most stable structures, the square antiprism and the dodecahedron, are considered in more detail below. There are only two dodecahedral structures, compared with ten square antiprismatic structures, in agreement with the above repulsion energy coefficients. The double triangular prism, hexagonal bipyramid, and bi-(end-capped) trigonal prism have prohibitively high repulsion energies, have not been experimentally observed, and will not be considered further.

B. The Square Antiprism

The square antiprism is shown in Fig. 12.1. All ligand sites are identical, and the stereochemistry is completely defined by α, the angle the eight metal-ligand bonds make with the fourfold axis. It should also be noted that there is a twofold axis passing through the midpoint of every edge connecting triangular faces, the symmetry being D_{4d}.

For a square antiprism based on equal edge lengths (the hard sphere model,

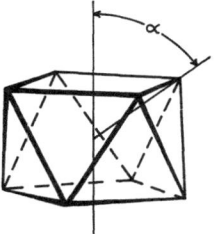

Fig. 12.1. The square antiprism

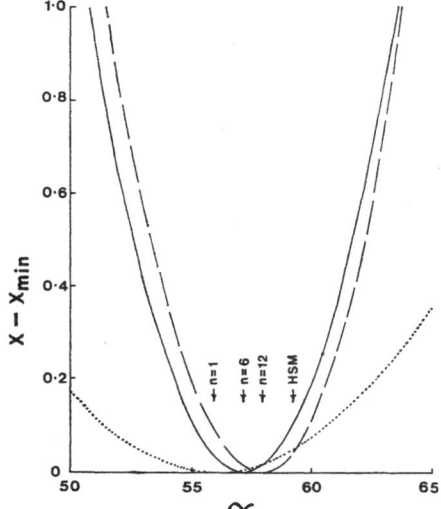

Fig. 12.2. Repulsion-energy coefficient, above that corresponding to the minimum, for the square antiprism as a function of α (in degrees). *Dotted line*, n = 1; *solid line*, n = 6; *broken line*, n = 12

HSM) and all metal-ligand bond lengths equal to r, $\alpha = 59.26°$ and the edge length is $1.2156r$. However, the stereochemistry of the polyhedron calculated by minimization of the repulsion energy is elongated along the major axis, $\alpha = 57.1°$ (n = 6), with a contraction of the larger square faces (Fig. 12.2).

This predicted small decrease in α from 59.3° to about 57° is observed in those square antiprismatic molecules whose structure has been precisely determined (Table 12.1).

C. The Dodecahedron

The dodecahedron is shown in Fig. 12.3. One set of four donor atoms is necessarily different from the other set of four donor atoms. The four A atoms constitute a tetrahedron that has been elongated along the fourfold inversion axis. The set of B atoms form a tetrahedron that has been squashed along the same axis, creating a

Table 12.1. Square antiprismatic [M(unidentate)$_8$]

Complex	α	Ref.
$H_4[W(CN)_8]6 H_2O$	57.6°	[82]
$Na_3[W(CN)_8]4 H_2O$	59.1°	[124]
$H_4[W(CN)_8]4 HCl \cdot 12 H_2O$	56.1°	[125]
$[Nd(ONC_5H_5)_8](ClO_4)_3$	56.1°	[34]
$[Sr(H_2O)_8](AgI_2)_2$	58.2°	[466]
$[Cu(H_2O)_6]_2[ZrF_8]$	57.1°	[439]
$Cs_4[U(NCS)_8]$	56.7°	[133]
$Na_3[TaF_8]$	59°	[564]
$Cd_2[Mo(CN)_8]2 N_2H_4 \cdot 4 H_2O$	58°	[233]
$[La\{OC(CHCMe)_2O\}_8](ClO_4)_3$		[219]

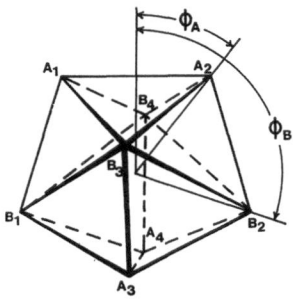

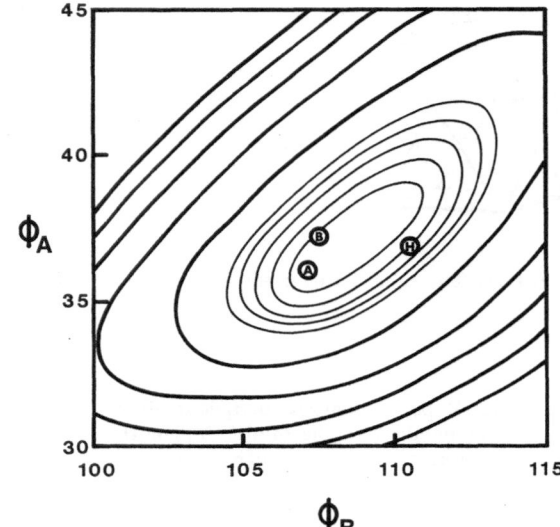

Fig. 12.3. The dodecahedron

Fig. 12.4. Projection of the potential energy surface for the dodecahedron as a function of ϕ_A and ϕ_B (in degrees). The five *faint contour lines* are for successive 0.02 increments above the minimum, and the five *heavy contour lines* are for successive 0.2 increments above the minimum. n = 6. A: $K_4[Mo(CN)_8]2\,H_2O$. B: $(Bu_4N)_3[Mo(CN)_8]$. H: hard sphere model

puckered square interposed between the two pairs of A atoms. A twofold axis passes through the midpoint of each B—B edge, the symmetry being D_{2d}.

An alternative useful way of viewing the dodecahedron is to consider it as two interpenetrating planar trapezoids, $B_1A_1A_2B_2$ and $B_3A_3A_4B_4$, which are at right angles to each other (Fig. 12.3).

The detailed geometry of the dodecahedron is described by two angular variables, ϕ_A and ϕ_B, the angles the M—A and M—B bonds make with the fourfold inversion axis, and the bond length ratio MA/MB (Fig. 12.3).

For a "hard sphere model", and all metal-ligand distances equal to r, A—A = A—B = 1.1993r, but B—B = 1.4986r = 1.2496 A—A, so that there are significant holes in the structure along the B—B edges. For this hard sphere model, $\phi_A = 36.85°$ and $\phi_B = 110.54°$.

The potential energy surface is shown in Fig. 12.4, projected onto $\phi_A - \phi_B$. The minimum at $\phi_A = 37.3°$, $\phi_B = 108.6°$ corresponds to a further flattening of the tetrahedron of B atoms, shortening the long B—B edges.

Figure 12.4 also shows that the structural parameters of the dodecahedral $K_4[Mo(CN)_8]2\,H_2O$ ($\phi_A = 36.0°$, $\phi_B = 107.1°$) [563] and $(Bu_4N)_3[Mo(CN)_8]$ ($\phi_A = 37.2°$, $\phi_B = 107.5°$) [261] lie closer to the minimum predicted from the repulsion calculations than to the hard sphere model.

The repulsion energy calculations also show that the A atoms experience a greater repulsion than the B atoms, $Y_A/Y_B = 1.14$ (n = 6). However the bond lengths in the above compounds are equal, MA/MB = 1.00, again showing that the bond lengths to π-bonding ligands do not follow predictions from simple repulsion theory.

D. Relations Between the Square Antiprism, Dodecahedron, and Cube

The relations between the three most stable polyhedra will be illustrated with potential energy surfaces, firstly projected onto the $\theta_A - \theta_B$ plane, and secondly projected onto the $\phi_A - \phi_B$ plane (as in Fig. 12.4).

The general stereochemistry is shown in Fig. 12.5. The points are arranged in two sets of four, with a D_2 axis passing through the metal atom.

The first type of potential energy surface is shown in Fig. 12.6. The center of the surface, at $\theta_A = \theta_B = 45°$, corresponds to the dodecahedron, labelled D_1. Distortion of this dodecahedron by decreasing θ_A to 30.9° and increasing θ_B to 55.3° produces a square antiprism (A_1) with $A_1B_1A_4B_4$ and $A_2B_2A_3B_3$ comprising the two square faces ($\phi_A = 38.9°$, $\phi_B = 108.8°$). The potential energy surface is symmetrical about D_1, since the A_1A_2 atoms can be twisted away from 90° relative to the A_3A_4 atoms in either a clockwise or anticlockwise direction.

Further continuation along the reaction coordinate in Fig. 12.6 generates an alternative dodecahedron D_2 at $\theta_A = 25.5°$ and $\theta_B = 61.7°$ (and $\phi_A = 47.9°$, $\phi_B = 115.4°$), which is identical to the original dodecahedron D_1, but with the A and B labels interchanged. This dodecahedron is then converted into a square antiprism A_2, at $\theta_A = 22.5°$ and $\theta_B = 67.5°$ ($\phi_A = 57.1°$, $\phi_B = 122.9°$), the square faces being $A_1B_3A_2B_4$ and $A_3B_2A_4B_1$.

Continuation up this potential energy valley in the same direction to $\theta_A = 0$ and $\theta_B = 90°$ leads to the cube, labelled C_1. Moving up a different potential energy valley from the first square antiprism leads to the cube C_2 at $\theta_A = 0$ and $\theta_B = 54.7°$. Direct distortion from the dodecahedron D_1 to a third cube, C_3, also situated at $\theta_A = \theta_B = 45°$, can be seen if the potential energy surface is projected onto the $\phi_A - \phi_B$ plane (Fig. 12.7), rather than the $\theta_A - \theta_B$ plane. In this case the cube can be considered as two interpenetrating regular tetrahedra, and is on the saddle at $\phi_A = 180° - \phi_B = 54.7°$ between two identical dodecahedra, depending on which tetrahedron is elongated and which is compressed.

Views of the polyhedra corresponding to D_1, A_1, D_2, A_2, C_1, C_2, and C_3 are shown in Fig. 12.8.

Potential energy surfaces of the type shown in Fig. 12.6 indicate that structures intermediate between a square antiprism and a dodecahedron might be expected,

(*Continued p. 158*)

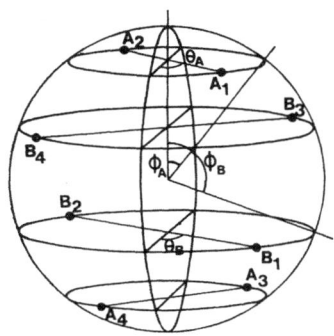

Fig. 12.5. General stereochemistry for [M(unidentate)$_8$]

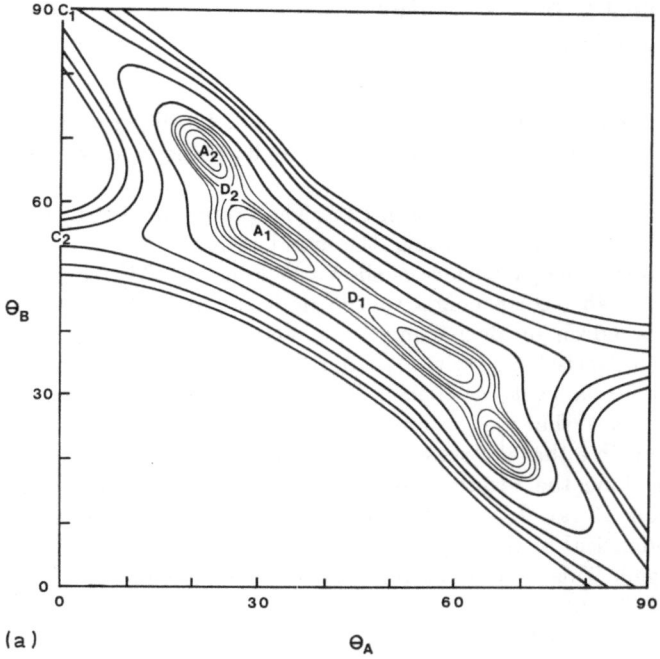

(a)

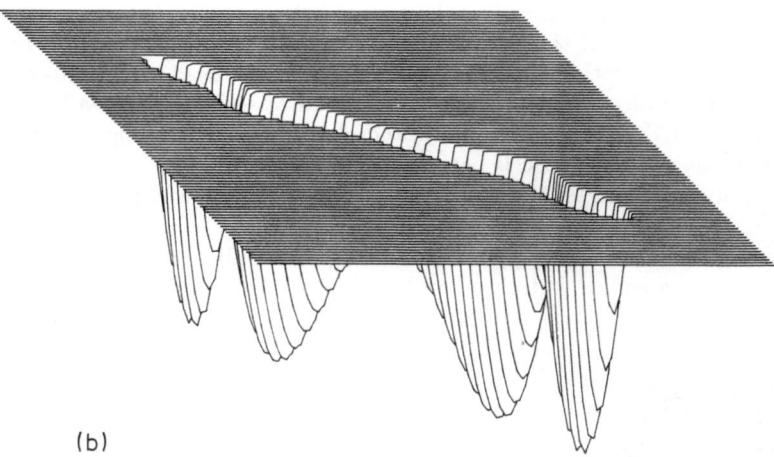

(b)

Fig. 12.6. **a** Projection of the potential energy surface for [M(unidentate)$_8$] onto the $\theta_A - \theta_B$ plane (in degrees). The five *faint contour lines* are for successive 0.02 increments above the minima, and the five *heavy contour lines* are for successive 0.2 increments above the minima. n = 6. The locations of the dodecahedra (D), square antiprisms (A), and cubes (C) are shown. **b** As in **a**, but with truncation at X = 0.1

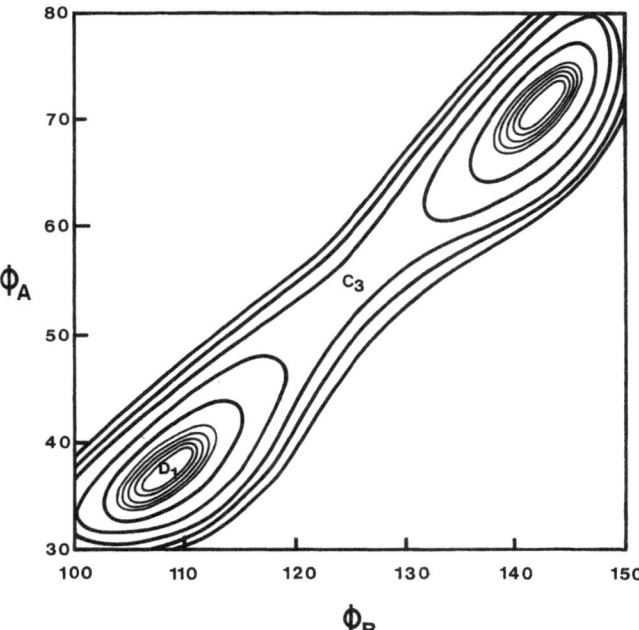

Fig. 12.7. Projection of the potential energy surface for [M(unidentate)$_8$] onto the $\phi_A - \phi_B$ plane (in degrees). The five *faint contour lines* are for successive 0.02 increments above the minima, and the five *heavy contour lines* are for successive 0.2 increments above the minima, at D$_1$. n = 6. The locations of the dodecahedra (D$_1$) and cube (C$_3$) are shown

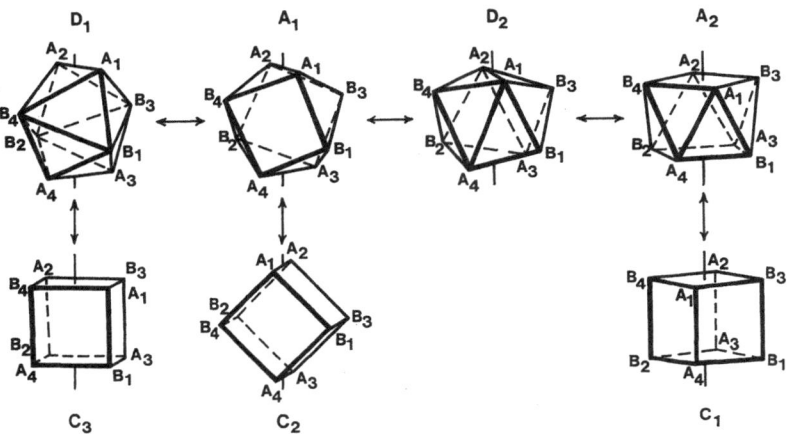

Fig. 12.8. The relation between the polyhedra corresponding to Figs. 12.6 and 12.7

with smaller distortions in the direction of a cube also permitted. An additional intermediate structure, not allowed by the symmetry assumed in Fig. 12.5, can be visualized if the square antiprism — dodecahedron interconversion is considered as a creasing of each square of the square antiprism by shorting one of the diagonals to form two triangular faces. For example, shortening B_1B_4 and B_2B_3 converts A_1 into D_1 (Fig. 12.8). If only one of these square faces is creased, for example by shortening B_1B_4 of square antiprism A_1, then a bicapped trigonal prism is formed with $B_1B_3A_3$—$B_4A_2B_2$ forming the prism and A_1 and A_4 the capping atoms.

Small distortions from the idealized stereochemistries have been observed in some of the above square antiprismatic and dodecahedral molecules, and have been discussed elsewhere [638]. However $(Et_3NH)_2(H_3O)_2[Mo(CN)_8]$ appears to be approximately midway between a square antiprism and dodecahedron [706]. In $Li_6(BeF_4)[ZrF_8]$ [977], the dodecahedral structure is distorted towards the C_3 cube, $\phi_A = 43.0$, $\phi_B = 115.0$, $\theta_A = \theta_B = 45.0°$. The most obvious reason for this distortion is interaction with lithium ions, which lie outside all the A—B edges, but not outside the A—A or B—B edges.

In contrast to $Cs_4[U(NCS)_8]$ which is close to the ideal square antiprism (Table 12.1), the structure of the tetraethylammonium salt, $(Et_4N)_4[U(NCS)_8]$ [271], is remarkable. This compound is centrosymmetric, and the structure is virtually a perfect cube. This cubic structure is not retained in solution. The structure must be attributed to the tetraethylammonium groups, which are disordered and sit outside each face of the cube. The ethyl groups are in two alternative arrangements at right angles to each other, so that half ethyl groups are projected between every pair of thiocyanate groups.

Cubic stereochemistry is also found in Na_3PaF_8 [156]. The structure is closely related to the fluorite structure, with both sodium and protactinium occupying the cubic sites.

The potential energy surface in Fig. 12.6 shows that there is no potential barrier between the square antiprism and the dodecahedron, and very rapid rearrangements may be expected [120]. The ions $[M(CN)_8]^{4-}$ and $[M(CN)_8]^{3-}$, where M is Mo or W, have been extensively studied [505]. As shown above, the structures in the solid state may be square antiprismatic, dodecahedral, or an intermediate stereochemistry. In solution, only a single ^{13}C-NMR signal is observed for $[Mo(CN)_8]^{4-}$ in CH_2Cl_2—$CHClF_2$ mixtures at temperatures as low as $-165°C$, consistent with very rapid intramolecular rearrangements [813].

E. [M(unidentate A)$_4$(unidentate B)$_4$]

It has been shown above that for eight equivalent unidentate ligands, the square antiprism is slightly more stable than the dodecahedron. However, the energy difference is not large, leading to the occurrence of both structures in the solid state.

It has also been pointed out that a major difference between the square antiprism and the dodecahedron is that the former has eight identical ligand sites,

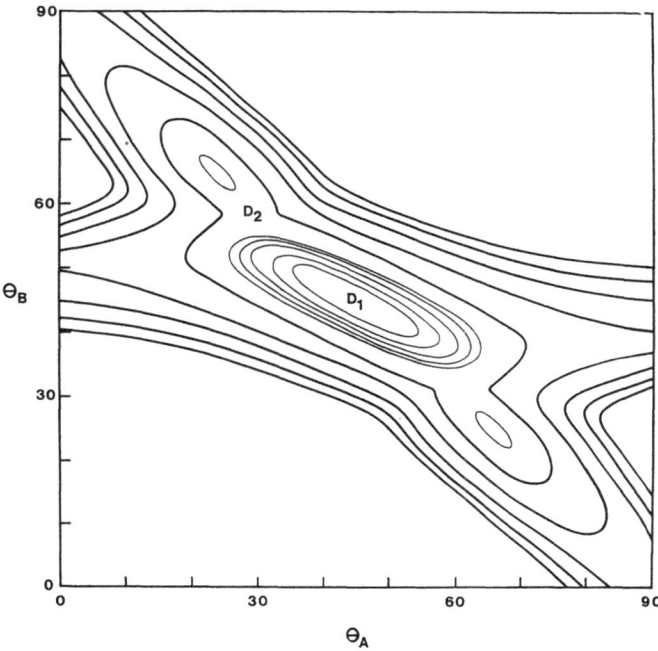

Fig. 12.9. Projection of the potential energy surface for [M(unidentate A)₄(unidentate B)₄] onto the $\theta_A - \theta_B$ plane (in degrees). The five *faint contour lines* are for successive 0.02 increments above the minimum, and the five *heavy contour lines* are for successive 0.2 increments above the minimum, at D_1. n = 6, R = 1.2. The position of the dodecahedra D_1 and D_2 are shown

whereas the dodecahedron has four A sites associated with higher repulsion energy than the four B sites. It follows that for compounds of the type [M(unidentate A)₄(unidentate B)₄], in which there are significant differences between the two types of ligand, the sorting into the appropriate A and B sites may stabilize the dodecahedron relative to the square antiprism.

This effect is shown by the potential energy surface in Fig. 12.9, calculated for the effective bond length ratio R(A/B) = 1.20. Figure 12.9 is directly comparable to Fig. 12.6, in which all metal-ligand bonds were equal. Fig. 12.9 clearly shows that the dodecahedron at $\theta_A = \theta_B = 45°$ has been stabilized relative to all other stereochemistries, and corresponds to the long bonds occupying the A sites of the dodecahedron. Conversely, the dodecahedron at $\theta_A \sim 29°$ and $\theta_B \sim 58°$, corresponding to the long bonds occupying the B sites, has been significantly destabilized.

The stereochemistry corresponding to the minimum in the potential energy surface changes smoothly from the square antiprism at R(A/B) = 1.0 to the dodecahedron at R(A/B) = 1.2, the angular parameters being shown in Fig. 12.10. As R(A/B) is increased further, the dodecahedron D_1 in turn is transferred into the "cube" C_3, with $\phi_A = 180 - \phi_B = 54.7°$ (Fig. 12.10). This C_3 stereochemistry

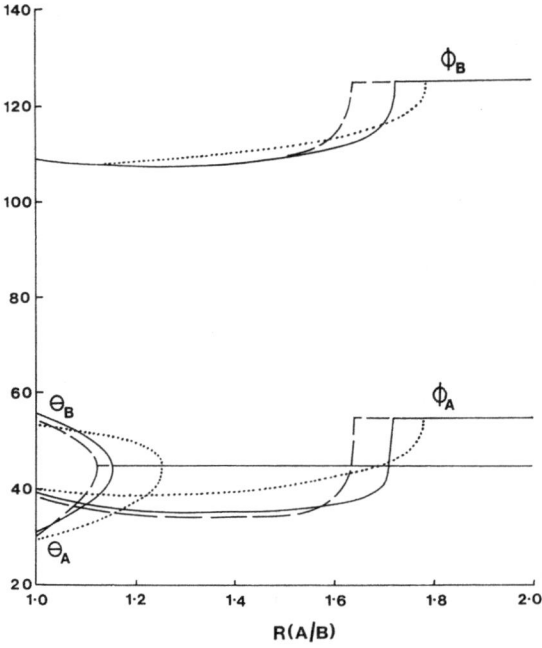

Fig. 12.10. Angular coordinates (in degrees) for [M(unidentate A)$_4$(unidentate B)$_4$], as a function of the bond-length ratio, R(A/B). *Dotted lines*, n = 1; *solid lines*, n = 6; *broken lines*, n = 12

consists of two different sized regular interpenetrating tetrahedra, or can alternatively be considered as a tetrahedron of four B atoms with an A atom outside each face. (This "cube" C$_3$ is not the same as the "cubes" C$_1$ and C$_2$, as the latter are composed of two interpenetrating A$_4$ and B$_4$ rectangles.)

The relative energies of all these stereochemistries are shown in Fig. 12.11.

In agreement with these predictions, the dodecahedral structure is now more common than for [M(unidentate)$_8$] compounds. Dodecahedral structures are found for K$_4$[Nd(NCS)$_4$(H$_2$O)$_4$](NCS)$_3$ [700], K$_4$[Eu(NCS)$_4$(H$_2$O)$_4$](NCS)$_3$·2H$_2$O [700], [ThCl$_4$(Me$_2$SO)$_4$] [931], [MoH$_4$(PPh$_2$Me)$_4$] [777], and [Mo(CN)$_4$(CNMe)$_4$] [214, 839], the last being significantly distorted towards a square antiprism. The square antiprismatic structure is retained in [U(NCS)$_4$(Me$_3$PO)$_4$] [930] and [U(NCS)$_4$(Ph$_3$PO)$_4$] [131].

F. [M(unidentate A)$_2$(unidentate B)$_6$]

Compounds of this stoichiometry are rare, but the stereochemistry is of some interest because of the relation to that of [M(unidentate)$_2$(bidentate)$_3$] (Sect. 13.2), which for bidentate ligands of small normalized bite and for short metal-unidentate ligand bond lengths, may be hexagonal bipyramidal with linear unidentate-

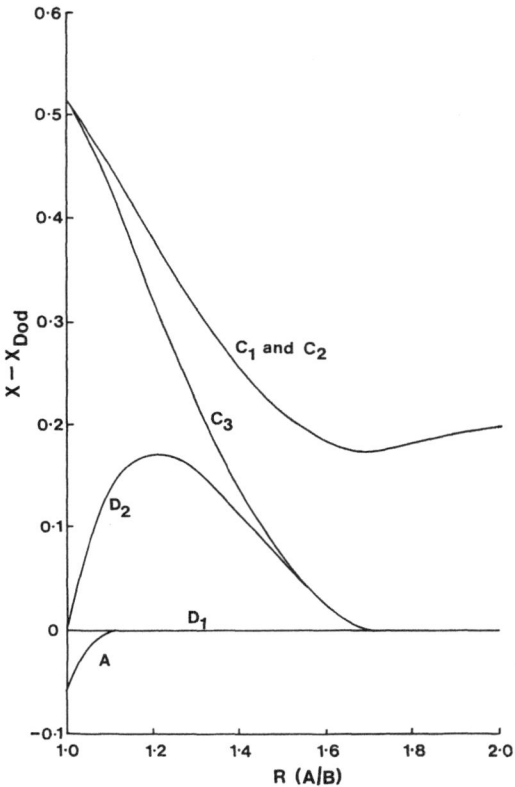

Fig. 12.11. Repulsion-energy coefficients, relative to that for the dodecahedron D_1, for [M(unidentate A)$_4$-(unidentate B)$_4$], as a function of the bond-length ratio, R(A/B). n = 6. C, cube; A, square antiprism

metal-unidentate groups. The best known examples are uranyl complexes such as [UO$_2$(NO$_3$)$_3$]$^-$.

The various square antiprismatic isomers of [M(unidentate A)$_2$(unidentate B)$_6$] which exist for R(A/B) $\sim$ 1.0 progressively distort and pass through distorted dodecahedral and bicapped trigonal prismatic isomers as R(A/B) is varied. However, a bicapped octahedron with linear (unidentate A)-M-(unidentate A) bonds is reached only for R(A/B) < 0.3, and is not expected to be experimentally found. In agreement with this prediction, the Cl—M—Cl groups in [UCl$_2$(Me$_2$SO)$_6$] ·[UCl$_6$] [127] and [LnCl$_2$(H$_2$O)$_6$]Cl$_2$ (Ln = Nd, Sm, Eu, Gd, Lu) [94, 510] are nonlinear.

CHAPTER 13

Eight-Coordinate Compounds Containing Chelate Groups

A. [M(bidentate)$_2$(unidentate)$_4$]

With values of the normalized bite from 1.20 to 1.25, which are appropriate for the bidentate ligands fitting along various edges of polyhedra calculated for [M(unidentate)$_8$] (Chap. 12), seven separate isomers have been located as potential energy minima. These are most readily envisaged as various ways of arranging ligands along the edges of distorted square antiprisms (isomers A—F), or a distorted dodecahedron (isomer G) (Fig. 13.1). The square antiprismatic isomers are loosely ordered as progressively changing from a "*trans*" arrangement of bidentate ligands (isomer A), to the most "*cis*" arrangement of ligands (isomers E and F).

These seven isomers all contain some elements of symmetry, namely three twofold axes in isomer A, one twofold axis in isomers B, E, and F, a twofold axis and two mirror planes in isomer D, and one mirror plane in isomers C and G. Both bidentate ligands are symmetrical and equivalent and all four unidentate ligands are equivalent only in isomer A.

Potential energy surfaces analogous to that shown in Fig. 12.6 for [M(unidentate)$_8$] may be calculated for isomer A if the chelate rings span the A_1A_2 and A_3A_4 edges shown in Fig. 12.5. At large normalized bites (Fig. 13.2), the minima correspond to the above square antiprism, but these minima progressively move

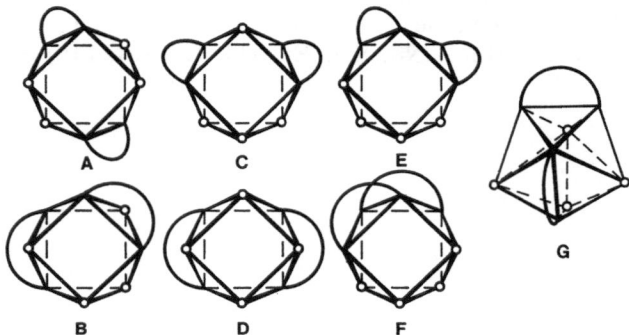

Fig. 13.1. Stereochemistries for [M(bidentate)$_2$(unidentate)$_4$]

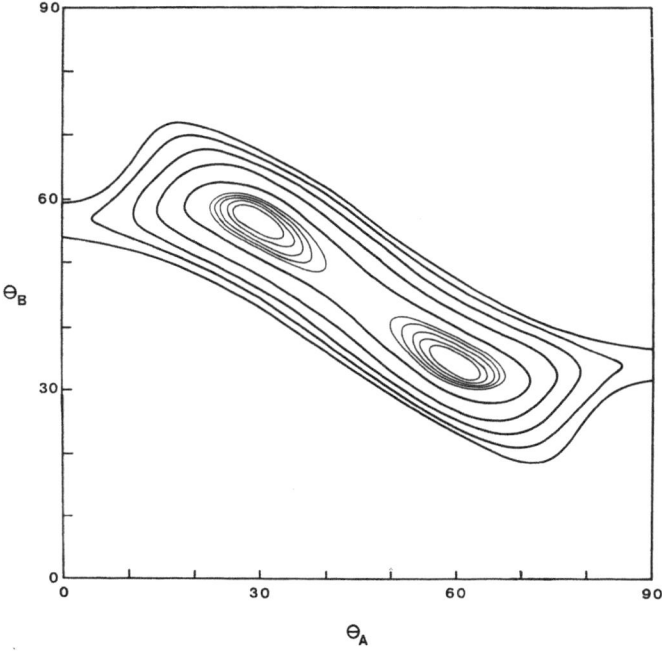

Fig. 13.2. Projection of the potential energy surface for isomer A of [M(bidentate)$_2$(uni-dentate)$_4$] onto the $\theta_A - \theta_B$ plane (in degrees). The five *faint contour lines* are for successive 0.02 increments above the minima, and the five *heavy contour lines* are for successive 0.2 increments above the minima. n = 6, b = 1.3

closer together as b is reduced, until a single minimum is obtained at $\theta_A = \theta_B = 45°$ (Fig. 13.3) corresponding to a dodecahedron with the bidentate ligands spanning the two A—A edges, and the four unidentate ligands occupying the B sites. The dodecahedron is additionally stabilized by the effect examined in Sect. 12.E, namely, by decreasing the effective bond length to the unidentate ligands, R(unidentate/bidentate).

Similar changes occur in the stereochemistries of the other isomers as b is reduced. Isomer B distorts towards a dodecahedron where one trapezoid is formed from the bidentate ligands, and one from the four unidentate ligands. At low b isomer C approximates to a bicapped trigonal prism, with the bidentate ligands spanning edges linking capping sites and prism sites. Isomer D also forms a bicapped trigonal prism, but with the two bidentate ligands forming the uncapped rectangular face of the prism. Isomers E and F are converted into isomer D below b = 1.0. The stereochemistry of isomer G is relatively insensitive to the value of b.

The relative energies of each of these isomers, and also its range of existence as a discrete potential energy minimum, are shown in Fig. 13.4. Isomers A, C and E are of equal stability at b = 1.26, which is the value appropriate for bidentate ligands spanning the non-square edges of a square antiprism, and isomers B, D and

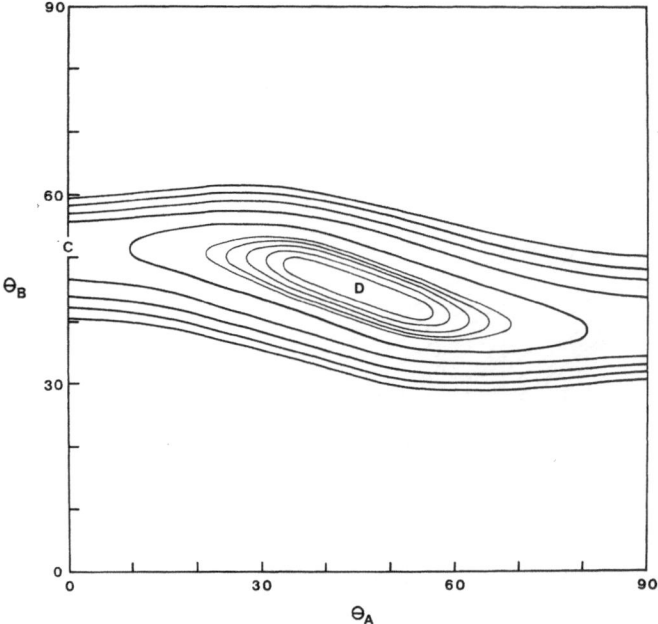

Fig. 13.3. Projection of the potential energy surface for isomer A of [M(bidentate)$_2$(unidentate)$_4$] onto the θ_A — θ_B plane (in degrees). The five *faint contour lines* are for successive 0.02 increments above the minimum, and the five *heavy contour lines* are for successive 0.2 increments above the minimum, at D. n = 6, b = 0.9. The locations of the dodecahedron (D) and "cube" (C) are indicated.

F are of equal stability at b = 1.19, corresponding to the square edges of a square antiprism.

This order of stability changes dramatically at R(unidentate/bidentate) < 1.0, where isomer A is greatly stabilized with respect to all other isomers (Fig. 13.5). This stereochemistry will therefore be strongly stabilized for charged unidentate ligands.

The stereochemistries observed for compounds of the type [M(bidentate)$_2$-(unidentate)$_4$] are listed in Table 13.1. All compounds [M{C$_6$H$_4$(AsMe$_2$)$_2$}$_2$X$_4$]$^{x+}$ are of isomer A as predicted, and the experimental bond angles may be fitted against those calculated to yield R(X$^-$/bidentate) $\sim$ 0.7 (n = 6). For all other compounds in Table 13.1, it is found that R(unidentate/bidentate) $\sim$ 1.0. The stabilization of isomers B and C for [Er(HOCH$_2$COO)$_2$(H$_2$O)$_4$][Er(HOCH$_2$COO)$_4$] and [Ca(picrate)$_2$(H$_2$O)$_4$] respectively is probably largely attributale to these structures being particularly suitable for the accommodation of unsymmetrical bidentate ligands.

The last two compounds in Table 13.1, [M(phen)$_2$(H$_2$O)$_4$](ClO$_4$)$_2$·2phen, have exceptional distorted cubic structures with the two phenanthroline ligands spanning opposite edges (Fig. 13.6). Both the coordinated and uncoordinated phenanthroline molecules are packed in parallel planes, which appears to be a

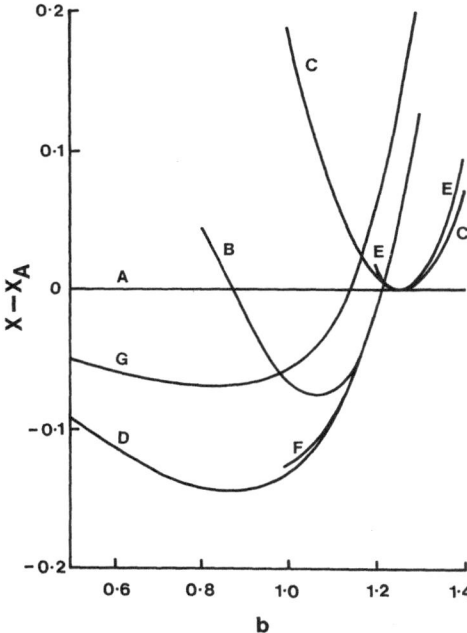

Fig. 13.4. Repulsion energy coefficients for [M(bidentate)$_2$(unidentate)$_4$] stereochemistries relative to those for stereochemistry A, as a function of normalized bite, b. R(unidentate/bidentate) = 1.0, n = 6.

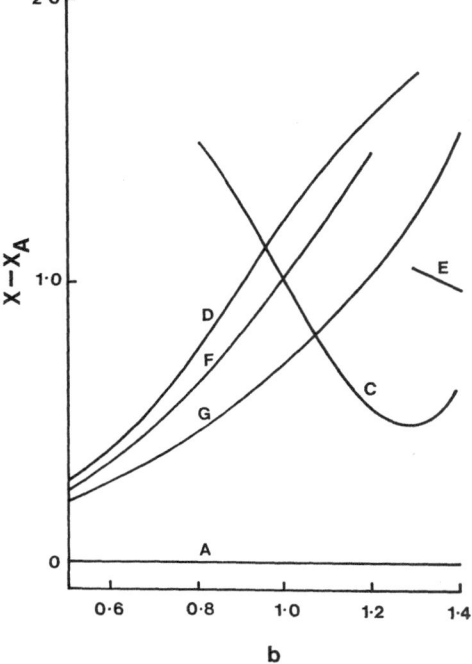

Fig. 13.5. Repulsion energy coefficients for [M(bidentate)$_2$(unidentate)$_4$] stereochemistries relative to those for stereochemistry A, as a function of normalized bite, b. R(unidentate/bidentate) = 0.75, n = 6

Table 13.1. Stereochemistries of [M(bidentate)$_2$(unidentate)$_4$]

Complex	b	Stereo-chemistry	Ref.
[Ta{C$_6$H$_4$(AsMe$_2$)$_2$}$_2$Cl$_4$][TaCl$_5$(OEt)]	1.18	A	[319]
[Nb{C$_6$H$_4$(AsMe$_2$)$_2$}$_2$Cl$_4$]$_2$[NbO$_2$Cl$_3$]	1.18	A	[319]
[Nb{C$_6$H$_4$(AsMe$_2$)$_2$}$_2$Cl$_4$][NbOCl$_4$]	1.19	A	[319]
[Nb{C$_6$H$_4$(AsMe$_2$)$_2$}$_2$Cl$_4$]	1.19	A	[644]
[Ti{C$_6$H$_4$(AsMe$_2$)$_2$}$_2$Cl$_4$]	1.19	A	[240]
[Tc{C$_6$H$_4$(AsMe$_2$)$_2$}$_2$Cl$_4$](PF$_6$)	1.21	A	[481]
[Ta{C$_6$H$_4$(AsMe$_2$)$_2$}$_2$Br$_4$][TaBr$_6$]	1.22	A	[367]
[Mo{C$_6$H$_4$(AsMe$_2$)$_2$}$_2$Cl$_4$](I$_3$)	1.23	A	[344]
[Zr(bipy)$_2$(NCS)$_4$]	1.12	A	[867]
[Nb(bipy)$_2$(NCS)$_4$]	1.15	A	[867]
[Cd(NO$_3$)$_2$(H$_2$O)$_4$]	0.83	A	[747]
[Er(HOCH$_2$COO)$_2$(H$_2$O)$_4$][Er(HOCH$_2$COO)$_4$]	1.10	B	[504]
[Ca(picrate)$_2$(H$_2$O)$_4$]H$_2$O	1.06	C	[322]
[W{ONN(Me)O}$_2$Me$_4$]	1.14	D	[442]
[Ca(NO$_3$)$_2$(MeOH)$_4$]	0.85	G	[703]
[Ba(phen)$_2$(H$_2$O)$_4$](ClO$_4$)$_2$ · 2phen	0.93	see text	[995]
[Sr(phen)$_2$(H$_2$O)$_4$]$_2$ · 2phen	0.98	see text	[995]

particularly favorable arrangement. The only stereochemistry listed above which allows the two bidentate ligands to be in the same plane is the dodecahedral version of isomer B, in which both bidentate ligands span edges of the same trapezoid. This arrangement is not possible with ligands such as 1,10-phenanthroline because of the clash of the α-hydrogen atoms (Fig. 13.7), but if the bidentate ligands move apart to prevent this clash, the observed distorted cubic structure is formed.

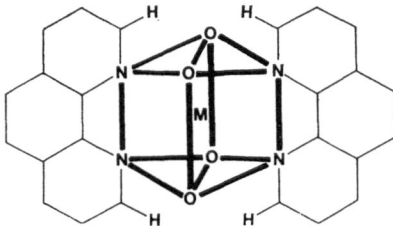

Fig. 13.6. [Ba(phen)$_2$(H$_2$O)$_4$](ClO$_4$)$_2$ · 2phen

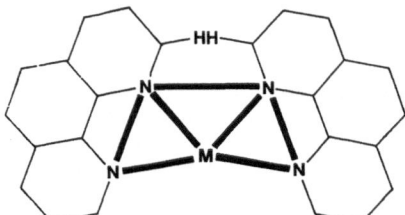

Fig. 13.7. A hypothetical M(phen)$_2$ trapezoid showing clash of α-hydrogen atoms

B. [M(bidentate)$_3$(unidentate)$_2$]

Eight-coordinate compounds containing three bidentate ligands are listed in Table 13.2. They can be classified into the following three types, depending on the normalized bite of the bidentate ligands:

1. $b < \sim 0.9$: These compounds are all of hexagonal bipyramidal stereochemistry, with a linear unidentate-metal-unidentate group. The hexagonal plane formed by the three bidentate ligands may be slightly puckered.

2. $b \sim 1.1$ to 1.2: These compounds are adequately described as distorted square antiprisms or distorted dodecahedra. There are 17 different ways (not counting optical isomers) three bidentate ligands can be arranged around the edges of a square antiprism or a dodecahedron (excluding the longer B—B edges).

3. $b \sim 1.0$: The only example of intermediate normalized bite is found in [Nb(O$_2$CNMe$_2$)$_5$], in which two of the carbamate ligands act only as unidentate ligands.

The results obtained from repulsion energy calculations for two of the simpler stereochemistries are described below. The first contains a twofold axis and is of particular relevance to the hexagonal bipyramidal molecules, while the second contains a mirror plane and is of particular relevance to the structure of [Nb(O$_2$CNMe$_2$)$_5$].

The first isomer is shown in Fig. 13.8, and contains a twofold axis through the midpoint of one bidentate ligand, labelled AB. For $b \sim 0.7$, a hexagonal bipyramid is formed, the hexagonal plane being formed by the three bidentate ligands. Above $b \sim 0.75$, the unidentate-M-unidentate group remains linear, but the

Table 13.2. Stereochemistries of [M(bidentate)$_3$(unidentate)$_2$]

Complex	b	Stereochemistry	Ref.
(Me$_4$N)[Pb(O$_2$CMe)$_3$Ph$_2$]	0.83	Hexagonal bipyramid	[24]
Ba[Np(O$_2$CMe)$_3$O$_2$]2 H$_2$O	0.86	Hexagonal bipyramid	[189]
Rb[U(NO$_3$)$_3$O$_2$]	0.87	Hexagonal bipyramid	[72]
Na[U(O$_2$CMe)$_3$O$_2$]	0.89	Hexagonal bipyramid	[1113]
(NH$_4$)$_4$[U(CO$_3$)$_3$O$_2$]	0.91	Hexagonal bipyramid	[496]
(Me$_4$N)[U(S$_2$CNEt$_2$)$_3$O$_2$]		Hexagonal bipyramid	[141]
[Nb(O$_2$CNMe$_2$)$_5$]	0.99	Intermediate	[232]
[La{S$_2$P(OEt)$_2$}$_3$(Ph$_3$PO)$_2$]	1.08	Distorted square antiprism	[882]
[Eu(MeCOCHCOMe)$_3$(H$_2$O)$_2$]H$_2$O	1.11	Distorted square antiprism	[605]
[La(MeCOCHCOMe)$_3$(H$_2$O)$_2$]	1.13	Distorted square antiprism	[873]
[Nd(C$_4$H$_3$SCOCHCHCF$_3$)$_3$(Ph$_3$PO)$_2$]	1.15	Distorted dodecahedron	[708]
[Eu(BuCOCHCOBu)$_3$(Me$_2$NCHO)$_2$]	1.16	Distorted square antiprism	[289]
[Nd(MeCOCHCOMe)$_3$(H$_2$O)$_2$]	1.17	Distorted square antiprism	[56]
[Eu(BuCOCHCOBu)$_3$(C$_5$H$_5$N)$_2$]	1.17	Distorted square antiprism	[282]
[Y(MeCOCHCOMe)$_3$(H$_2$O)$_2$]H$_2$O	1.18	Distorted square antiprism	[287]
[Ho(MeCOCHCOMe)$_3$(H$_2$O)$_2$]2 H$_2$O	1.18	Distorted dodecahedron	[55]
[Ho(BuCOCHCOBu)$_3$(MeC$_5$H$_4$N)$_2$]		Distorted square antiprism	[574]
[Eu(C$_4$H$_3$SCOCHCOCF$_3$)$_3$(H$_2$O)$_2$]	1.19	Distorted square antiprism	[1088]

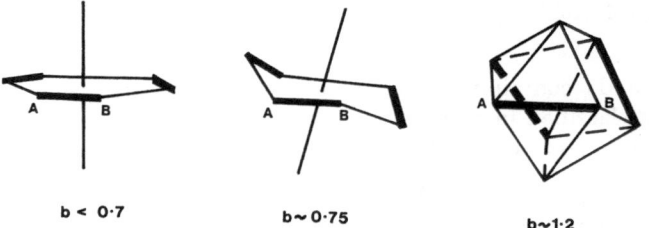

b < 0·7   b ~ 0·75   b ~ 1·2

Fig. 13.8. The first isomer for [M(bidentate)$_3$(unidentate)$_2$]. b < 0.7: hexagonal bipyramid; b $\sim$ 0.75: puckered hexagonal bipyramid; b $\sim$ 1.2: square antiprism

 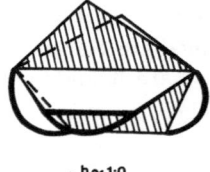

b ~ 1·2   b ~ 1·0

Fig. 13.9. The second isomer for [M(bidentate)$_3$(unidentate)$_2$]. b $\sim$ 1.2: square antiprism; b $\sim$ 1.0: interpenetrating pentagons

hexagonal plane puckers. The commencement of puckering as the normalized bite is increased is very much delayed for R(unidentate/bidentate) < 1.0[638]. As the normalized bite is further increased, the stereochemistry continuously changes until at b $\sim$ 1.2 a distorted square antiprism is attained, with two of the bidentate ligands spanning edges of different squares (Fig. 13.8). This particular isomer is found for [Eu(BuCOCHCOBu)$_3$(C$_5$H$_5$N)$_2$] and [Ho(BuCOCHCOBu)$_3$(MeC$_5$H$_4$N)$_2$].

In the second isomer, a mirror plane bisects one bidentate ligand and also contains the two unidentate ligands (Fig. 13.9). At b $\sim$ 1.2, the structure is a distorted square antiprism with only one bidentate ligand spanning a square edge. This isomer is observed for [Eu(C$_4$H$_3$SCOCHCOCF$_3$)$_3$(H$_2$O)$_2$]. The stereochemistry formed by reduction to b $\sim$ 1.0 can be described as two interpenetrating pentagons that are almost perpendicular to each other (Fig. 13.9). It is this structure which can be used to describe the structure of [Nb(O$_2$CNMe$_2$)$_5$], which contains three bidentate and two unidentate ligands.

C. [M(bidentate)$_4$]

1. The Theoretical Stereochemistries

The general stereochemistry is the same as that shown previously (Fig. 12.5), with the added condition that bidentate ligands span the A$_1$B$_1$, A$_2$B$_2$, A$_3$B$_3$ and A$_4$B$_4$ edges. Representative potential energy surfaces are shown in Figs. 13.10 to 13.13, and are directly comparable to those calculated above (Figs. 12.6 and 12.9). The minima on the potential energy surfaces correspond to three different

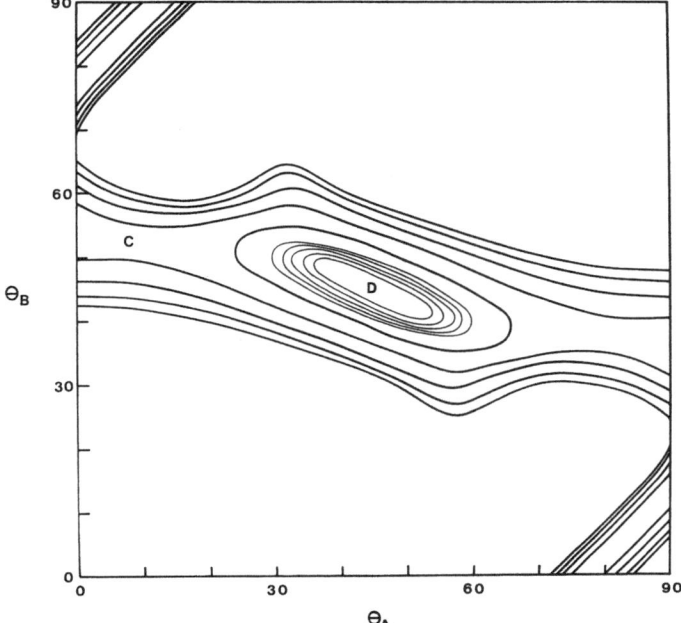

Fig. 13.10. Projection of the potential energy surface for [M(bidentate)$_4$] onto the $\theta_A - \theta_B$ plane (in degrees). The five *faint contour lines* are for successive 0.02 increments above the minimum, and the five *heavy contour lines* are for successive 0.2 increments above the minimum, at D. n = 6, b = 1.0. The locations of the dodecahedron (D) and "cube" (C) are indicated

stereochemistries, the relative stabilities of which depend on the choice of normalized bite b.

At low normalized bites (Fig. 13.10) the minimum at $\theta_A = \theta_B = 45°$ corresponds to the dodecahedron of overall molecular symmetry D$_{2d}$ (Fig. 13.14).

As the normalized bite is progressively increased (Fig. 13.11), this central minimum becomes increasingly shallow and then divides into two identical minima. These minima then move further apart as b is increased further (Fig. 13.12). These minima correspond to the intermediate stereochemistry, and the polyhedron chosen to best describe the structure depends upon the value of b. At the low end of the range, b $\sim$ 1.1, the stereochemistry is adequately represented as a distorted D$_{2d}$ dodecahedron. At b = 1.19 (for n = 6), the structure is a square antiprism with the bidentate ligands along the square edges (Fig. 13.15a), the polyhedron being identical to that calculated for [M(unidentate)$_8$]. At other values of the normalized bite, these square faces are creased, and one of the dodecahedral isomers shown in Fig. 13.15 b, c may be the preferred description.

Figure 13.12 shows that at b = 1.2 an additional feature has appeared on the potential energy surface, which at b = 1.3 has formed a deep minimum (Fig. 13.13). The two minima correspond to the two optical isomers of the D$_4$ square antiprism

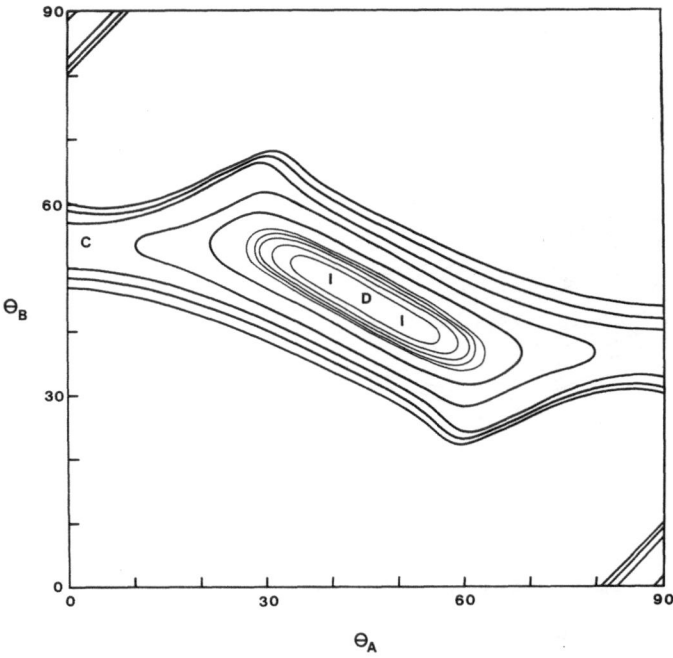

Fig. 13.11. Projection of the potential energy surface for [M(bidentate)₄] onto the $\theta_A - \theta_B$ plane (in degrees). The five *faint contour lines* are for successive 0.02 increments above the minima, and the five *heavy contour lines* are for successive 0.2 increments above the minima, at I. n = 6, b = 1.1. The locations of the dodecahedron (D), intermediate stereochemistries (I), and "cube" (C) are indicated

(Fig. 13.16). It may be noted that in contrast to the "three-bladed propeller" which is the dominant stereochemistry for tris(bidentate) complexes (Chap. 8), this "four-bladed propeller" is only expected in tetrakis(bidentate) complexes where the bidentate ligands have exceptionally large normalized bites. At b = 1.26, the angle of twist $\theta = 22.5°$ and the two square faces are staggered with respect to each other (Fig. 13.16). In an analogous way to the behaviour of tris(bidentate) complexes, a decrease in b leads to a decrease in θ.

2. Comparison with Experiment

The above predictions, that the stereochemistry is largely controlled by the normalized bite of the bidentate ligands, is in general agreement with the known structures (Table 13.3).

Compounds with normalized bites less than ~ 1.1 generally have the D_{2d} dodecahedral structure, with some distorted structures occurring at the higher end of the range. The stereochemistry of these compounds is discussed in more detail below. As b is increased from ~ 1.1 to ~ 1.2, the D_{2d} dodecahedral structures are progressively replaced by the intermediate stereochemistries (Table 13.3).

Table 13.3. Stereochemistries of [M(bidentate)₄]

Complex	b	Stereochemistry	Ref.
$K_3[Cr(O_2)_4]$	0.78	D_{2d} dodecahedral	[1017]
$[Zn(NH_3)_4][Mo(O_2)_4]$	0.79	D_{2d} dodecahedral	[1021]
$K_3[Hg(NO_2)_4](NO_3)$	0.84	see text	[516, 895]
$(Ph_4As)_2[Mn(NO_3)_4]$	0.90	D_{2d} dodecahedral	[371]
$(Cd(naphthyridine)_4)(ClO_4)_2$	0.90	D_{2d} dodecahedral	[414]
$[Fe(naphthyridine)_4](ClO_4)_2$	0.95	D_{2d} dodecahedral	[991]
$[Sn(NO_3)_4]$	0.99	D_{2d} dodecahedral	[463]
$[Sn(O_2CMe)_4]$	1.00	D_{2d} dodecahedral	[31]
$(Ph_4As)[Fe(NO_3)_4]$	1.00	D_{2d} dodecahedral	[650]
$(NO)_3(NO_3)[Fe(NO_3)_4]_2$	1.01	D_{2d} dodecahedral	[119]
$(Et_4N)[Bi(S_2COEt)_4]$	1.02	D_{2d} dodecahedral	[917]
$(Et_4N)[Np(S_2CNEt_2)_4]$	1.02	Distorted D_{2d} dodecahedral	[158]
$[Ti(NO_3)_4]$	1.03	D_{2d} dodecahedral	[464]
$[Th(S_2CNEt_2)_4]$	1.03	Distorted D_{2d} dodecahedral	[157]
$[U(bipy)_4]$	1.05	Cubic, see text	[312]
$[Te(S_2CNEt_2)_4]$	1.07	D_{2d} dodecahedral	[599]
$[Te(S_2CNC_4H_8O)_4]3\,C_6H_6$	1.07	D_{2d} dodecahedral	[417]
$[Te(S_2CNMeC_2H_4OH)_4]$	1.08	D_{2d} dodecahedral	[597]
$[Zr(S_2CNEt_2)_4]$	1.09	D_{2d} dodecahedral	[254]
$[Mo(S_2CPh)_4]$	1.10	D_{2d} dodecahedral	[137]
$[Mo(S_2CMe)_4]$	1.10	Distorted D_{2d} dodecahedral	[314]
$[V(S_2CCH_2Ph)_4]$	1.10	D_{2d} dodecahedral	[136]
$Na_4[Th(O_2C_6H_4)_4]21\,H_2O$	1.10	D_{2d} dodecahedral	[1001]
$[V(S_2CMe)_4]^a$	1.10	Distorted D_{2d} dodecahedral	[424]
^{a}	1.11	D_{2d} dodecahedral	[424]
$[V(S_2CPh)_4]$	1.11	D_{2d} dodecahedral	[136]
$Na_4[U(O_2C_6H_4)_4]21\,H_2O$	1.11	D_{2d} dodecahedral	[1001]
$[Ti(S_2CNEt_2)_4]$	1.11	D_{2d} dodecahedral	[255]
$[Ta(S_2CNMe_2)_4]Cl \cdot CH_2Cl_2$	1.11	D_{2d} dodecahedral	[717]
$[Ta(S_2CNMe_2)_4][TaCl_6]\tfrac{1}{2}CH_2Cl_2$	1.11	D_{2d} dodecahedral	[717]
$[Mo(S_2CNEt_2)_4]Cl$	1.11	D_{2d} dodecahedral	[462]
$[Mo(S_2CNEt_2)_4]_2(Mo_6O_{19})$	1.12	D_{2d} dodecahedral	[462]
$[W(S_2CNEt_2)_4]Br$	1.12	D_{2d} doedcahedral	[1098]
$Na_4[Ce(O_2C_6H_4)_4]21\,H_2O$	1.12	D_{2d} dodecahedral	[1002]
$[La(bipyO_2)_4](ClO_4)_3$	1.12	Cubic, see text	[32]
$(Ph_4P)[Pr(S_2PMe_2)_4]$	1.12	Intermediate	[881]
$[Co(MeCONHCOMe)_4](ClO_3)_3 \cdot H_2O$	1.12	Intermediate	[468]
$(C_9H_8N)[Ce(CF_3COCHCOC_4H_3S)_4]$	1.13	Distorted D_{2d} dodecahedral	[751]
$H[Sc(O_2C_7H_5)_4]$	1.13	Distorted D_{2d} dodecahedral	[47, 302]
$[Zr(O_2C_7H_5)_4]2.25\,CHCl_3$	1.14	D_{2d} dodecahedral	[304]
$[Hf(O_2C_7H_5)_4]HCONMe_2$	1.14	D_{2d} dodecahedral	[1053]
$[Th(MeCOCHCOMe)_4]\tfrac{1}{2}C_6H_6$	1.14	Intermediate	[711]
$[Th(C_4H_3SCOCHCOCF_3)_4]$	1.15	D_{2d} dodecahedral	[712]
$(C_5H_{12}N)[Gd(MeCOCHCOPh)_4]C_5H_{11}N$	1.15	Intermediate	[200]
$H[Eu(MeCOCHCOPh)_4]Et_2NH$	1.15	Intermediate	[606]
$K_4[Zr(C_2O_4)_4]5\,H_2O$	1.16	D_{2d} dodecahedral	[666]
$K_4[Hf(C_2O_4)_4]5\,H_2O$	1.16	D_{2d} dodecahedral	[1052]
$[Nb(O_2C_7H_5)_4]_2[(H_3O)Cl_3]MeCN$	1.16	Distorted D_{2d} dodecahedral	[303]
$[Th(MeCOCHCOMe)_4]$	1.16	Intermediate	[36]
$[U(MeCOCHCOMe)_4]$	1.16	Intermediate	[1048]

ᵃ Two different stereochemistries in unit cell.

(Continued)

Table 13.3 (continued)

Complex	b	Stereochemistry	Ref.
[U(CF₃COCHCOPh)₄]	1.16	Intermediate	[831]
(NH₄)[Pr(CF₃COCHCOC₄H₃S)₄]H₂O	1.16	Intermediate	[686]
(C₅H₆N)[Nd(CF₃COCHCOC₄H₃S)₄]	1.17	Intermediate	[707]
Na₄[Hf(O₂C₆H₄)₄]21 H₂O	1.17	D₂d dodecahedral	[1002]
Na₄[Zr(C₂O₄)₄]3 H₂O	1.17	Distorted D₂d dodecahedral	[482]
[Ce(MeCOCHCOMe)₄]	1.17	Intermediate	[771, 1047]
[Np(MeCOCHCOMe)₄]	1.17	Intermediate	[35]
[Nb{S₂P(OPr)₂}₄]	1.18	D₂d dodecahedral	[1050]
(C₅H₁₂N)[Eu(MeCOCHCOPh)₄]	1.18	Intermediate	[606]
Cs[Eu(CF₃COCHCOCF₃)₄]	1.18	Intermediate	[187]
Cs[Am(CF₃COCHCOCF₃)₄]	1.18	Intermediate	[187]
Cs[Y(CF₃COCHCOCF₃)₄]	1.19	Intermediate	[96]
[Zr{S₂P(OPr)₂}₄]	1.19	D₂d dodecahedral	[1050]
[Zr(PhCOCHCOPh)₄]	1.21	Intermediate	[237]
[Zr(MeCOCHCOMe)₄]	1.22	Intermediate	[990]
[Nb(BuCOCHCOBu)₄]	1.28	D₄ square antiprism	[883]

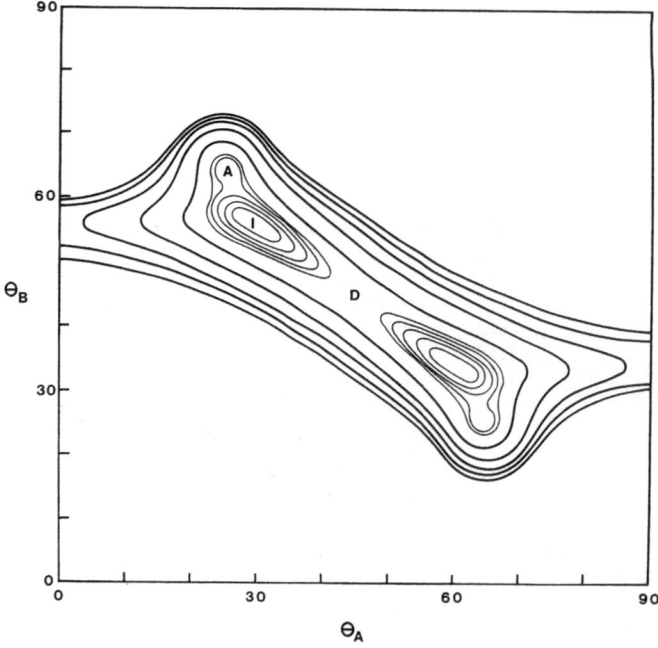

Fig. 13.12. Projection of the potential energy surface for [M(bidentate)₄] onto the $\theta_A - \theta_B$ plane (in degrees). The five *faint contour lines* are for successive 0.02 increments above the minima, and the five *heavy contour lines* are for successive 0.2 increments above the minima, at I. n = 6, b = 1.2. The locations of the dodecahedron (D), intermediate stereochemistries (I), and square antiprism (A) are indicated

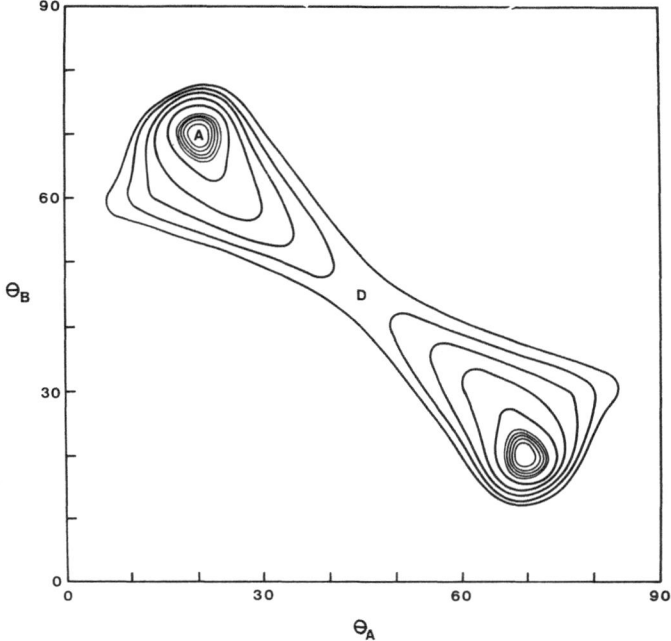

Fig. 13.13. Projection of the potential energy surface for [M(bidentate)$_4$] onto the $\theta_A - \theta_B$ plane (in degrees). The five *faint contour lines* are for successive 0.02 increments above the minima, and the five *heavy contour lines* are for successive 0.2 increments above the minima, at A. n = 6, b = 1.3. The locations of the dodecahedron (D) and square antiprism (A) are indicated

The only compound with the bidentate ligand having a large normalized bite, [Nb(BuCOCHCOBu)$_4$] with b = 1.28, has the D$_4$ square antiprismatic structure as predicted. As would be expected from the potential energy surfaces, [M(bidentate)$_4$] molecules may be rigid or fluxional in solution, both types of behavior being observed. Some exceptional structures in Table 13.3, those of K$_3$[Hg(NO$_2$)$_4$](NO$_3$), [U(bipy)$_4$], and [La(bipyO$_2$)$_4$](ClO$_3$)$_3$, are discussed in the next section.

The D$_{2d}$ dodecahedron consists of two interlocking trapezoids each formed from two bidentate ligands (Fig. 13.14), and the stereochemistry is defined by ϕ_A, ϕ_B,

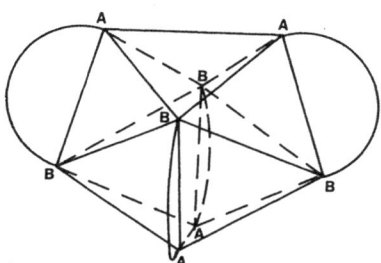

Fig. 13.14. The D$_{2d}$ dodecahedral isomer of [M(bidentate)$_4$]

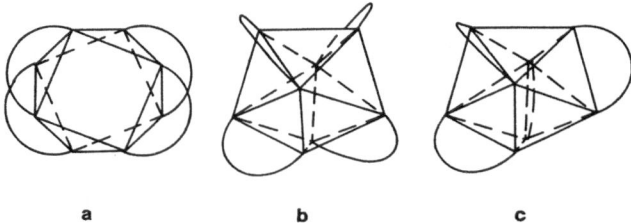

a   **b**   **c**

Fig. 13.15. The intermediate stereochemistry of [M(bidentate)$_4$] represented as a square antiprism (**a**) and as dodecahedra (**b** and **c**)

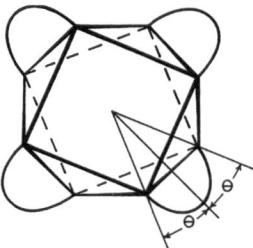

Fig. 13.16. The D$_4$ square antiprismatic isomer of [M(bidentate)$_4$]

and the bond length ratio (M—A)/(M—B). The dependence of the calculated values of ϕ_A and ϕ_B on the normalized bite is shown in Fig. 13.17. It can be seen that the main effect of reducing b is to decrease ϕ_B, until at b $\sim$ 0.8 the four B atoms form a square plane.

The stereochemical parameters for D$_{2d}$ dodecahedral molecules are given in Table 13.4, and the values of ϕ_A and ϕ_B are also plotted in Fig. 13.17. The agreement between prediction and experiment is good, although it may be noted that at

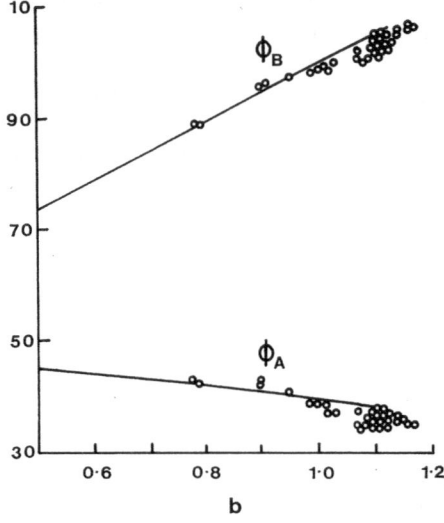

Fig. 13.17. Angular coordinates (in degrees) for D$_{2d}$ dodecahedral complexes of the type [M(bidentate)$_4$]; theoretical curve for n = 6 also shown

Table 13.4. D_{2d} dodecahedral [M(bidentate)$_4$]

Complex	b	ϕ_A	ϕ_B	MA/MB
K$_3$[Cr(O$_2$)$_4$]	0.78	43.2°	89.1°	0.95
[Zn(NH$_3$)$_4$][Mo(O$_2$)$_4$]	0.79	42.3°	88.6°	0.97
(Ph$_4$As)$_2$[Mn(NO$_3$)$_4$]	0.90	43.0°	96.3°	0.97
[Cd(naphthyridine)$_4$](ClO$_4$)$_2$	0.90	42.2°	96.0°	1.07
[Fe(naphthyridine)$_4$](ClO$_4$)$_2$	0.95	41.1°	97.6°	1.12
[Sn(NO$_3$)$_4$]	0.99	38.9°	98.3°	1.01
[Sn(O$_2$CMe)$_4$]	1.00	39.0°	98.8°	1.01
(NO)$_3$(NO$_3$)[Fe(NO$_3$)$_4$]$_2$	1.01	38.8°	99.2°	1.02
(Et$_4$N)[Bi(S$_2$COEt)$_4$]	1.02	36.9°	98.6°	1.02
[Ti(NO$_3$)$_4$]	1.03	37.2°	100.2°	1.01
[Te(S$_2$CNEt$_2$)$_4$]	1.07	35.0°	100.1°	1.02
[Te(S$_2$CNC$_4$H$_8$O)$_4$]3 C$_6$H$_6$	1.07	37.4°	102.2°	1.04
[Te(S$_2$CNMeC$_2$H$_4$OH)$_4$]	1.08	35.0°	100.0°	1.02
[Zr(S$_2$CNEt$_2$)$_4$]	1.09	35.4°	100.7°	
[Mo(S$_2$CPh)$_4$]	1.10	37.9°	104.7°	1.03
[V(S$_2$CCH$_2$Ph)$_4$]	1.10	37.5°	104.0°	1.02
Na$_4$[Th(O$_2$C$_5$H$_4$)$_4$]21 H$_2$O	1.10	37.9°	104.6°	1.00
[V(S$_2$CMe)$_4$]	1.11	35.4°	103.1°	1.02
[V(S$_2$CPh)$_4$]	1.11	38.0°	105.7°	1.04
Na$_4$[U(O$_2$C$_6$H$_4$)$_4$]21 H$_2$O	1.11	37.1°	104.8°	1.01
[Ti(S$_2$CNEt$_2$)$_4$]	1.11	35.2°	102.5°	1.03
[Ta(S$_2$CNMe$_2$)$_4$]Cl·CH$_2$Cl$_2$	1.11	35.0°	102.6°	1.03
[Ta(S$_2$CNMe$_2$)$_4$][TaCl$_6$]½ CH$_2$Cl$_2$	1.11	35.4°	103.4°	1.03
[Mo(S$_2$CNEt$_2$)$_4$]Cl	1.11	34.6°	102.2°	1.01
[Mo(S$_2$CNEt$_2$)$_4$]$_2$(Mo$_6$O$_{19}$)	1.12	35.7°	103.6°	1.02
[W(S$_2$CNEt$_2$)$_4$]Br	1.12	34.7°	102.4°	1.01
Na$_4$[Ce(O$_2$C$_6$H$_4$)$_4$]21 H$_2$O	1.12	36.9°	105.1°	1.00
[Zr(O$_2$C$_7$H$_5$)$_4$]2.25 CHCl$_3$	1.14	35.8°	105.6°	1.01
[Hf(O$_2$C$_7$H$_5$)$_4$]HCONMe$_2$	1.14	36.0°	105.6°	1.01
K$_4$[Zr(C$_2$O$_4$)$_4$]5 H$_2$O	1.16	35.4°	106.3°	1.02
K$_4$[Hf(C$_2$O$_4$)$_4$]5 H$_2$O	1.16	35.3°	107.0°	1.02
Na$_4$[Hf(O$_2$C$_6$H$_4$)$_4$]21 H$_2$O	1.17	35.1°	106.6°	1.01

high values of b, both ϕ_A and ϕ_B appear to be consistently about 2° too low. A ready explanation for this observation is given below.

The three tellurium(IV) dithiocarbamate compounds, [Te(S$_2$CNR$_2$)$_4$], and the bismuth(III) complex, [Bi(S$_2$COEt)$_4$]$^-$, in Table 13.4 possess a nonbonding pair of electrons, but these do not noticeably distort the structures.

The other structural feature apparent from Table 13.4 is the slight asymmetry of the bidentate ligands. For normalized bites greater than about 0.9, the M—A bonds are about 2% longer than the M—B bonds, which reflects the greater repulsion experienced by the A atoms compared with the B atoms (Fig. 13.18). This lengthening of the M—A bonds allows a small decrease in ϕ_A, which must be accompanied by the same decrease in ϕ_B, and apparently amounts to about 2° (see above). In contrast, for normalized bites less than about 0.9, it is predicted that the M—A bonds will be *shorter* than the M — B bonds (Fig. 13.18), which is again in agreement with the experiment (Table 13.4). These structural changes

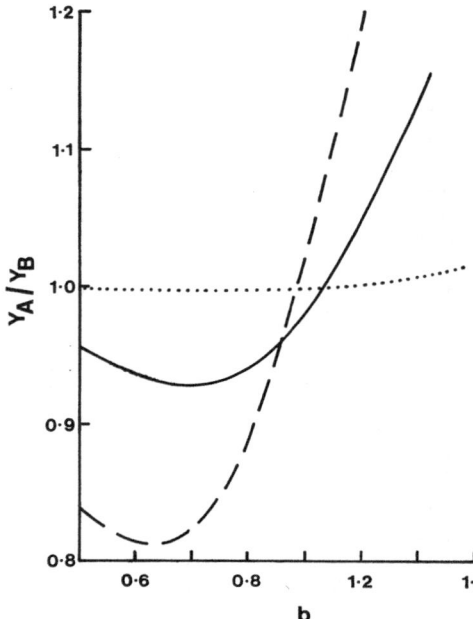

Fig. 13.18. Ratio of the individual atom repulsion coefficients for the two ends of each bidentate ligand in the D_{2d}-dodecahedral [M(bidentate)$_4$] as a function of normalized bite b. *Dotted line*, n = 1; *solid line*, n = 6, *broken line*, n = 12

are emphasized in Fig. 13.19, which show the trapezoidal plane for two typical compounds.

This slight asymmetry must not be confused with the gross asymmetry found in $(Ph_4As)_2[Zn(NO_3)_4]$ [93] (MA/MB = 0.88), $(Ph_4As)_2[Co(NO_3)_4]$ [100] (MA/MB = 0.84), and $(Ph_4As)_2[Co(CF_3COO)_4]$ [101] (MA/MB = 0.64), which can alternatively be regarded as approaching four-coordinate and tetrahedral. The opposite asymmetry is found in $Ca[Cu(CH_3COO)_4]6\,H_2O$ [158] (MA/MB = 1.42), $(Ph_4As)_2[Cu(NO_3)_4]CH_2Cl_2$ [651] (MA/MB = 1.40) and $K[Au(NO_3)_4]$ [465], which can be regarded as four-coordinate and square planar.

3. Some Exceptional Structures

There are three structures in Table 13.3 which do not fit into any of the above stereochemical types.

The first is the uranium(0) complex [U(bipy)$_4$], which has the C_2 cubic structure (Fig. 13.20). The bipyridyl ligand in this complex has a normalized bite of 1.05,

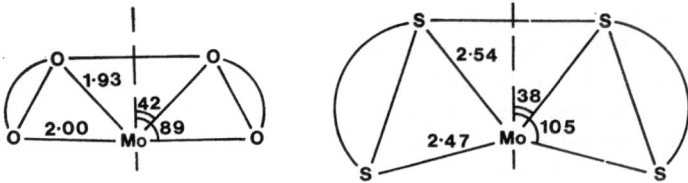

Fig. 13.19. Trapezoid geometries, degrees and Å, in $[Zn(NH_3)_4][Mo(O_2)_4]$ and $[Mo(S_2CPh)_4]$

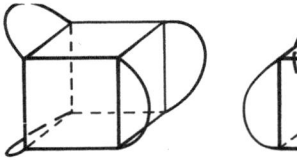 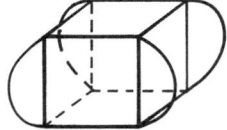

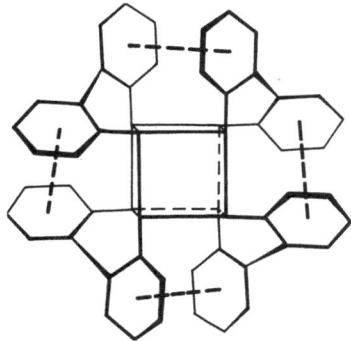

Fig. 13.20. [U(bipy)₄] and [La(bipyO₂)₄](ClO₄)₃

Fig. 13.21. [La(bipyO₂)₄](ClO₄)₃. The stacking of parallel C_5H_4N planes between different ligands is shown by the *broken lines*

which leads to the expectation of a D_{2d} dodecahedral structure, with $\phi_A \sim 38°$ and $\phi_B \sim 101°$. However such a structure is not possible, as two bipyridyl ligands cannot form such a planar trapezium because of the clash of the α-hydrogen atoms. A similar argument was used in Sect. 13.A to explain the cubic structures of $[M(phen)_2(H_2O)_4]^{2+}$ (Fig. 13.7).

A cubic arrangement of ligands is also found in $[La(bipyO_2)_4](ClO_4)_3$ (b = 1.12) (Fig. 13.20), which can also be attributed to steric interactions. The pyridine rings from different ligands are stacked parallel to each other above four of the cube faces (Fig. 13.21) and this interaction prevents twisting into, for example, a square antiprism.

The structure of $K_3[Hg(NO_2)_4](NO_3)$ is remarkable. The structure was represented as a grossly distorted square antiprism, with one bidentate ligand lying on a mirror plane spanning a diagonal of the square face (Fig. 13.22). The eight mercury-oxygen bond lengths lie within the range 2.39 to 2.57 Å.

4. Further Comments on Ligand Design

The distance between the donor atoms of any chelate group is relatively constant, depending to only a small extent upon the coordination number and the size of the metal atom. For example [638]:

β-Diketonates (RCOCHCOR)⁻: $O\text{---}O \sim 2.8 \pm 0.1$ Å

Dithiocarbamates, xanthates, etc. (S₂CX)⁻: $S\text{---}S \sim 2.8 \pm 0.1$ Å

Nitrate (NO₃)⁻: $O\text{---}O \sim 2.1 \pm 0.05$ Å

Any particular ligand will tend to form higher coordination numbers with larger metal atoms, although there may be some overlap between different groups. For example [638]:

(RCOCHCOR)⁻: $[M(bidentate)_3]^{x\pm}$ for M—O $< \sim 2.1$ Å
$[M(bidentate)_4]^{x\pm}$ for M—O $> \sim 2.1$ Å

(S₂CX)⁻: $[M(bidentate)_3]^{x\pm}$ for M—S $< \sim 2.5$ Å
$[M(bidentate)_4]^{x\pm}$ for M—S $> \sim 2.5$ Å

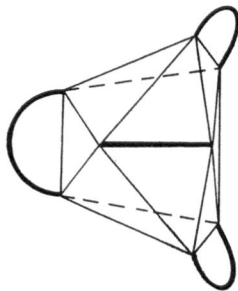

Fig. 13.22. $K_3[Hg(NO_2)_4](NO_3)$

(NO_3^-):

$[M(bidentate)_3]^{x\pm}$ for M—O $< \sim 2.0$ Å

$[M(bidentate)_4]^{x\pm}$ for M—O ~ 2.0 to ~ 2.4 Å

$[M(bidentate)_5]^{x\pm}$ for M—O ~ 2.4 to ~ 2.6 Å

$[M(bidentate)_6]^{x\pm}$ for M—O $> \sim 2.6$ Å

Large metal atoms with unusually low coordination numbers are sterically uncrowded, and may have exceptional structures where additional effects such as crystal packing play an unusually important role. On the other hand, molecules with unusually high coordination numbers are often observed to contain bidentate ligands which are bonded in a grossly unsymmetrical manner, as one end of the chelate ring is squeezed out of the coordination sphere.

Because of this change in coordination number with size of the metal atom, the *normalized* bite of any particular ligand, defined as the distance between the donor atoms divided by the metal-ligand bond length, is lower in eight-coordinate compounds than in six-coordinate compounds. The β-diketonates are of particular interest. Small metal atoms form six-coordinate complexes having M—O bond lengths of 1.9 to 2.1 Å and a normalized bite of 1.4, which is an appropriate fit for a fairly regular octahedron (ideal, b = 1.41). Large metal atoms, namely the early transition metals, lanthanoids and actinoids, form eight-coordinate β-diketonate complexes having M—O bond lengths of 2.1—2.5 Å, but the corresponding normalized bite of approximately 1.2 is, coincidently, the appropriate value for a fairly regular square antiprism or dodecahedron.

In Sect. 8.G, it was noted that for compounds of the type [M(bidentate)₃], the normalized bite decreased as the number of atoms in the chelate ring decreased, and as the size of those atoms decreased. A parallel effect is observed for compounds of the type [M(bidentate)₄]:

Chelate ring	b for C.N. = 6	b for C.N. = 8
6-membered (e.g. acetylacetonate)	~ 1.4	~ 1.2
5-membered (e.g. tropolonate)	~ 1.3	~ 1.1
4-membered: 3 large atoms (e.g. dithiophosphate)	~ 1.3	~ 1.1
2 large atoms (e.g. dithiocarbamate)	~ 1.2	~ 1.1
0 large atoms (e.g. nitrate)	~ 1.1	~ 1.0
3-membered (e.g. peroxide)	—	~ 0.8

CHAPTER 14

Nine-Coordinate Compounds

A. [M(unidentate)$_9$]

Five polyhedra with nine vertices were described in Chap. 2. All these polyhedra will distort, by placing all vertices on the surface of a sphere, and by decreasing the sizes of the larger faces. The repulsion energy coefficients X, calculated for n = 6, are listed [427, 429, 508, 520, 636, 649, 936]:

	X
Tricapped trigonal prism	8.105
Capped square antiprism	8.117
Capped cube	9.435
Tridiminished icosahedron	11.219
Triangular cupola	12.459

The tricapped trigonal prism is slightly more stable than the capped square antiprism, but the other structures have prohibitively high repulsion energies and need not be considered further.

The tricapped trigonal prism is shown in Fig. 14.1, where atoms ABCGHI form the trigonal prism, capped by atoms DEF. If all ligands lie on the surface of a sphere of radius r, the stereochemistry is defined by ϕ_A, the angle the six "prismatic" metal-ligand bonds make with the threefold axis. For a "hard sphere model" each unidentate ligand is in contact with four neighbouring unidentate ligands, that is, $\phi_A = \arcsin (2/3) = 41.81°$, the polyhedral edge lengths being $AB = AD = 2/3^{1/2}r = 1.1547r$ and $AG = (2/3)5^{1/2}r = 1.4907r$.

The most favourable polyhedron (MFP) formed by minimization of the total repulsion energy is squashed along the threefold axis (Fig. 14.2), ϕ_A being increased to 44.7° (n = 6). This results in a decrease in all four ligand-ligand distances to each capping atom, increasing the repulsion experienced by these atoms, $Y_D/Y_A = 1.138$ (n = 6). The resultant increase in metal-capping atom bond length will lead to a further increase in ϕ_A, for example, for R(D/A) = 1.0 and 1.1, $\phi_A = 44.7°$ and 45.6° respectively [636].

The capped square antiprism is closely related to the tricapped trigonal prism. The general structure, with C$_{2v}$ symmetry, is shown in Fig. 14.3. The symmetry axis is coincident with the M−D bond, the other atoms being labelled as above

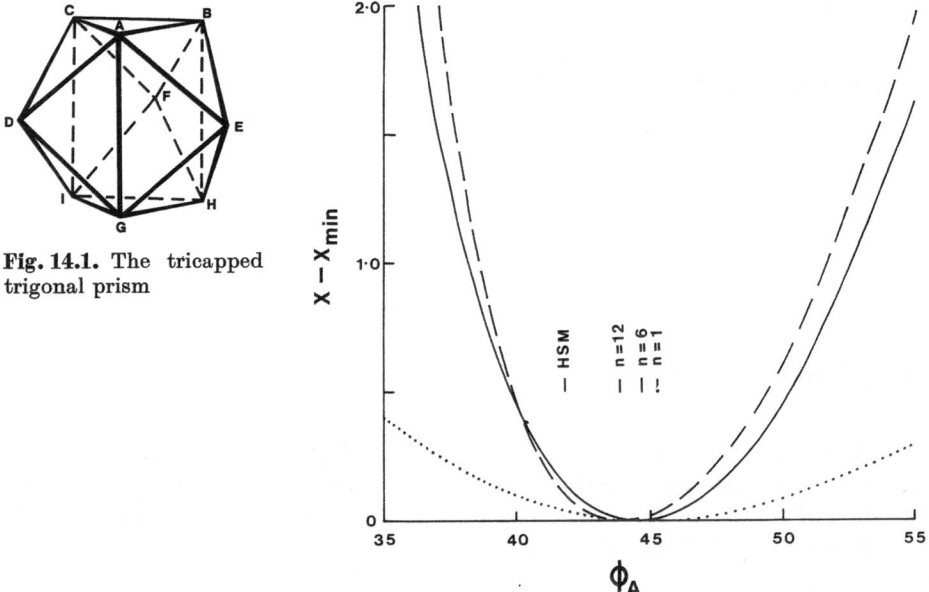

Fig. 14.1. The tricapped trigonal prism

Fig. 14.2. Repulsion energy coefficient, above that corresponding to the minimum, for the tricapped trigonal prism as a function of ϕ_A (in degrees). *Dotted line*, n = 1; *solid line*, n = 6; *broken line*, n = 12

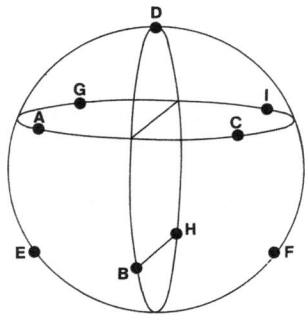

Fig. 14.3. General stereochemistry for [M(unidentate)$_9$]

for the tricapped trigonal prism. The capped square antiprism is defined by $\phi_B = \phi_F$, creating a square BFHE face, and $\theta_C = 45°$. The two structures may be compared by examination of the angular coordinates (calculated for n = 6):

	ϕ_B	ϕ_F	ϕ_C	θ_C
Capped square antiprism	127.0°	127.0°	68.9°	45.0°
Tricapped trigonal prism	134.7°	120.0°	69.4°	40.6°

The potential energy surface for [M(unidentate)$_9$] is shown in Fig. 14.4, projected onto the $\phi_B - \phi_F$ plane. There is no potential energy barrier between the two tricapped trigonal prisms and the capped square antiprism. Movement along the

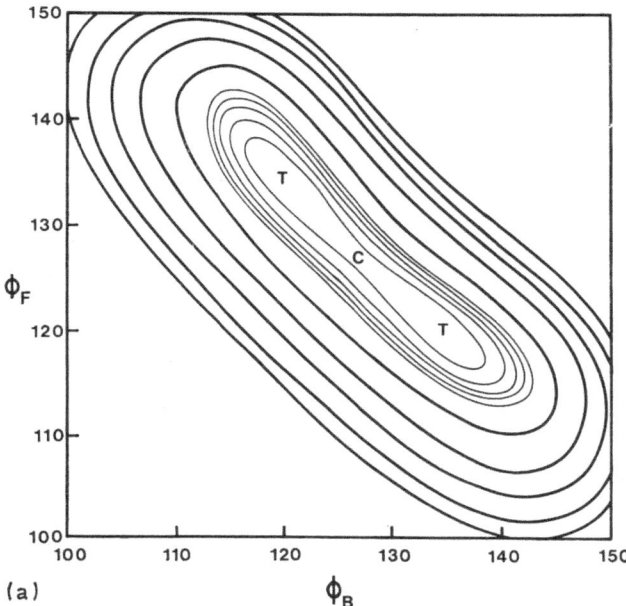

(a)

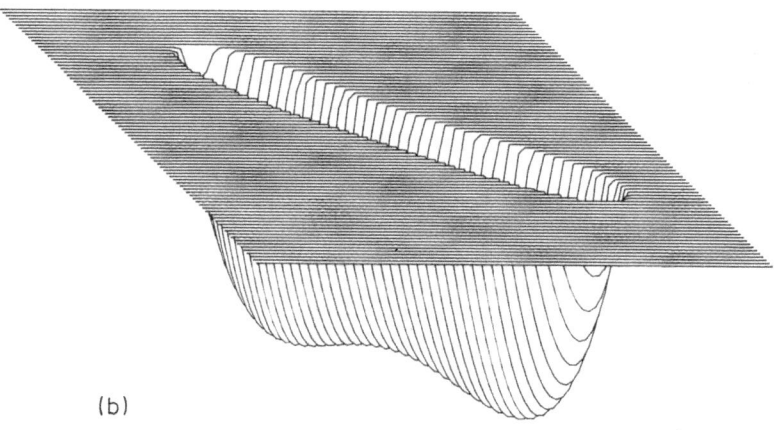

(b)

Fig. 14.4. a Projection of the potential energy surface for [M(unidentate)$_9$] onto the $\phi_B - \phi_F$ plane (in degrees). The five faint contour lines are for successive 0.02 increments above the minima, and the five heavy contour lines are for 0.2 increments above the minima, at T. n = 6. The location of the tricapped trigonal prisms (**T**) and the capped square antiprism (C) are indicated. **b** As in (**a**), but with truncation at X = 0.1

valley between the two stereochemistries necessarily scrambles all nine atoms, as any one of the three capping atoms of the tricapped trigonal prism can be selected to become the unique atom of the capped square antiprism.

All compounds containing nine unidentate ligands have the tricapped trigonal

Table 14.1. Tricapped trigonal prismatic [M(unidentate)$_9$]

Complex	ϕ_A	$\dfrac{M-D}{M-A}$	Ref.
[Sm(H$_2$O)$_9$](BrO$_3$)$_3$	—	1.04	[989]
[Pr(H$_2$O)$_9$](BrO$_3$)$_3$	47.4°	1.01	[19]
[Yb(H$_2$O)$_9$](BrO$_3$)$_3$	47.5°	1.05	[19]
[Y(H$_2$O)$_9$](EtSO$_4$)$_3$	45.0°	1.06	[151]
[Pr(H$_2$O)$_9$](EtSO$_4$)$_3$	45.1°	1.05	[19]
[Ho(H$_2$O)$_9$](EtSO$_4$)$_3$	45.4°	1.04	[586]
[Yb(H$_2$O)$_9$](EtSO$_4$)$_3$	45.1°	1.08	[19]
[Yb(H$_2$O)$_9$](CF$_3$SO$_3$)$_3$	—	—	[533]
K$_2$[ReH$_9$]	43.2°	0.99	[3, 659]

prismatic structure, and the structural parameters of $\phi_A \sim 46°$ and $(M-D)/(M-A) \sim 1.05$ (Table 14.1 and Fig. 14.1), are in good agreement with predictions.

B. [M(bidentate)$_3$(unidentate)$_3$]

A number of stereochemistries may be envisaged for complexes of the type [M(bidentate)$_3$(unidentate)$_3$], which are based on the tricapped trigonal prism and the capped square antiprism. The most important of these contains a threefold axis (Fig. 14.5), and becomes the same as the tricapped trigonal prismatic structure calculated for [M(unidentate)$_9$], when b = 1.1387 (for n = 6). The main effect of reducing the normalized bite is to twist the lower GHI triangular face relative to the ABC and DEF triangles. For example, the angle of twist between the DEF and GHI triangles decreases from 60° at b = 1.1387, to 44° at b = 0.87 (as in the complexes below) (Fig. 14.6). Complete twisting of the GHI face so that it is eclipsed with respect to DEF and staggered with respect to ABC would form the tridiminished icosahedron, but this stereochemistry is not achieved, reflecting the high repulsion energy of this polyhedron.

The tricapped trigonal prismatic structure is observed in a number of trinitrato complexes, [Er(NO$_3$)$_3$(OSMe$_2$)$_3$][57], [Lu(NO$_3$)$_3$(OSMe$_2$)$_3$][58], [Yb(NO$_3$)$_3$(OSMe$_2$)$_3$] [109], [Tl(NO$_3$)$_3$(H$_2$O)$_3$] [423], and [Nd(NO$_3$)$_3$(OC$_3$HN$_2$Me$_2$Ph)$_3$] [110]. The observed normalized bites of $\sim$0.87 and the angles of twist between the DEF and

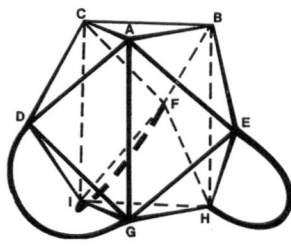

Fig. 14.5. Tricapped trigonal prismatic [M(bidentate)$_3$(unidentate)$_3$]

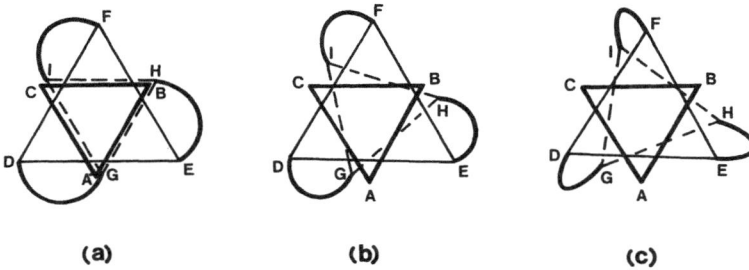

Fig. 14.6a—c. Tricapped trigonal prismatic [M(bidentate)₃(unidentate)₃]. **a** b = 1.1387; **b** b = 0.87; **c** b = 0.50

GHI triangles of $\sim 46°$ are in good agreement with the results obtained from repulsion energy calculations (Fig. 14.6b).

A notable feature of this stereochemistry is that the dihedral angle between any two bidentate ligands is close to 90° [427]. For example in the above trinitrato complexes it is $\sim 88°$. The three bidentate ligands can therefore be considered to form half an icosahedral [M(bidentate)₆] unit (Chap. 16), whereas the three unidentate ligands form half an octahedral [M(unidentate)₆] unit.

The repulsion energy calculations also predict that each bidentate ligand will be unsymmetrically bonded, the capping atom experiencing more repulsion than the prismatic atom. This is experimentally observed, $(M—D)/(M—G) \sim 1.06$ for the above trinitrato complexes.

Different structural isomers of [M(bidentate)₃(unidentate)₃] are found for [Eu(NO₃)₃{OC(NMe₂)₂}₃] [229], [Sm{NH₂NHC(C₅H₄N)O}₃(H₂O)₃](NO₃)₃ [1122], and [C(NH₂)₃]₅[Th(CO₃)₃F₃] [1070].

C. [M(bidentate)₄(unidentate)]

Four equivalent bidentate ligands can be wrapped around a nine-coordinate metal atom by forming a capped square antiprism, with the bidentate ligands spanning edges linking the two square faces of the antiprism (Fig. 14.7a). This

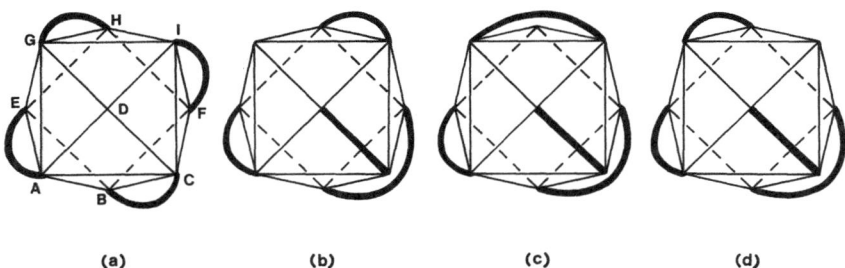

Fig. 14.7a—d. Capped square antiprismatic isomers of [M(bidentate)₄(unidentate)]

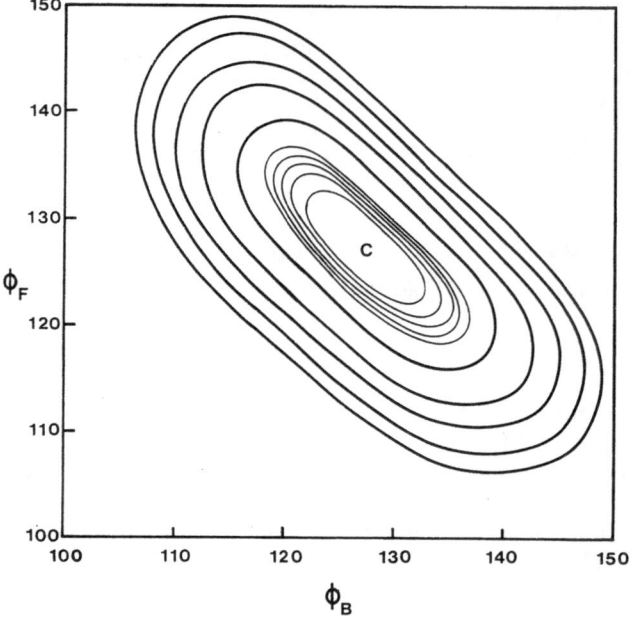

Fig. 14.8. Projection of the potential energy surface for [M(bidentate)$_4$(unidentate)] onto the $\phi_B - \phi_F$ plane (in degrees). The five *faint contour lines* are for successive 0.02 increments above the minimum, and the five *heavy contour lines* are for 0.2 increments above the minimum, at C. n = 6. The location of the capped square antiprism (C) is indicated

structure is observed for [Th(CF$_3$COCHCOCH$_3$)$_4$(H$_2$O)] [520]. A distorted form is found for [Pb(phen)$_4$(ClO$_4$)](ClO$_4$) [1].

Repulsion energy calculations carried out over the range b = 0.5—1.4 confirm that this isomer exists as a discrete potential energy minimum. The stabilization of the capped square antiprism compared with the tricapped trigonal prism is shown by the typical potential energy surface in Fig. 14.8, and should be contrasted with that calculated for [M(unidentate)$_9$] (Fig. 14.4) which shows the stabilization of the tricapped trigonal prism.

The ACIG and BFHE square faces of the antiprism are staggered at b = 1.1746, and the angle of twist between these faces decreases from 45° at b = 1.1746 to 0° at b = 0.75 (n = 6). The angle of twist of 42° at b = 1.15 observed for [Th(CF$_3$COCHCOCH$_3$)$_4$(H$_2$O)] is in excellent agreement with the calculated value of 43°. This twisting distortion of axially-symmetric structures containing bidentate ligands has been noted previously for [M(bidentate)$_3$] (Chap. 8), [M(bidentate)$_3$(unidentate)] (Sect. 11.C), [M(bidentate)$_4$] (Sect. 13.C), and [M(bidentate)$_3$(unidentate)$_3$] (Sect. 14.B).

The repulsion energy calculations predict that the bidentate ligands in capped square antiprismatic molecules will be unsymmetrically bonded, the atoms forming the capped face experiencing considerably less repulsion than the atoms forming the uncapped face, $Y_C/Y_B \sim 0.88$ for b ~ 0.8—1.4 (n = 6). This is in

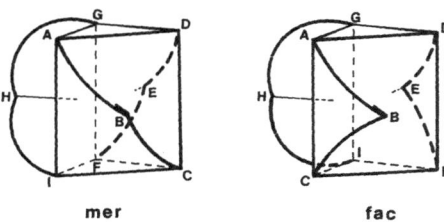

mer   fac

Fig. 14.9. The two tricapped trigonal prismatic isomers of [M(tridentate)₃]

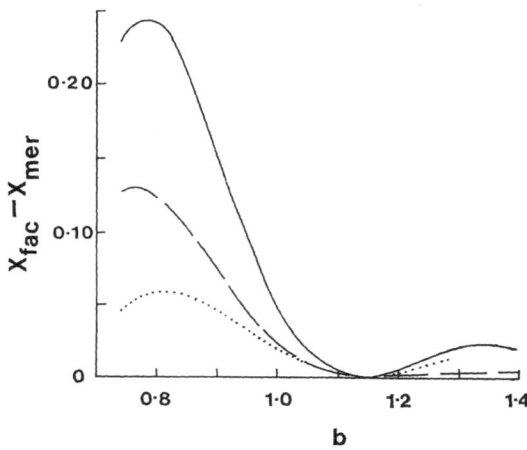

Fig. 14.10. Relative repulsion energy coefficients for the *fac*- and *mer*-isomers of [M(tridentate)₃] as a function of b. *Dotted line*, n = 1; *solid line*, n = 6; *broken line*, n = 12

agreement with the observed Th−O bond lengths of 2.39 and 2.46 Å respectively in [Th(CF₃COCHCOCH₃)₄(H₂O)].

Three other isomers of [M(bidentate)₄(unidentate)] based on distorted capped square antiprisms are shown in Fig. 14.7 b−d. The first contains a mirror plane, and is observed for [Sr(MeCONHCOMe)₄(H₂O)](ClO₄) [467] and [Th(O₂C₇H₅)₄(Me₂NCHO)] [307]. The observed structures are again in good agreement with those calculated by minimization of the repulsion energy [428]. For example, if the molecules are

Table 14.2. Capped trigonal prismatic *mer*-[M(tridentate)₃] complexes

Complex	b	ABC	ϕ_A	θ_B	Ref.
Na₃[Yb{C₅H₃N(COO)₂}₃]13 H₂O	1.08	114.6°	45.9°	54.6°	[14]
Na₃[Yb{C₅H₃N(COO)₂}₃]14 H₂O	1.08	114.7°	45.7°	54.2°	[12]
Na₄[Yb{C₅H₃N(COO)₂}₃](ClO₄)10 H₂O	1.07	115.4°	45.9°	53.3°	[15]
Na₃[Nd{C₅H₃N(COO)₂}₃]15 H₂O	1.04	117.3°	44.2°	48.8°	[16]
(Ph₄As)₂[U{C₅H₃N(COO)₂}₃]3 H₂O	1.02	−	−	−	[70]
Na₅[Yb{O(CH₂COO)₂}₃](ClO₄)₂·6 H₂O	1.07	115.5°	46.2°	53.4°	[13]
Na₅[Gd{O(CH₂COO)₂}₃](ClO₄)₂·6 H₂O	1.06	116.0°	44.7°	51.4°	[11]
Na₅[Nd{O(CH₂COO)₂}₃](ClO₄)₂·6 H₂O	1.04	117.2°	46.0°	50.6°	[13]
Na₅[Ce{O(CH₂COO)₂}₃](ClO₄)₂·6 H₂O	1.03	117.7°	46.3°	49.8°	[17]
Na₅[Ce{O(CH₂COO)₂}₃]9 H₂O	1.02	118.6°	46.3°	48.4°	[18, 399]
[Eu(terpy)₃](ClO₄)₃	−	−	−	−	[455]

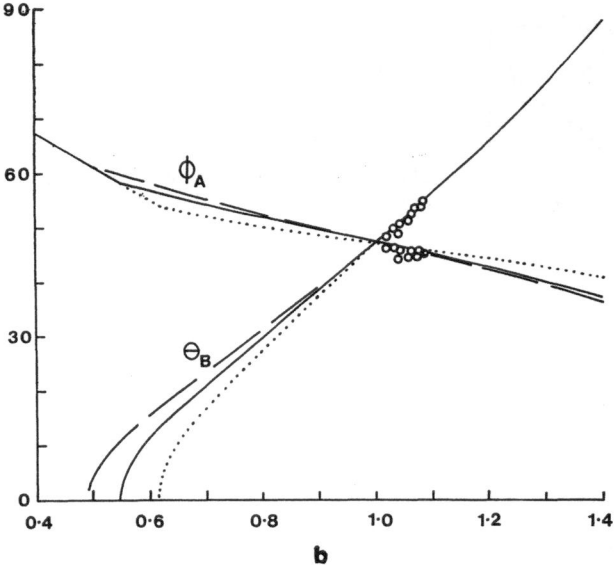

Fig. 14.11. Angular coordinates (degrees) for *mer*-[M(tridentate)$_3$] as a function of b. $\theta_A = 0$ *Dotted lines*, n = 1; *solid lines*, n = 6; *broken lines*, n = 12. Experimental values are marked

viewed down the metal-capping atom bond, the angles of twist of the two symmetry related bidentate ligands are 33° and 36° respectively, compared with calculated values for b = 1.07 and 1.04 of 37° and 36° respectively.

The isomer shown in Fig. 14.7c contains no elements of symmetry, and can also be represented as a tricapped trigonal prism. The structural details of [Th(C$_4$H$_3$SCOCHCOCF$_3$)$_4${(C$_8$H$_{17}$)$_3$PO}] [710] are again in agreement with those calculated. For example, viewing down the metal-capping atom bond, the twist angles of the three bidentate ligands are 49, 74 and 82°, compared with those calculated of 50, 74, and 85° respectively [428].

The isomer shown in 14.7d also contains no elements of symmetry, but is known for [Th(O$_2$C$_7$H$_4$Pr)$_4$(H$_2$O)]H$_2$O [587]. Again viewing down the metal-capping atom bond, the observed angles of twist of the other three bidentate ligands are 36, 37 and 80°, compared with those calculated for b = 1.04 of 35, 36 and 80°.

Other compounds which are known, but have not yet been subjected to detailed structural analysis are [Th(O$_2$C$_7$H$_5$)$_4$(H$_2$O)] [558], and (NO)$_2$[Sc(NO$_3$)$_4$(ONO$_2$)] [5].

D. [M(tridentate)$_3$]

There are two ways three equivalent symmetrical tridentate ligands can be wrapped around a tricapped trigonal prism. These are referred to as the *mer*- and *fac*-isomers, and are shown in Fig. 14.9. Repulsion energy calculations based on a flexible tridentate ligand show that the *mer*-isomer exists as a minimum at all values of the normalized bite, but the *fac*-isomer exists as a separate minimum only above b = 0.74 (for n = 6). The *mer*-isomer is the more stable, particularly at low b (Fig. 14.10).

An important difference between the two isomers is the value of the tridentate angle ABC. For example, for the tricapped trigonal prism calculated for [M(unidentate)$_9$], that is, $\phi_A = 44.7°$, the tridentate angle ABC in [M(tridentate)₃] is 110.6° for the *mer*-isomer and 77.3° for the *fac*-isomer.

All compounds of the type [M(tridentate)₃] contain dipicolinate or oxydiacetate as the symmetrical tridentate ligand. They all have the *mer*-stereochemistry. The structural parameters are given in Table 14.2, and compared with calculations in Fig. 14.11. Agreement between calculated and experimental angular parameters is again excellent.

CHAPTER 15

Ten-Coordinate Compounds

A. [M(unidentate)$_{10}$]

There is as yet no structurally characterized ten-coordinate compound containing ten unidentate ligands. Nevertheless the geometry of these hypothetical compounds is a useful introduction to the stereochemistry of known ten-coordinate complexes such as [M(bidentate ligand)$_5$].

Two semiregular polyhedra and five non-uniform polyhedra with ten vertices are described in Chap. 2. An additional convex chemical polyhedron is created if the tetracapped trigonal prism, in which the three square faces and one triangular face of the prism are capped, is distorted so that all vertices lie on the surface of a sphere. Minimization of the total repulsion energy and/or placing all atoms on the surface of a sphere leads to distortion in all cases, and the elimination of the bidiminished icosahedron from the list, as it becomes identical to the sphenocorona. The repulsion energy coefficients X, calculated for n = 6, are listed for the remaining structures [427, 723, 936]:

	X
Bicapped square antiprism	12.337
Sphenocorona	12.362
Tetracapped trigonal prism	12.363
Pentagonal antiprism	14.637
Bicapped cube	15.375
Pentagonal prism	15.522
Capped tridiminished icosahedron	21.692

The first three structures are substantially more stable than the other four, and are shown in Fig. 15.1. The sphenocorona can alternatively be considered as a *cis*-bicapped cube, the capping atoms I and J lying outside the CDHG and FEHG faces respectively, with the ACGF and BDHE faces each being converted into two triangular faces by creasing along AG and BH respectively. The sphenocorona can be considered as a distorted bicapped square antiprism if atoms C and E are the capping atoms, and AGID and BFJH the two square faces of the antiprism. Similarly the tetracapped trigonal prism can be considered as a distorted bicapped square antiprism if atoms B and F are taken as the two capping atoms, the two square faces being AGHE and CDJI.

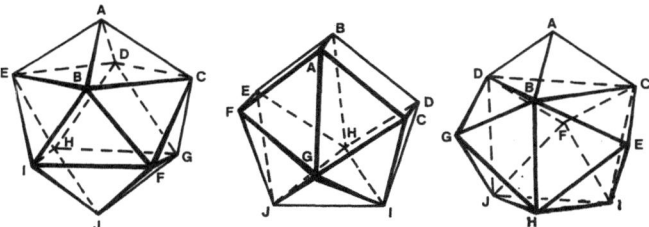

Fig. 15.1. The bicapped square antiprism, sphenocorona, and tetracapped trigonal prism

The differences in energy between these three polyhedra are small and all three stereochemistries, or intermediates lying along various reaction coordinates connecting them, would be possible. The pentagonal antiprism and pentagonal prism are among the less stable ten-coordinate polyhedra, but they do allow the positioning of five equivalent bidentate ligands, and hence are of relevance to complexes of the type [M(bidentate)$_5$].

B. [M(bidentate)$_5$]

1. The Theoretical Stereochemistries

Five minima have so far been located on potential energy surfaces calculated for [M(bidentate)$_5$], all of which show some relationship to the structures calculated for [M(unidentate)$_{10}$]. Isomers I—IV contain a twofold axis, while isomer V contains a mirror plane.

The bicapped square antiprism is the most stable stereochemistry for complexes of the type [M(unidentate)$_{10}$]. There are 16 edges of length approximately 1.085r, along which five bidentate ligands can be arranged in two different ways (Fig. 15.2). In isomer I, the two bidentate ligands with atoms at the capping sites have their other ends on two adjacent vertices of the square antiprism. This structure can accordingly be called the *cis*-isomer. The other three bidentate ligands form a three bladed propeller. As the normalized bite is reduced from b = 1.085, the two square faces of the antiprism are rotated towards one another. Isomer I exists as a discrete minimum over the whole range of normalized bite.

In isomer II, Fig. 15.2, the two bidentate ligands attached to the capping atoms of the bicapped square antiprism are also attached to two nearly opposite

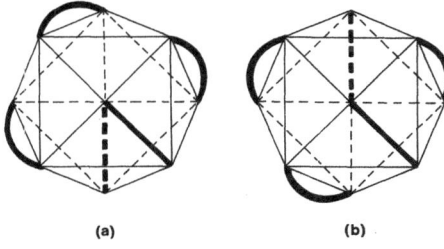

(a)   (b)

Fig. 15.2a, b. Bicapped square antiprismatic isomers of [M(bidentate)$_5$]. **a** Isomer I (*cis*). **b** Isomer II (*trans*).

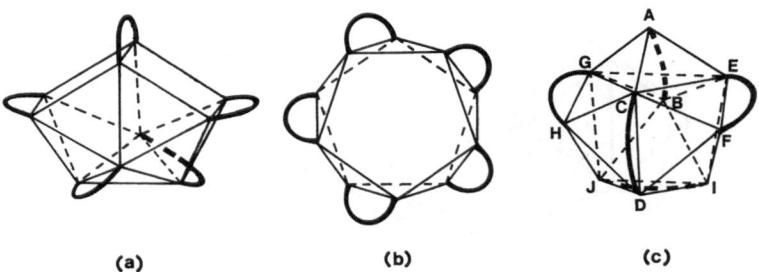

Fig. 15.3a—c. Additional isomers of [M(bidentate)$_5$]. **a** Isomer III, sphenocorona. **b** Isomer IV, pentagonal antiprism. **c** Isomer V, tetracapped trigonal prism

vertices of the antiprism, and this can be called the *trans*-isomer. Of the remaining three bidentate ligands, one is twisted in the opposite direction to the other two when viewed down the fourfold axis of the antiprism, and variation of the normalized bite cannot be accommodated by rotation of the square faces of the antiprism relative to each other. Isomer II exists as a separate minimum only over the range b = 0.97—1.16 (for n = 6).

Isomer III is shown in Fig. 15.3a, and is based upon the sphenocorona structure of [M(unidentate)$_{10}$]. It exists as a minimum on the potential energy surface only above b = 1.13.

At large values of the normalized bite, b > 1.16, a twisted pentagonal prism of symmetry D_5 is formed, the bidentate ligands being arranged as a five-bladed propeller (Fig. 15.3b).

Stereochemistry V exists as a minimum on the potential energy surface only above b = 1.10. There is a mirror plane through ABCD, and the stereochemistry is related to a tetracapped trigonal prism, with CEG being the capped triangular face and DIJ the uncapped face of the prism (Fig. 15.3c).

The difference between the repulsion energy coefficient for the five [M(bidentate)$_5$] stereochemistries are shown in Fig. 15.4. Compounds of stereochemistry II are expected only over the range b ~ 0.95—1.15, whereas stereochemistry I is stable at all values of the normalized bite. The remaining isomers only occur above b ~ 1.1, which is probably not experimentally possible for ten-coordination.

The relation between the two important stereochemistries, isomers I and II, is also shown by the potential energy surfaces in Figs. 15.5 and 15.6 [427]. The surface for b = 0.8 (Fig. 15.5) shows only the presence of isomer I, but at b = 1.05 (Fig. 15.6) both isomers exist as minima, separated by a small potential energy barrier.

2. Comparison with Experiment

Table 15.1 lists the known compounds of the type [M(bidentate)$_5$]. The only monomeric molecule in which the isomer II structure is observed is the barium diacetamide complex [Ba(MeCONHCOMe)$_5$](ClO$_4$)$_2$. The normalized bite of 0.98 is

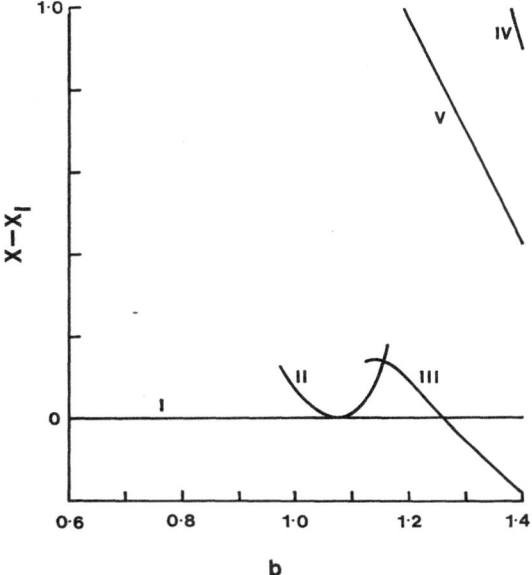

Fig. 15.4. Repulsion energy coefficients, relative to those for isomer I, for the various isomers of [M(bidentate)$_5$], as a function of b. n = 6

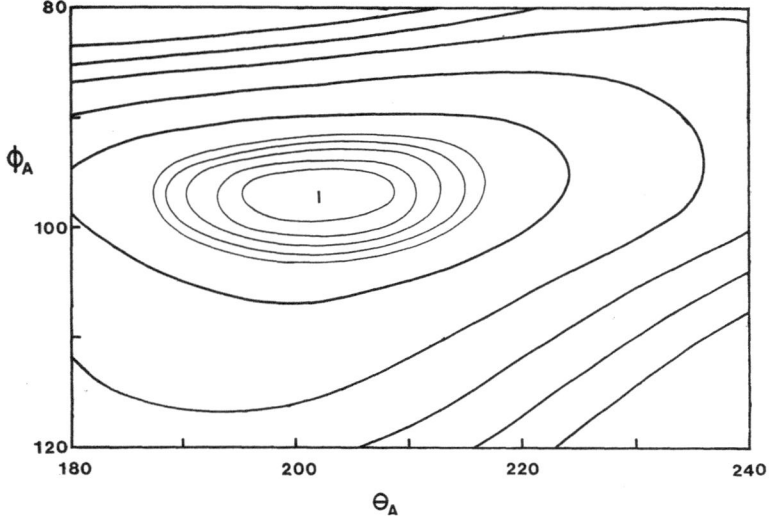

Fig. 15.5. Projection of the potential energy surface for [M(bidentate)$_5$] onto the $\phi_A - \theta_A$ plane (in degrees). The five *faint contour lines* are for successive 0.02 increments above the minimum, and the five *heavy contour lines* are for 0.2 increments above the minimum, at I. n = 6, b = 0.8. The location of isomer I is indicated

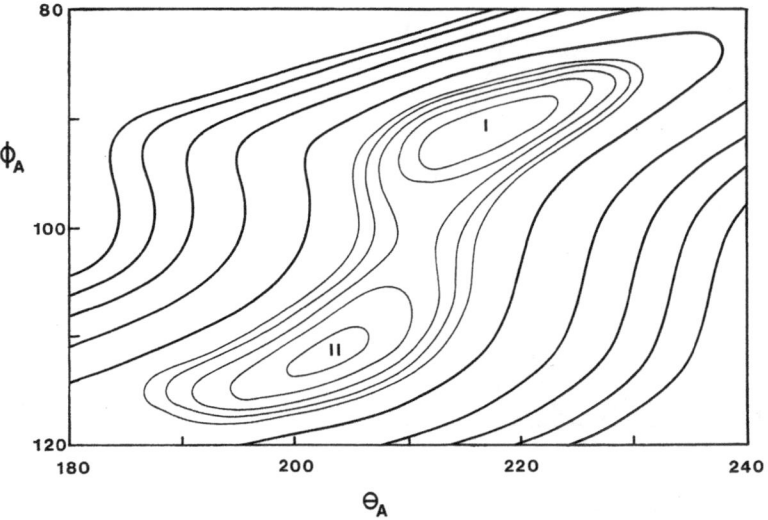

Fig. 15.6. Projection of the potential energy surface for [M(bidentate)₅] onto the $\phi_A - \theta_A$ plane (in degrees). The five *faint contour lines* are for successive 0.02 increments above the minima, and the five *heavy contour lines* are for 0.2 increments above the minima, at I and II. n = 6, b = 1.053. The location of isomers I and II are indicated

within the range where this stereochemistry is expected. Although polymeric structures have not generally been included in this work, it may be noted that this stereochemistry is also observed in $K_4[Th(C_2O_4)_4]4\,H_2O$ [9], in which the anion is a linear polymer with bridging oxalate groups. The polymeric uranium analogue $K_4[U(C_2O_4)_4]6\,H_2O$ is the only example of stereochemistry III [856].

All other compounds in Table 15.1 have four-membered nitrate or carbonate chelate rings, with normalized bites in the range 0.8—0.9. As predicted from the calculations, they are all of stereochemistry I. The agreement between observed and calculated bond angles is within about 3° [427].

Table 15.1. Structures of [M(bidentate)₅]

Complex	b	Structure	Ref.
$(Ph_3EtP)_2[Ce(NO_3)_5]$	0.84	I	[33]
$K_2[Er(NO_3)_5]$	0.88	I	[984]
$(NO)_2[Ho(NO_3)_5]$	0.88	I	[1051]
$Na_6[Th(CO_3)_5]12\,H_2O$	0.88	I	[1072]
$Na_6[Ce(CO_3)_5]12\,H_2O$	0.89	I	[1073]
$[C(NH_2)_3]_6[Th(CO_3)_5]4\,H_2O$	0.88	I	[1071]
$[C(NH_2)_3]_6[Ce(CO_3)_5]4\,H_2O$	0.90	I	[1074]
$[Ba(MeCONHCOMe)_5](ClO_4)_2$	0.98	II	[470]

The appreciation of such complicated stereochemistries is not always easy, and the problem has sometimes been simplified by discarding some of the stereochemical information and considering only the arrangment of the midpoints of the bidentate ligands [427]. For stereochemistry I, to a reasonable approximation there are six "bond angles" between these metal-to-midpoint vectors of 90°, three of 120°, and one of 180°, as required for a trigonal bipyramidal arrangement of "midpoints" about the metal atom. This pattern of angles is much less regular for isomer II, and does not correspond to either a trigonal bipyramidal or square pyramidal arrangement of midpoints.

CHAPTER 16

Twelve-Coordinate Compounds

A. [M(unidentate)$_{12}$]

There are no known compounds of the type [M(unidentate)$_{12}$], and it does not appear likely that any could be prepared. The geometry of these hypothetical compounds is nevertheless a useful introduction to the stereochemistry of known twelve-coordinate complexes such as [M(bidentate ligand)$_6$].

The repulsion energy coefficients for the relevant polyhedra with twelve vertices on the surface of a sphere (Chap. 2) are listed below, calculated for n = 6 (427):

	X		X
Icosahedron	23.531	Hexagonal antiprism	36.755
Bicapped pentagonal prism	25.928	Hexagonal prism	38.095
Anticuboctahedron	26.280	Truncated tetrahedron	48.692
Cuboctahedron	26.483	Square cupola	60.012

The four most stable structures are shown in Fig. 16.1. The icosahedron is clearly the most stable structure, and is expected to be by far the most chemically important. The icosahedron is a regular polyhedron with twelve identical vertices, 30 identical edges, and 20 identical equilateral triangular faces. Each vertex is linked to five other vertices at a distance of AB = 1.0515r (the five next nearest neighbours being at a distance $AG = \left[(\sqrt{5} + 1)/2\right]AB = 1.6180\,AB$, the "divine proportion" of classical times). The bond angle subtended by a polyhedral edge at the central metal atom is AMB = 63.43°.

The three next most stable structures have a mixture of triangular and quadrilateral faces. The bicapped pentagonal prism (Fig. 16.1 b) is formed by capping the two pentagonal faces of the prism, or can alternatively be formed by rotating one half of an icosahedron (ABCDEF in Fig. 16.1 a) by 36° relative to the other half (GHIJKL).

It should be noted that the anticuboctahedron (Fig. 16.1 c) is a fragment of hexagonal close packing, whereas the cuboctahedron (Fig. 16.1 d) is a fragment of cubic close packing, the close packed layers being ABC, DEFGHI, and JKL. For discrete molecules and ions the anticuboctahedron is more stable than the cuboctahedron, as it can distort by shorting the DE, FG, and HI edges which join two square faces, converting them into smaller trapezoids. Such distortions are not

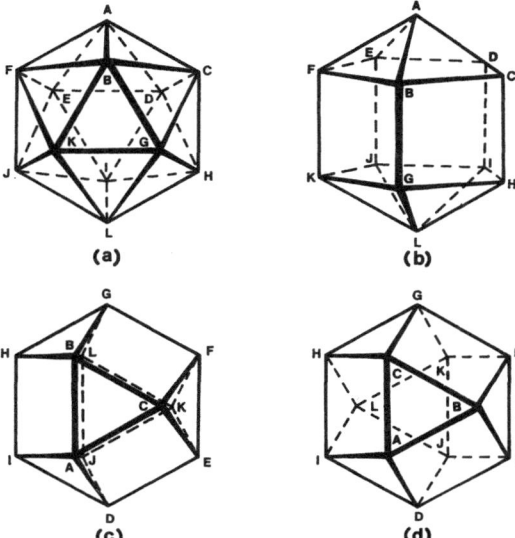

Fig. 16.1. a The icosahedron. **b** The bicapped pentagonal prism. **c** The anticuboctahedron. **d** The cuboctahedron

possible for the cuboctahedron, as every edge is part of both a triangle and a square.

The remaining four polyhedra listed above have hexagonal or octagonal faces, and very high repulsion energies.

B. [M(bidentate)$_6$]

Just as compounds of the type [M(unidentate)$_{12}$] are not observed, compounds of the type [M(bidentate)$_6$] are not expected unless the normalized bite is significantly reduced from b = 1.0515, the value corresponding to the edge length of a regular icosahedron. Nevertheless the icosahedron remains a convenient starting point for considering the stereochemistries of these complexes.

Three structural isomers may be formed by wrapping six bidentate ligands along the edges of a regular icosahedron (Fig. 16.2).

Stereochemistry I is highly symmetrical, with the very rare point group symmetry T$_h$. There are four C$_3$—S$_6$ axes through the ACE-HJL, AFK-DGJ, BCG-FIL, and the DEI-BHK faces, an inversion centre, three twofold axes passing through the midpoints of the bidentate ligands, and three mirror planes each containing a pair of bidentate ligands. On reducing the normalized bite the symmetry remains unaltered and all vertices remain identical (Fig. 16.3a). All polyhedral edge lengths not spanned by a bidentate ligand are equal, and of length a:

$$a = [2 - (b^2 - b^4/4)^{1/2}]^{1/2}.$$

Stereochemistry II (Fig. 16.2b), of symmetry D$_3$, is defined by the angular coordinates of each end of the bidentate ligand. The ϕ_A and ϕ_B coordinates are

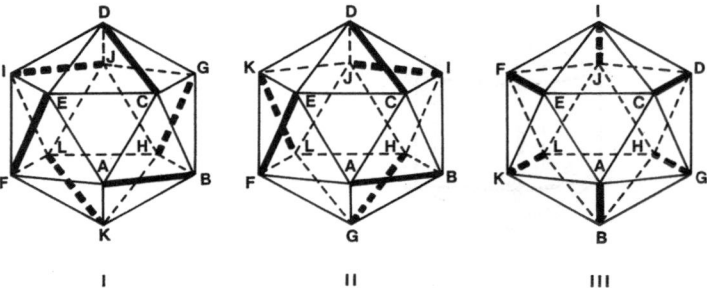

Fig. 16.2. Regular icosahedral isomers of [M(bidentate)₆]

the angles the M—A and M—B bonds make with the threefold axis through ACE-HJL, while the θ coordinates are defined relative to the twofold axis passing through the midpoints of BG and DK (Fig. 16.2 b). At b = 1.0515 the icosahedron is regular, ϕ_A = 37.38, ϕ_B = 79.19°, θ_A = −30.00 and θ_B = 30.00°. The main change on decreasing the normalized bite is to progressively increase θ_A (Fig. 16.4). At b = 0.62 (for n = 6), θ_A = 0° and the upper ACE triangular face becomes eclipsed relative to the lower HJL face, while θ_B remains approximately 30° (Fig. 16.3 b). This structure, however, differs from the anticuboctahedron (Fig. 16.1 d) in that $\phi_B \neq 90.0°$, resulting in a substantial distortion and creasing of the square faces of the anticuboctahedron. The A end of the bidentate ligand experiences slightly less repulsion than does the B end [427].

A clear distinction between isomers I and II is shown by the angles between vectors joining the central metal atom and the midpoint of each bidentate ligand. For isomer I there are twelve angles of 90.0° and three of 180.0° at all values of the normalized bite, with the midpoint of each bidentate ligand situated at the apex of a regular octahedron about the metal atom. For isomer II of the regular icosahedron, b = 1.0515, there are three angles of 72.0°, six of 90°, three of 120.0°, and three of 144.0°. The first twelve angles approach 90°, and the last three approach 180°, as b is progressively decreased to zero.

Stereochemistry III (Fig. 16.2 c) of D$_{3d}$ symmetry exists as a minimum on the potential energy surface only above b ∼ 0.8.

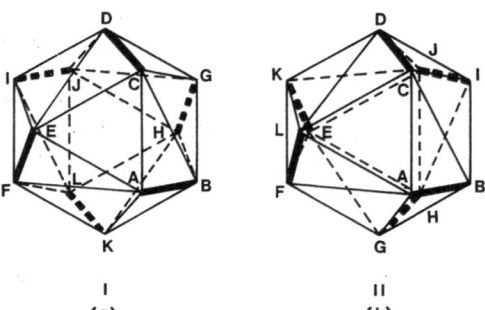

Fig. 16.3. Isomers of [M(bidentate)₆] for b = 0.62

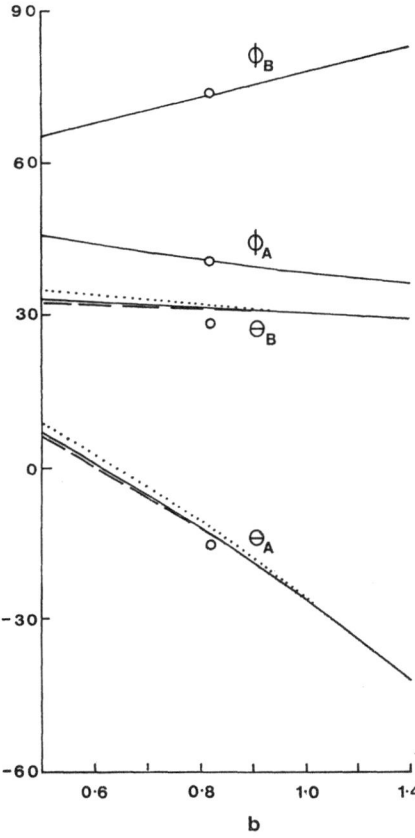

Fig. 16.4. Angular coordinates (degrees) for isomer II of [M(bidentate)$_6$] as a function of b. *Dotted lines*, n = 1; *solid lines*, n = 6; *broken lines*, n = 12. The angular coordinates for [Pr(naphthyridine)$_6$](ClO$_4$)$_3$ are marked

The relative energies (n = 6) of these three isomers are shown in Fig. 16.5 Isomer III is unstable for b < 1.0, and the existence of compounds with this structure is not expected. Isomers I and II are both expected to be possible, the latter being more stable.

Structure I is observed for the hexanitrato complexes of cerium(III), lanthanum(III), and thorium(IV) listed in Table 16.1.

Table 16.1. Structural parameters for [M(bidentate)$_6$] complexes, isomer I

Complex	b	a	Ref.
[Mg(H$_2$O)$_6$]$_3$[Ce(NO$_3$)$_6$]$_2$·6 H$_2$O	0.82	1.12	[1115]
[Mg(H$_2$O)$_6$]$_3$[La(NO$_3$)$_6$]$_2$·6 H$_2$O	0.80	1.13	[40]
[La(NO$_3$)$_2$(C$_{18}$H$_{36}$N$_2$O$_6$)][La(NO$_3$)$_6$]2 MeOH	0.82	1.12	[534]
[Mg(H$_2$O)$_6$][Th(NO$_3$)$_6$]2 H$_2$O	0.80	1.13	[965]
[Th(NO$_3$)$_3$(OPMe$_3$)$_4$]$_2$[Th(NO$_3$)$_6$]	0.81	1.12	[27]

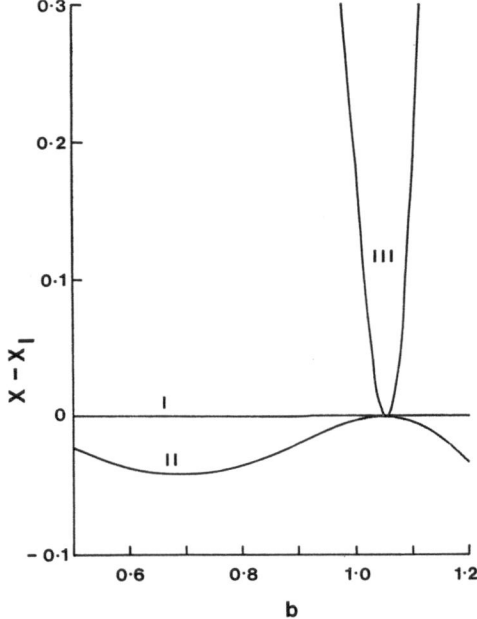

Fig. 16.5. Repulsion energy coefficients, relative to those for isomer I, for the various isomers of [M(bidentate)$_6$] as a function of b. n = 6

Structure II is observed only for the praseodymium(III) naphthyridine complex, [Pr(napy)$_6$](ClO$_4$)$_3$ [243]. The experimental angular parameters, b = 0.82, ϕ_A = 40.6°, ϕ_B = 73.9°, θ_A = −15.5° and θ_B = 28.2°, are marked on Fig. 16.4, the agreement with those calculated being reasonable.

References

1 Ablov, A. V., Kon, A. Yu., Burshtein, I. F., Malinovskii, T. I., Levitskaya, Z. G.: Dokl. Akad. Nauk S. S. S. R. *217*, 1317 (1974)
2 Abrahams, S. C., Bernstein, J. L.: Acta Crystallogr. *B33*, 3601 (1977)
3 Abrahams, S. C., Ginsberg, A. P., Knox, K.: Inorg. Chem. *3*, 558 (1964)
4 Abrahamson, H., Heiman, J. R., Pignolet, L. H.: Inorg. Chem. *14*, 2070 (1975)
5 Addison, C. C., Greenwood, A. J., Haley, M. J., Logan, N.: J. Chem. Soc., Chem. Commun., 580 (1978)
6 Ahmed, A., Schwarz, W., Weidlein, J., Hess, H.: Z. anorg. Chem. *434*, 207 (1977)
7 Akabori, K., Kushi, Y.: J. Inorg. Nucl. Chem. *40*, 625 (1978)
8 Akhtar, M., Ellis, P. D., MacDiarmid, A. B., Odom, J. D.: Inorg. Chem. *11*, 2917 (1972)
9 Akhtar, M. N., Smith, A. J.: Acta Crystallogr. *B 31*, 1361 (1975)
10 Albano, V. G., Bellon, P. L.: J. Organomet. Chem. *37*, 151 (1972)
11 Albertsson, J.: Acta Chem. Scand. *22*, 1563 (1968)
12 Albertsson, J.: Acta Chem. Scand. *24*, 1213 (1970)
13 Albertsson, J.: Acta Chem. Scand. *24*, 3527 (1970)
14 Albertsson, J.: Acta Chem. Scand. *26*, 985 (1972)
15 Albertsson, J.: Acta Chem. Scand. *26*, 1005 (1972)
16 Albertsson, J.: Acta Chem. Scand. *26*, 1023 (1972)
17 Albertsson, J., Elding, I.: Acta Chem. Scand. *A 31*, 21 (1977)
18 Albertsson, J., Elding, I.: Acta Crystallogr. *B32*, 3066 (1976)
19 Albertsson, J., Elding, I.: Acta Crystallogr. *B33*, 1460 (1977)
20 Albertsson, J., Elding, I., Oskarsson, A.: Acta Chem. Scand. *A 33*, 703 (1979)
21 Albertsson, J., Oskarsson, A.: Acta Crystallogr. *B33*, 1871 (1977)
22 Albertsson, J., Oskarsson, A., Nygren, M.: Acta Crystallogr. *B35*, 1473 (1979)
23 Albright, J. O., Clardy, J. C., Verkade, J. G.: Inorg. Chem. *16*, 1575 (1977)
24 Alcock, N. W.: J. Chem. Soc., Dalton Trans., 1189 (1972)
25 Alcock, N. W., Countryman, R. M., Esperas, S., Sawyer, J. F.: J. Chem. Soc., 854 (1979)
26 Alcock, N. W., Esperas, S.: J. Chem. Soc., Dalton Trans., 893 (1977)
27 Alcock, N. W., Esperas, S., Bagnall, K. W., Hsian-Yun, W.: J. Chem. Soc., Dalton Trans., 638 (1978)
28 Alcock, N. W., de Meester, P., Kemp, T. J.: Acta Crystallogr. *B34*, 3367 (1978)
29 Alcock, N. W., Sawyer, J. F.: J. Chem. Soc., Dalton Trans., 1090 (1977)
30 Alcock, N. W., Tracy, V. L.: J. Chem. Soc., Dalton Trans., 2246 (1976)
31 Alcock, N. W., Tracy, V. L.: Acta Crystallogr. *B35*, 80 (1979)
32 Al-Karaghouli, A. R., Day, R. O., Wood, J. S.: Inorg. Chem. *17*, 3702 (1978)
33 Al-Karaghouli, A. R., Wood, J. S.: J. Chem. Soc., Dalton Trans., 2318 (1973)
34 Al-Karaghouli, A. R., Wood, J. S.: Inorg. Chem. *18*, 1177 (1979)
35 Allard, B., Acta Chem. Scand. *26*, 3492 (1972)
36 Allard, B., Acta Chem. Scand. *A 30*, 461 (1976)
37 Allcock, H. R., Bissell, E. C.: J. Am. Chem. Soc. *95*, 3154 (1973)
38 Almenningen, A., Gundersen, G., Haaland, A.: Acta Chem. Scand. *22*, 328 (1968)

39 Alyea, A. C., Costin, A., Ferguson, G., Fey, G. T., Goel, R. G., Restivo, R. J.: J. Chem. Soc., Dalton Trans., 1294 (1975)
40 Anderson, M. R., Jenkin, G. T., White, J. W.: Acta Crystallogr. *B33*, 3933 (1977)
41 Anderson, O., Husebye, S.: Acta Chem. Scand. *24*, 3141 (1970)
42 Anderson, O. P.: J. Chem. Soc., Dalton Trans., 2597 (1972)
43 Anderson, O. P.: J. Chem. Soc., Dalton Trans., 1237 (1973)
44 Anderson, O. P.: Inorg. Chem. *14*, 730 (1975)
45 Anderson, T. J., Neuman, M. A., Melson, G. A.: Inorg. Chem. *12*, 927 (1973)
46 Anderson, T. J., Neuman, M. A., Melson, G. A.: Inorg. Chem. *13*, 159 (1974)
47 Anderson, T. J., Neuman, M. A., Melson, G. A.: Inorg. Chem. *13*, 1884 (1974)
48 Antti, B.-M.: Acta Chem. Scand. *A 30*, 24 (1976)
49 Antti, B.-M.: Acta Chem. Scand. *A 30*, 103 (1976)
50 Antti, B.-M., Lundberg, B. K. S., Ingri, N.: Acta Chem. Scand. *26*, 3984 (1972)
51 Appelman, E. H., Beagley, B., Cruickshank, D. W. J., Foord, A., Rustad, S., Ulbrecht, V.: J. Mol. Struct. *35*, 139 (1976)
52 Archer, C., Durand, J., Cot, L., Galigne, J. L.: Can. J. Chem. *57*, 899 (1979)
53 Arutyunyan, E. G., Porai-Koshits, M. A.: J. Struct. Chem. *4*, 96 (1963)
54 Ashworth, T. V., Singleton, E., Laing, M.: J. Chem. Soc., Chem. Commun., 875 (1976)
55 Aslanov, L. A., Korytnyi, E. F., Porai-Koshits, M. A.: J. Struct. Chem. *12*, 600 (1971)
56 Aslanov, L. A., Porai-Koshits, M. A., Dekaprilevich, M. O.: J. Struct. Chem. *12*, 431 (1971)
57 Aslanov, L. A., Soleva, L. I., Porai-Koshits, M. A.: J. Struct. Chem. *13*, 1021 (1972)
58 Aslanov, L. A., Soleva, L. I., Porai-Koshits, M. A.: J. Struct. Chem. *14*, 998 (1973)
59 Atwood, J. L., Hrncir, D. C., Newberry, W. R.: Cryst. Struct. Comm. *3*, 615 (1974)
60 Atwood, J. L., Hunter, W. E., Carmona-Guzman, E., Wilkinson, G.: J. Chem. Soc., Dalton Trans., 467 (1980)
61 Ault, H. K., Husbye, S.: Acta Chem. Scand. *A 32*, 157 (1978)
62 Avdeef, A., Costamagna, J. A., Fackler, J. P.: Inorg. Chem. *13*, 1854 (1974)
63 Baenziger, N. C., Schultz, R. J.: Inorg. Chem. *10*, 661 (1971)
64 Bailey, N. A., Bird, P. H., Wallbridge, M. G. H.: Inorg. Chem. *7*, 1575 (1968)
65 Baker, J., Engelhardt, L. M., Figgis, B. N., White, A. H.: J. Chem. Soc., Dalton Trans., 530 (1975)
66 Baker, R. W., Pauling, P.: J. Chem. Soc., Chem. Commun., 1495 (1969)
67 Baker, W. A., Williams, D. E.: Acta Crystallogr. *B34*, 3739 (1978)
68 Bancroft, G. M., Davies, B. W., Payne, N. C., Sham, T. K.: J. Chem. Soc., Dalton Trans., 973 (1975)
69 Bandoli, G., Bortolozzo, G., Clemente, D. A., Croatto, U., Panattoni, C.: J. Chem. Soc. (A), 2778 (1970)
70 Baracco, L., Bombieri, G., Degetto, S., Forsellini, E., Graziani, R., Marongoni, G.: Inorg. Nucl. Chem. Letters *10*, 1045 (1974)
71 Barclay, G. A., Hoskins, B. F., Kennard, C. H. L.: J. Chem. Soc., 5691 (1963)
72 Barclay, G. A., Sabine, T. M., Taylor, J. C.: Acta Crystallogr. *19*, 205 (1965)
73 Barclay, G. A., Vagg, R. S., Watton, E. C.: Acta Crystallogr. *B33*, 3777 (1977)
74 Bartell, L. S., Clippard, F. B., Jacob, E. J.: Inorg. Chem. *15*, 3009 (1976)
75 Bartell, L. S., Gavin, R. M.: J. Chem. Phys. *48*, 2466 (1968)
76 Bartlett, N., Beaton, S., Reeves, L. W., Wells, E. J.: Can. J. Chem. *42*, 2531 (1964)
77 Bartlett, N., De Boer, B. G., Hollander, F. J., Sladky, F. O., Templeton, D. H., Zalkin, A.: Inorg. Chem. *13*, 780 (1974)
78 Bartlett, N., Einstein, F., Stewart, D. F., Trotter, J.: J. Chem. Soc. (A), 1190 (1967)
79 Bartlett, N., Gennis, M., Gibler, D. D., Morrell, B. K., Zalkin, A.: Inorg. Chem. *12*, 1717 (1973)
80 Bassett, J. M., Berry, D. E., Barker, G. K., Green, M., Howard, J. A. K., Stone, F. G. A.: J. Chem. Soc., Dalton Trans., 1003 (1979)
81 Bassett, J. M., Green, M., Howard, J. A. K., Stone, F. G. A.: J. Chem. Soc., Chem. Commun., 853 (1977)
82 Basson, S. S., Bok, L. D. C., Leipoldt, J. G.: Acta Crystallogr. *B26*, 1209 (1970)
83 Bats, J. W.: Acta Crystallogr. *B34*, 1679 (1978)

84 Beauchamp, A. L., Bennett, M. J., Cotton, F. A.: J. Am. Chem. Soc. *90*, 6675 (1968)
85 Beauchamp, A. L., Bennett, M. J., Cotton, F. A.: J. Am. Chem. Soc. *91*, 297 (1969)
86 Beauchamp, A. L., Saperas, B., Rivest, R.: Can. J. Chem. *49*, 3579 (1971)
87 Becker, G., Beck, H. P.: Z. anorg. Chem. *430*, 91 (1977)
88 Beckett, R., Heath, G. A., Hoskins, B. F., Kelly, B. P., Martin, R. L., Roos, I. A. G., Weickhardt, P. L.: Inorg. Nucl. Chem. Letters *6*, 257 (1970)
89 Begley, M. J., Haley, M. J., King, T. J., Morris, A., Pike, R., Smith, B.: Inorg. Nucl. Chem. Letters *12*, 99 (1976)
90 Begley, M. J., Sowerby, D. B., Haiduc, I.: J. Chem. Soc., Chem. Commun., 64 (1980)
91 Bellard, S., Rivera, A. V., Sheldrick, G. M.: Acta Crystallogr. *B34*, 1034 (1978)
92 Bellavance, P. L., Corey, E. R., Corey, J. Y., Hey, G. W.: Inorg. Chem. *16*, 462 (1977)
93 Bellitto, C., Gastaldi, L., Tomlinson, A. A. G.: J. Chem. Soc., Dalton Trans., 989 (1976)
94 Bel'skii, N. K., Struchkov, Yu. T.: Sov. Phys. Crystallogr. *10*, 15 (1965)
95 Beltram, F. G., Capilla, A. V., Aranda, R. A.: Cryst. Struct. Commun. *8*, 87 (1979)
96 Bennett, M. J., Cotton, F. A., Legzdins, P., Lippard, S. J.: Inorg. Chem. *7*, 1770 (1968)
97 Beran, G., Dymock, K., Patel, H. A., Carty, A. J., Boorman, P. M.: Inorg. Chem. *11*, 896 (1972)
98 Bereman, R. D., Churchill, M. R., Nalewajek, D.: Inorg. Chem. *18*, 3112 (1979)
99 Berg, G. C. van den, Oskam, A., Olie, K.: J. Organomet. Chem. *80*, 363 (1974)
100 Bergman, J. G., Cotton, F. A.: Inorg. Chem. *5*, 1208 (1966)
101 Bergman, J. G., Cotton, F. A.: Inorg. Chem. *5*, 1420 (1966)
102 Bernal, I., Elliott, N., Lalancette, R.: J. Chem. Soc., Chem. Commun., 803 (1971)
103 Bernstein, P. K., Rodley, G. A., Marsh, R., Gray, H. B.: Inorg. Chem. *11*, 3040 (1972)
104 Berry, R. S.: J. Chem. Phys. *32*, 933 (1960)
105 Bertini, I., Dapporto, P., Gatteschi, D., Scozzafava, A.: Inorg. Chem. *14*, 1639 (1975)
106 Bertini, I., Dapporto, P., Gatteschi, D., Scozzafava, A.: J. Chem. Soc., Daton Trans., 1409 (1979)
107 Bertrand, J. A., Carpenter, D. A.: Inorg. Chem. *5*, 514 (1966)
108 Bertrand, J. A., Plymale, M. A.: Inorg. Chem. *5*, 879 (1966)
109 Bhandary, K. K., Manohar, H., Vankatesan, K.: J. Chem. Soc., Dalton Trans., 288 (1975)
110 Bhandary, K. K., Manohar, H., Vankatesan, K.: Acta Crystallogr. *B32*, 861 (1976)
111 Biagini, S., Cannas, M.: J. Chem. Soc. (A), 2398 (1970)
112 Biagini Cingi, M., Guistini, C., Musatti, A., Nardelli, M.: Acta Crystallogr. *B28*, 667 (1972)
113 Bird, P. H., Fraser, A. R., Lau, C. F.: Inorg. Chem. *12*, 1322 (1973)
114 Birker, P. J. M. W. L., Prick, P., Beurskens, P. T.: Cryst. Struct. Commun. *5*, 135 (1976)
115 Biscarini, P., Fusina, L., Nivellini, G., Pelizzi, G.: J. Chem. Soc., Dalton Trans., 664 (1977)
116 Bishop, E. O., Butler, G., Chatt, J., Dilworth, J. R., Leigh, G. J., Orchard, D., Bishop, M. W.: J. Chem. Soc., Dalton Trans., 1654 (1978)
117 Bishop, M. W., Chatt, J., Dilworth, J. R., Hursthouse, M. B., Motevalli, M.: J. Chem. Soc., Chem. Commun., 780 (1976)
118 Bjorvatten, T., Hassel, O., Lindheim, A.: Acta Chem. Scand. *17*, 689 (1963)
119 Blackwell, L. J., Nunn, E. K., Wallwork, S. C.: J. Chem. Soc., Dalton Trans., 2068 (1975)
120 Blight, D. G., Kepert, D. L.: Theor. Chim. Acta *11*, 51 (1968)
121 Boese, R., Schmid, G.: J. Chem. Soc., Chem. Commun., 349 (1979)
122 Boeyens, J. C. A., de Villiers, J. P. P.: J. Cryst. Mol. Struct. *1*, 297 (1971)
123 Bois, C., Nguyen Quy Dao, Rodier, N.: Acta Crystallogr. *B32*, 1541 (1976)
124 Bok, L. D. C., Leipoldt, J. G., Basson, S. S.: Acta Crystallogr. *B26*, 684 (1970)
125 Bok, L. D. C., Leipoldt, J. G., Basson, S. S.: Z. anorg. Chem. *392*, 303 (1972)
126 Boldrini, P., Gillespie, R. J., Ireland, P. R., Schrobilgen, G. J.: Inorg. Chem. *13*, 1690 (1974)
127 Bombieri, G., Bagnall, K. W.: J. Chem. Soc., Chem. Commun., 188 (1975)
128 Bombieri, G., Brown, D., Graziani, R.: J. Chem. Soc., Dalton Trans., 1873 (1975)

129   Bombieri, G., Brown, D., Mealli, C.: J. Chem. Soc., Dalton Trans., 2025 (1976)
130   Bombieri, G., Croatto, U., Forsellini, E., Zarli, B., Graziani, R.: J. Chem. Soc.,
      Dalton Trans., 560 (1972)
131   Bombieri, G., De Paoli, G., Forsellini, E., Brown, D.: J. Inorg. Nucl. Chem. *41*, 1315
      (1979)
132   Bombieri, G., Forsellini, E., Brown, D., Whittaker, B.: J. Chem. Soc., Dalton Trans.,
      735 (1976)
133   Bombieri, G., Moseley, P. T., Brown, D.: J. Chem. Soc., Dalton Trans., 1520 (1975)
134   Bonamico, M., Dessy, G., Fares, V., Flamini, A., Scaramuzza, L.: J. Chem. Soc.,
      Dalton Trans., 1743 (1976)
135   Bonamico, M., Dessy, G., Fares, V., Scaramuzza, L.: J. Chem. Soc., Dalton Trans., 876
      (1973)
136   Bonamico, M., Dessy, G., Fares, V., Scaramuzza, L.: J. Chem. Soc., Dalton Trans.,
      1258 (1974)
137   Bonamico, M., Dessy, G., Fares, V., Scaramuzza, L.: J. Chem. Soc., Dalton Trans.,
      2079 (1975)
138   Bone, S. P., Sowerby, D. B.: J. Chem. Soc., Dalton Trans., 718 (1979)
139   Bortin, O.: Acta Chem. Scand. *A 30*, 657 (1976)
140   Bottomley, F., White, P. S.: Acta Crystallogr. *B 35*, 2193 (1979)
141   Bowman, K., Dori, Z.: J. Chem. Soc., Chem. Commun., 636 (1968)
142   Brabant, C., Blanck, B., Beauchamp, A. L.: J. Organomet. Chem. *82*, 231 (1974)
143   Brabant, C., Hubert, J., Beauchamp, A. L.: Can. J. Chem. *51*, 2952 (1973)
144   Bradbury, J. R., Mackay, M. F., Wedd, A. G.: Aust. J. Chem. *31*, 2423 (1978)
145   Bradley, D. C., Hursthouse, M. B., Malik, K. M. A., Vuru, G. B. C.: Inorg. Chim.
      Acta *44*, L5 (1980)
146   Brant, P., Cotton, F. A., Sekutowski, J. C., Wood, T. E., Walton, R. A.: J. Am. Chem.
      Soc. *101*, 6588 (1979)
147   Brawner, S. A., Lin, I. J. B., Kim, J. H., Everett, G. W.: Inorg. Chem. *17*, 1304 (1978)
148   Brennan, T., Bernal, I.: J. Phys. Chem. *73*, 443 (1969)
149   Brennan, T. F., Bernal, I.: Inorg. Chim. Acta *7*, 283 (1973)
150   Bright, D., Ibers, J. A.: Inorg. Chem. *8*, 709 (1969)
151   Broach, R. W., Williams, J. M., Felcher, G. P., Hinks, D. G.: Acta Crystallogr. *B 35*,
      2317 (1979)
152   Brock, C. P., Webster, D. F.: Acta Crystallogr. *B 32*, 2089 (1976)
153   Brotherton, P. D., Epstein, J. M., White, A. H., Wild, S. B.: Aust. J. Chem. *27*, 2667
      (1974)
154   Brotherton, P. D., Kepert, D. L., White, A. H., Wild, S. B.: J. Chem. Soc., Dalton
      Trans., 1870 (1976)
155   Brown, B. W., Lingafelter, E. C.: Acta Crystallogr. *16*, 753 (1963)
156   Brown, D., Easey, J. F., Rickard, C. E. F.: J. Chem. Soc. (A), 1161 (1969)
157   Brown, D., Holah, D. G., Rickard, C. E. F.: J. Chem. Soc. (A), 423 (1970)
158   Brown, D., Holah, D. G., Rickard, C. E. F.: J. Chem. Soc. (A), 786 (1970)
159   Brown, D., Reynolds, C. T., Moseley, P. T.: J. Chem. Soc., Dalton Trans., 857 (1972)
160   Brown, D. S., Bushnell, G. W.: Acta Crystallogr. *22*, 296 (1967)
161   Brown, G. F., Stiefel, E. I.: Inorg. Chem. *12*, 2140 (1973)
162   Brown, G. M., Walker, L. A.: Acta Crystallogr. *20*, 220 (1966)
163   Brown, I. D., Lock, C. J. L., Wan, C.: Can. J. Chem. *51*, 2073 (1973)
164   Brown, K. L.: Cryst. Struct. Commun. *3*, 493 (1974)
165   Brown, L. D., Greig, D. R., Raymond, K. N.: Inorg. Chem. *14*, 645 (1975)
166   Brown, L. D., Raymond, K. N.: Inorg. Chem. *14*, 2590 (1975)
167   Brown, R. K., Holmes, R. R.: J. Am. Chem. Soc. *99*, 3326 (1977)
168   Brownlee, G. S., Walker, A., Nyburg, S. C., Szymanski, J. T.: J. Chem. Soc., Chem.
      Commun., 1073 (1971)
169   Browty, C., Spinat, P., Whuler, A.: Acta Crystallogr. *B 33*, 3453 (1977)
170   Browty, C., Spinat, P., Whuler, A., Herpin, P.: Acta Crystallogr. *B 32*, 2153 (1976)
171   Browty, C., Spinat, P., Whuler, A., Herlin, P.: Acta Crystallogr. *B 33*, 1913 (1977)
172   Browty, C., Spinat, P., Whuler, A., Herpin, P.: Acta Crystallogr. *B 33*, 1920 (1977)

173 Browty, C., Whuler, A., Spinat, P., Herpin, P.: Acta Crystallogr. *B33*, 2563 (1977)
174 Bryan, R. F., J. Chem. Soc. (A), 172 (1967)
175 Bryan, R. F., J. Chem. Soc. (A), 192 (1967)
176 Buchanan, R. M., Kessel, S. L., Downs, H. H., Pierpont, C. G., Hendrickson, D. N.: J. Am. Chem. Soc. *100*, 7894 (1978)
177 Bullen, G. J.: Acta Crystallogr. *12*, 703 (1959)
178 Bullock, J. I., Parrett, F. W., Post, M. L., Taylor, N. J.: J. Chem. Soc., Dalton Trans., 1536 (1978)
179 Burbank, R. D.: Acta Crystallogr. *15*, 1207 (1962)
180 Burbank, R. D., Bartlett, N.: J. Chem. Soc., Chem. Commun., 645 (1968)
181 Burbank, R. D., Bensey, F. N.: J. Chem. Phys. *21*, 602 (1953)
182 Burbank, R. D., Jones, G. R.: Inorg. Chem. *3*, 1071 (1974)
183 Burbank, R. D., Jones, G. R.: J. Am. Chem. Soc. *96*, 43 (1974)
184 Burgi, H. B., Stedman, D., Bartell, L. S.: J. Mol. Struct. *10*, 31 (1971)
185 Burke-Laing, M. E., Trueblood, K. N.: Acta Crystallogr. *B33*, 2698 (1977)
186 Burns, J. H., Agron, P. A., Levy, H. A.: Science *139*, 1208 (1963)
187 Burns, J. H., Danford, M. D.: Inorg. Chem. *8*, 1780 (1969)
188 Burns, J. H., Ellison, R. D., Levy, H. A.: Acta Crystallogr. *18*, 11 (1965)
189 Burns, J. H., Musikas, C.: Inorg. Chem. *16*, 1619 (1977)
190 Butcher, R. J., O'Connor, C. J., Sinn, E.: Inorg. Chem. *18*, 492 (1979)
191 Butcher, R. J., Penfold, B. R.: J. Cryst. Mol. Struct. *6*, 13 (1976)
192 Butcher, R. J., Penfold, B. R., Sinn, E.: J. Chem. Soc., Dalton Trans., 668 (1979)
193 Butcher, R. J., Sinn, E.: J. Chem. Soc., Dalton Trans., 2517 (1975)
194 Butcher, R. J., Sinn, E.: J. Am. Chem. Soc. *98*, 2440 (1976)
195 Butcher, R. J., Sinn, E.: J. Am. Chem. Soc. *98*, 5159 (1976)
196 Butcher, R. J., Sinn, E.: Inorg. Chem. *16*, 2334 (1977)
197 Butler, K. R., Snow, M. R.: J. Chem. Soc. (A), 565 (1971)
198 Butler, K. R., Snow, M. R.: Acta Crystallogr. *B31*, 354 (1975)
199 Butler, K. R., Snow, M. R.: J. Chem. Soc., Dalton Trans., 251 (1976)
200 Butman, L. A., Aslanov, L. A., Porai-Koshits, M. A.: J. Struct. Chem. *11*, 40 (1970)
201 Buttenshaw, A. J., Duchene, M., Webster, M.: J. Chem. Soc., Dalton Trans., 2230 (1975)
202 Bye, E.: Acta Chem. Scand. *A28*, 731 (1974)
203 Caira, M. R., Nassimbeni, L. R.: Acta Crystallogr. *B30*, 2332 (1974)
204 Calogero, S., Ganis, P., Peruzzo, V., Tagliavini, G.: J. Organomet. Chem. *179*, 145 (1979)
205 Calvo, C., Krishnamachari, N., Lock, C. J. L.: J. Cryst. Mol. Struct. *1*, 161 (1971)
206 Cameron, A. F., Taylor, D. W., Nuttall, R. H.: J. Chem. Soc., Dalton Trans., 1603 (1972)
207 Cameron, A. F., Taylor, D. W., Nutall, R. H.: J. Chem. Soc., Dalton Trans., 1608 (1972)
208 Cameron, A. F., Taylor, D. W., Nuttall, R. H.: J. Chem. Soc., Dalton Trans., 2130 (1973)
209 Cameron, T. S., Dahlen, B.: J. Chem. Soc., Perkin II, 1737 (1975)
210 Cameron, T. S., Howlett, K. D., Miller, K.: Acta Crystallogr. *B34*, 1639 (1978)
211 Cameron, T. S., Howlett, K. D., Shaw, R. A., Woods, M.: Phosphorus *3*, 71 (1973)
212 Cannas, M., Carta, G., Cristini, A., Marongiu, G.: J. Chem. Soc., Dalton Trans., 210 (1976)
213 Cannas, M., Carta, G., Marongiu, G.: J. Chem. Soc., Dalton Trans., 550 (1974)
214 Cano, F. H., Cruickshank, D. W. J.: J. Chem. Soc., Chem. Commun., 1617 (1971)
215 Carola, J. M., Freedman, D. D., McLaughlin, K. L., Reim, P. C., Schmidt, W. J., Haas, R. G., Broome, W. J., DeCarlo, E. A., Lawton, S. L.: Cryst. Struct. Commun. *5*, 393 (1976)
216 Carrondo, M. A. A. F. de C. T., Shakir, R., Skapski, A. C.: J. Chem. Soc., Dalton Trans., 844 (1978)
217 Carter, R. P., Holmes, R. R.: Inorg. Chem. *4*, 738 (1965))

218   Carty, A. J., Taylor, N. J., Coleman, A. W., Lappert, M. F.: J. Chem. Soc., Chem. Commun., 639 (1979)

219   Castellani Bisi, C., Della Giusta, A., Coda, A., Tazzoli, V.: Cryst. Struct. Commun. *3*, 381 (1974)

220   Castellani Bisi, C., Gori, M., Cannillo, E., Coda, A., Tazzoli, V.: Acta Crystallogr. *A 31*, S 134 (1975)

221   Castellano, E. E., Piro, O. E., Rivero, B. E.: Acta Crystallogr. *B 33*, 1725 (1977)

222   Castellano, E. E., Piro, O. E., Rivero, B. E.: Acta Crystallogr. *B 33*, 1728 (1977)

223   Cazzoli, G.: J. Mol. Spectrosc. *53*, 37 (1974)

224   Cazzoli, G., Favero, P. G., Dal Borgo, A.: J. Mol. Spectrosc. *50*, 82 (1974)

225   Chao, G. K.-J., Sime, R. L., Sime, R. J.: Acta Crystallogr. *B 29*, 2845 (1973)

226   Chapps, G. E., McCann, S. W., Wickmann, H. H., Sherwood, R. C.: J. Chem. Phys. *60*, 990 (1974)

227   Chen, H. W., Feckler, J. P.: Inorg. Chem. *17*, 22 (1978)

228   Chiang, C. C., Epps, L. A., Marzilli, L. G., Kistenmacher, T. J.: Inorg. Chem. *18*, 791 (1979)

229   Chieh, C., Toogood, G. E., Boyle, T. D., Burgess, C. M.: Acta Crystallogr. *B 32*, 1008 (1976)

230   Chiesi Villa, A., Gaetani Manfredotti, A., Guastini, C., Nardelli, M.: Acta Crystallogr. *B 28*, 2231 (1972)

231   Chisholm, M. H., Cotton, F. A., Extine, M. W.: Inorg. Chem. *17*, 2000 (1978)

232   Chisholm, M. H., Extine, M.: J. Am. Chem. Soc. *97*, 1623 (1975)

233   Chojnacki, J., Grochowski, J., Lebioda, L., Oleksyn, B., Stadnicka, K.: Roczn. Chem. *43*, 273 (1969)

234   Chomnilpan, S., Tellgren, R., Liminga, R.: Acta Crystallogr. *B 33*, 2108 (1977)

235   Christidis, P. C., Rentzeperis, P. J.: Acta Crystallogr. *B 34*, 2141 (1978)

236   Chui, K. M., Powell, H. M.: J. Chem. Soc., Dalton Trans., 1879, 2117 (1974)

237   Chun, H. K., Steffen, W. L., Fay, R. C.: Inorg. Chem. *18*, 2458 (1979)

238   Ciani, G., Manassero, M., Sansoni, M.: J. Inorg. Nucl. Chem. *34*, 1760 (1972)

239   Clark, A. H., Beagley, B., Cruickshank, D. W. J., Hewitt, T. G.: J. Chem. Soc. (A), 872 (1970)

240   Clark, R. J. H., Lewis, J., Nyholm, R. S., Pauling, P., Robertson, G. B.: Nature *192*, 222 (1961)

241   Clark, T. E., Day, R. O., Holmes, R. R.: Inorg. Chem. *18*, 1660 (1979)

242   Clark, T. E., Day, R. O., Holmes, R. R.: Inorg. Chem. *18*, 1668 (1979)

243   Clearfield, A., Gopal, R., Olsen, R. W.: Inorg. Chem. *16*, 91 (1977)

244   Clegg, W.: Acta Crystallogr. *B 34*, 3328 (1978)

245   Clegg, W., Greenhalg, D. A., Straghan, B. P.: J. Chem. Soc., Dalton Trans., 2591 (1975)

246   Clegg, W., Wheatley, P. J.: J. Chem. Soc., Dalton Trans., 90 (1973)

247   Clippard, F. B., Bartell, L. S.: Inorg. Chem. *9*, 805 (1970)

248   Codding, P. W., Kerr, K. A.: Acta Crystallogr. *B 34*, 3785 (1978)

249   Codding, P. W., Kerr, K. A.: Acta Crystallogr. *B 35*, 1261 (1979)

250   Coghi, L., Nardelli, M., Pelizzi, G.: Acta Crystallogr. *B 32*, 842 (1976)

251   Coghi, L., Pelizzi, C., Pelizzi, G.: J. Organomet. Chem. *114*, 53 (1976)

252   Cogne, A., Grand, A., Laugier, J., Robert, J. B., Wiesenfeld, L.: J. Am. Chem. Soc. *102*, 2238 (1980)

253   Colapietro, M., Domenicano, A., Scaramuzza, L., Vaciago, A., Zambonelli, L.: J. Chem. Soc., Chem. Commun., 583 (1967)

254   Colapietro, M., Vaciago, A.: unpublished, quoted in Ref. 137

255   Colapietro, M., Vaciago, A., Bradley, D. C., Hursthouse, M. B., Rendall, I. F.: J. Chem. Soc., Dalton Trans., 1052 (1972)

256   Collins, R. K., Drew, M. G. B., Rodgers, J.: J. Chem. Soc., Dalton Trans., 899 (1972)

257   Connor, J. A., McEwen, G. K., Rix, C. J.: J. Chem. Soc., Dalton Trans., 589 (1974)

258   Cook, P. M., Dahl, L. F., Hopgood, P., Jenkins, R. A.: J. Chem. Soc., Dalton Trans., 294 (1973)

259   Corbett, M., Hoskins, B. F.: Aust. J. Chem. *27*, 665 (1974)

260   Corbridge, D. E. C., Cox, E. G.: J. Chem. Soc., 594 (1956)
261   Corden, B. J., Cunningham, J. A., Eisenberg, R.: Inorg. Chem. *9*, 356 (1970)
262   Corradi, A. B., Palmieri, C. G., Nardelli, M., Pellinghelli, M. A., Tani, M. E. V.: J. Chem. Soc., Dalton Trans., 655 (1973)
263   Cotton, F. A., Davison, A., Day, V. W., Gage, L. D., Trop, H. S.: Inorg. Chem. *18*, 3024 (1979)
264   Cotton, F. A., Dunne, T. G., Wood, J. S.: Inorg. Chem. *4*, 318 (1965)
265   Cotton, F. A., Elder, R. C.: Inorg. Chem. *3*, 397 (1964)
266   Cotton, F. A., Hardcastle, K. I., Rusholme, G. A.: J. Coord. Chem. *2*, 217 (1973)
267   Cotton, F. A., Legzdins, P.: Inorg. Chem. *7*, 1777 (1968)
268   Cotton, F. A., Soderberg, R. H.: J. Am. Chem. Soc. *85*, 2402 (1963)
269   Cotton, F. A., Troup, J. M.: J. Am. Chem. Soc. *96*, 3438 (1974)
270   Coucouvanis, D., Hollander, F. J., Pedelty, R.: Inorg. Chem. *16*, 2691 (1977)
271   Countryman, R., McDonald, W. S.: J. Inorg. Nucl. Chem. *33*, 2213 (1971)
272   Cowie, M., Bennett, M. J.: Inorg. Chem. *15*, 1584 (1976)
273   Cowie, M., Bennett, M. J.: Inorg. Chem. *15*, 1589 (1976)
274   Cowie, M., Bennett, M. J.: Inorg. Chem. *15*, 1595 (1976)
275   Coyle, B. E., Ibers, J.: Inorg. Chem. *9*, 767 (1970)
276   Cradwick, E. N., Hall, D.: J. Organomet. Chem. *25*, 91 (1970)
277   Cramer, R. E., Cramer, S. W., Chudyk, M. A., Seff, K.: Inorg. Chem. *16*, 219 (1977)
278   Cramer, R. E., Huneke, J. T.: Inorg. Chem. *17*, 365 (1978)
279   Cramer, R. E., Lindsey, R. V., Prewitt, C. T., Stolberg, U. G.: J. Am. Chem. Soc. *78*, 658 (1965)
280   Cramer, R. E., Van Doorne, W., Huneke, J. T.: Inorg. Chem. *15*, 529 (1976)
281   Croatto, U.: Atti Accad. Peloritana Pericolanti, Classe Sci. Fis. Mat. Nat. *51*, 99 (1971)
282   Cromer, R. E., Seff, K.: Acta Crystallogr. *B28*, 3281 (1972)
283   Cullen, D. L., Lingafelter, E. C.: Inorg. Chem. *9*, 1858 (1970)
284   Cullen, D. L., Lingafelter, E. C.: Inorg. Chem. *9*, 1865 (1970)
285   Cullen, D. L., Lingafelter, E. C.: Inorg. Chem. *10*, 1264 (1971)
286   Cullen, W. R., Mihichuk, L. M.: Can. J. Chem. *54*, 2548 (1976)
287   Cunningham, J. A., Sands, D. E., Wagner, W. F.: Inorg. Chem. *6*, 499 (1967)
288   Cunningham, J. A., Sands, D. E., Wagner, W. F., Richardson, M. F.: Inorg. Chem. *8*, 22 (1969)
289   Cunningham, J. A., Sievers, R. E.: Inorg. Chem. *19*, 595 (1980)
290   Dahan, F., Jeannin, Y.: J. Organomet. Chem. *136*, 251 (1977)
291   Dahlen, B.: Acta Crystallogr. *B30*, 647 (1974)
292   Dahlen, B., Lindgren, B.: Acta Chem. Scand. *A33*, 403 (1979)
293   Daly, J. J.: J. Chem. Soc., 3799 (1964)
294   Daly, J. J., Sanz, F.: J. Chem. Soc., Dalton Trans., 2584 (1972)
295   Daly, J. J., Sanz, F., Sneedon, R. P. A., Zeiss, H. H.: J. Chem. Soc., Dalton Trans., 73 (1973)
296   Dapporto, P., DiVaira, M.: J. Chem. Soc. (A), 1891 (1971)
297   Dapporto, P., Mani, F., Mealli, C.: Inorg. Chem. *17*, 1323 (1978)
298   Darensbourg, D. J.: Inorg. Chem. *18*, 14 (1979)
299   Darensbourg, M. Y., Darensbourg, D. J., Barros, H. L. C.: Inorg. Chem. *17*, 297 (1978)
300   Dartiguenave, M., Dartiguenave, Y., Gleizes. A., Saint-Joly, C., Galy, J., Meier, P., Merbach, A. E.: Inorg. Chem. *17*, 3503 (1978)
301   Davies, G. R., Jarvis, J. A. J., Kilbourn, B. T., Mais, R. H. B., Owston, P. G.: J. Chem. Soc. (A), 1275 (1970)
302   Davis, A. R., Einstein, F. W. B.: Inorg. Chem. *13*, 1880 (1974)
303   Davis, A. R., Einstein, F. W. B.: Inorg. Chem. *14*, 3030 (1975)
304   Davis, A. R., Einstein, F. W. B.: Acta Crystallogr. *B34*, 2110 (1978)
305   Dawson, J. W., McLennan, T. J., Robinson, W., Merle, A., Dartiguenave, M., Dartiguenave, Y., Gray, H. B.: J. Am. Chem. Soc. *96*, 4428 (1974)
306   Day, V. W., Hoard, J. L.: J. Am. Chem. Soc. *90*, 3374 (1968)

307   Day, V. W., Hoard, J. L.: J. Am. Chem. Soc. *92*, 3626 (1970)
308   Deacon, G. B., Raston, C. L., Tunaley, D., White, A. H.: Aust. J. Chem. *32*, 2195 (1979)
309   De Almeida Santos, R. H., Mascarenhas, Y.: J. Coord. Chem. *9*, 59 (1979)
310   Degetto, S., Marangoni, G., Bombieri, G., Forsellini, E., Baracco, L., Graziani, R.: J. Chem. Soc., Dalton Trans., 1933 (1974)
311   Delaunay, J., Kappenstein, C., Hugel, R.: Acta Crystallogr. *B32*, 2341 (1976)
312   Del Piero, G., Perego, G., Zazzetta, A., Brandi, G.: Cryst. Struct. Commun. *4*, 521 (1975)
313   De Simone, R. E., Stucky, G. D.: Inorg. Chem. *10*, 1808 (1971)
314   Dessy, G., Fares, V., Scaramuzza, L.: Acta Crystallogr. *B34*, 3066 (1978)
315   Dettlaf, G., Behrens, U., Weiss, E.: J. Organomet. Chem. *152*, 95 (1978)
316   Dewan, J. C., Edwards, A. J., Calves, J. Y., Guerchais, J. E.: J. Chem. Soc., Dalton Trans., 981 (1977)
317   Dewan, J. C., Henrick, K., Kepert, D. L., Trigwell, K. R., White, A. H., Wild, S. B.: J. Chem. Soc., Dalton Trans., 546 (1975)
318   Dewan, J. C., Kepert, D. L., Maslen, E. N., Raston, C. L., Taylor, D., White, A. H.: J. Chem. Soc., Dalton Trans., 2082 (1973)
319   Dewan, J. C., Kepert, D. L., Raston, C. L., White, A. H.: J. Chem. Soc., Dalton Trans., 2031 (1975)
320   Dewan, J. C., Lippard, S. J., et al.: personal communication (1980)
321   De Wet, J. F., Du Preez, J. G. H.: J. Chem. Soc., Dalton Trans., 592 (1978)
322   Diakiw, V., Hambley, T. W., Kepert, D. L., Raston, C. L., White, A. H.: Aust. J. Chem. *32*, 301 (1979)
323   Diamantis, A. A., Snow, M. R., Vanzo, J. A.: J. Chem. Soc., Chem. Commun., 264 (1976)
324   Dietzsch, W., Sieler, J., Glowiak, T.: Z. anorg. Chem. *435*, 225 (1977)
325   Dirand, J., Ricard, L., Weiss, R.: J. Chem. Soc., Dalton Trans., 278 (1976)
326   Di Sipio, L., Tondello, E., Pelizzi, G., Ingletto, G., Montenero, A.: Cryst. Struct. Commun. *3*, 297, 301, 527, 731 (1974); *6*, 723 (1977)
327   Di Vaira, M.: J. Chem. Soc. (A), 148 (1971)
328   Di Vaira, M., Orioli, P. L.: Inorg. Chem. *8*, 2729 (1969)
329   Dixon, D. A., Marsh, R. E., Schaefer, W. P.: Acta Crystallogr. *B34*, 807 (1978)
330   Doddrell, D. M., Bendall, M. R., Healy, P. C., Smith, G., Kennard, C. H. L., Raston, C. L., White, A. H.: Aust. J. Chem. *32*, 1219 (1979)
331   Dodge, R. P., Templeton, D. H., Zalkin, A.: J. Chem. Phys. *35*, 55 (1961)
332   Domiano, P., Tiripicchio, A.: Cryst. Struct. Commun. *1*, 107 (1972)
333   Donaldson, J. D., Donoghue, M. T., Smith, C. H.: Acta Crystallogr. *B32*, 2098 (1976)
334   Donohue, J.: Acta Crystallogr. *18*, 1018 (1965)
335   Donohue, J., Caron, A.: Acta Crystallogr. *17*, 663 (1964)
336   Dori, Z., Eisenberg, R., Gray, H. B.: Inorg. Chem. *6*, 483 (1967)
337   Drager, M.: Chem. Ber. *108*, 1723 (1975)
338   Drager, M.: Z. anorg. Chem. *423*, 53 (1976)
339   Drager, M., Engler, M.: Chem. Ber. *108*, 17 (1975)
340   Drager, M., Engler, M.: Z. anorg. Chem. *413*, 229 (1975)
341   Drew, M. G. B.: J. Chem. Soc., Dalton Trans., 626 (1972)
342   Drew, M. G. B.: J. Chem. Soc., Dalton Trans., 1329 (1972)
343   Drew, M. G. B., Davis, K. M., Edwards, D. A., Marshalsea, J.: J. Chem. Soc., Dalton Trans., 1098 (1972)
344   Drew, M. G. B., Egginton, G. M., Wilkins, J. D.: Acta Crystallogr. *B30*, 1895 (1974)
345   Drew, M. G. B., Eve, D. J.: Acta Crystallogr. *B33*, 2919 (1977)
346   Drew, M. G. B., Goodgame, D. M. L., Hitchman, M. A., Rogers, D.: Proc. Chem. Soc., 363 (1964)
347   Drew, M. G. B., Hollis, S.: Acta Crystallogr. *B34*, 2853 (1978)
348   Drew, M. G. B., Mandyczewsky, R.: J. Chem. Soc., Chem. Commun., 292 (1970)
349   Drew, M. G. B., Mandyczewsky, R.: J. Chem. Soc. (A), 2815 (1970)
350   Drew, M. G. B., Mitchell, M. G. B., Pygall, C. F.: J. Chem. Soc., Dalton Trans., 1071 (1977)

351 Drew, M. G. B., Pygall, C. F.: Acta Crystallogr. *B33*, 2838 (1977)
352 Drew, M. G. B., Rix, C. J.: J. Organomet. Chem. *102*, 467 (1975)
353 Drew, M. G. B., Wilkins, J. D.: J. Chem. Soc., Dalton Trans., 1830 (1973)
354 Drew, M. G. B., Wilkins, J. D.: J. Chem. Soc., Dalton Trans., 2664 (1973)
355 Drew, M. G. B., Wilkins, J. D.: J. Chem. Soc., Dalton Trans., 198 (1974)
356 Drew, M. G. B., Wilkins, J. D.: J. Chem. Soc., Dalton Trans., 1579 (1974)
357 Drew, M. G. B., Wilkins, J. D.: J. Chem. Soc., Dalton Trans., 1973 (1974)
358 Drew, M. G. B., Wilkins, J. D.: J. Organomet. Chem. *69*, 271 (1974)
359 Drew, M. G. B., Wilkins, J. D.: Acta Crystallogr. *B31*, 177 (1975)
360 Drew, M. G. B., Wilkins, J. D.: Acta Crystallogr. *B31*, 2642 (1975)
361 Drew, M. G. B., Wilkins, J. D.: J. Chem. Soc., Dalton Trans., 2611 (1975)
362 Drew, M. G. B., Wilkins, J. D., Wolters, A. P.: J. Chem. Soc., Chem. Commun.,1278 (1972)
363 Drew, M. G. B., Wolters, A. P., J. Chem. Soc., Chem. Commun., 457 (1972)
364 Drew, M. G. B., Wolters, A. P.: Acta Crystallogr. *B33*, 205 (1977)
365 Drew, M. G. B., Wolters, A. P.: Acta Crystallogr. *B33*, 1027 (1977)
366 Drew, M. G. B., Wolters, A. P., Tomkins, I. M.: J. Chem. Soc., Dalton Trans., 974 (1977)
367 Drew, M. G. B., Wolters, A. P., Wilkins, J. D.: Acta Crystallogr. *B31*, 324 (1975)
368 Drew, R. E., Einstein, F. W. B.: Inorg. Chem. *11*, 1079 (1972)
369 Drew, R. E., Einstein, F. W. B., Gransden, S. E.: Can. J. Chem. *52*, 2184 (1974)
370 Drummond, J., Wood, J. S.: J. Chem. Soc., Chem. Commun., 1373 (1969)
371 Drummond, J., Wood, J. S.: J. Chem. Soc. (A), 226 (1970)
372 Durand, J., Cot, L., Galigne, J. L.: Acta Crystallogr. *B34*, 388 (1978)
373 Durand, J., Galigne, J. L., Cot, L.: Acta Crystallogr. *B33*, 1414 (1977)
374 Dymock, K., Palenik, G. J.: J. Chem. Soc., Chem. Commun., 884 (1973)
375 Dymock, K., Palenik, G. J.: Acta Crystallogr. *B30*, 1364 (1974)
376 Dymock, K., Palenik, G. J., Slezak, J., Raston, C. L., White, A. H.: J. Chem. Soc., Dalton Trans., 28 (1976)
377 Eaton, S. S., Eaton, G. R., Holm, R. H., Muetterties, E. L.: J. Am. Chem. Soc. *95*, 1116 (1973)
378 Eaton, S. S., Hutchison, J. R., Holm, R. H., Muetterties, E. L.: J. Am. Chem. Soc. *94*, 6411 (1972)
379 Edwards, A. J.: J. Chem. Soc. (A), 2751 (1970)
380 Edwards, A. J.: J. Chem. Soc., Dalton Trans., 582 (1972)
381 Edwards, A. J.: J. Chem. Soc., Dalton Trans., 1723 (1978)
382 Edwards, A. J., Jones, G. R.: J. Chem. Soc. (A), 1491 (1970)
383 Edwards, A. J., Jones, G. R.: J. Chem. Soc. (A), 1891 (1970)
384 Einstein, F. W. B., Enwall, E., Morris, D. M., Sutton D.: Inorg. Chem. *10*, 678 (1971)
385 Einstein, F. W. B., Field, J. S.: J. Chem. Soc., Chem. Commun., 1628 (1975)
386 Einstein, F. W. B., Huang, C. H.: Acta Crystallogr. *B34*, 1486 (1978)
387 Einstein, F. W. B., Penfold, P. R.: Acta Crystallogr. *20*, 924 (1966)
388 Einstein, F. W. B., Penfold, P. R.: J. Chem. Soc. (A), 3019 (1968)
389 Eisenberg, R., Gray, H. B.: Inorg. Chem. *6*, 1844 (1967)
390 Eisenberg, R., Ibers, J. A.: Inorg. Chem. *5*, 411 (1966)
391 Eisenhut, M., Mitchelle, H. L., Traficante, D. D., Kaufman, R. J., Deutsch, J. M., Whitesides, G. M.: J. Am. Chem. Soc., *96*, 5385 (1974)
392 Eisenhut, M., Schmutzler, R., Sheldrick, W. S.: J. Chem. Soc., Chem Commun., 144 (1973)
393 Elder, R. C.: Inorg. Chem. *7*, 1117 (1968)
394 Elder, R. C.: Inorg. Chem. *7*, 2316 (1968)
395 Elder, R. C., Heeg, M. J., Deutsch, E.: Inorg. Chem. *17*, 427 (1978)
396 Elder, R. C., Heeg, M. J., Payne, M. D., Trkula, M., Deutsch, E.: Inorg. Chem. *17*, 431 (1978)
397 Elder, R. C., Koran, D., Mark, H. B.: Inorg. Chem. *13*, 1644 (1974)
398 Elder, R. C., Marcuso, T., Boolchand, P.: Inorg. Chem. *16*, 2700 (1977)
399 Elding, I.: Acta Chem. Scand. *A 30*, 649 (1976)

400   Elema, R. J., de Boer, J. L., Vos, A.: Acta Crystallogr. *16*, 243 (1963)
401   Eller, P. G., Corfield, P. W. R.: J. Chem. Soc., Chem. Commun., 105 (1971)
402   Emerson, K., Ireland, P. R., Robinson, W. T.: Inorg. Chem. *9*, 436 (1970)
403   Enckevort, W. J. P. van, Hendriks, H. M., Beurskens, P. T.: Cryst. Struct. Commun.
      6, 531 (1977)
404   Enemark, J. H., Feltham, R. D.: J. Chem. Soc., Dalton Trans., 718 (1972)
405   Enemark, J. H., Feltham, R. D., Huie, B. T., Johnson, P. L., Swedo, K. B.: J. Am.
      Chem. Soc. *99*, 3285 (1977)
406   Enemark, J. H., Feltham, R. D., Riker-Nappier, J., Bizot, K. F.: Inorg. Chem. *14*,
      624 (1975)
407   Enemark, J. H., Quinby, M. S., Reed, L. L., Steuck, M. J., Walthers, K. K.: Inorg.
      Chem. *9*, 2397 (1970)
408   Engberts, J. B. F. N., Morssink, H., Vos, A.: J. Am. Chem. Soc. *100*, 799 (1978)
409   English, A. D., Ittel, S. D., Tolman, C. A., Meakin, P., Jesson, J. P.: J. Am. Chem.
      Soc. *99*, 117 (1977)
410   Enjalbert, R., Galy, J.: Compt. rend. *C287*, 259 (1978)
411   Enjalbert, R., Galy, J.: Acta Crystallogr. *B35*, 546 (1979)
412   Epstein, E. F., Bernal, I.: J. Chem. Soc. (A), 3628 (1971)
413   Epstein, E. F., Bernal, I.: Inorg. Chim. Acta *25*, 145 (1977)
414   Epstein, J. M., Dewan, J. C., Kepert, D. L., White, A. H.: J. Chem. Soc., Dalton
      Trans., 1949 (1974)
415   Erasmus, C. S., Boeyens, J. C. A.: J. Cryst. Mol. Struct. *1*, 83 (1971)
416   Esperas, S., Husebye, S.: Acta Chem. Scand. *26*, 3293 (1972)
417   Esperas, S., Husebye, S.: Acta Chem. Scand. *A29*, 185 (1975)
418   Estes, E. D., Hodgson, D. J.: Inorg. Chem. *12*, 2932 (1973)
419   Ewings, P. F. R., Harrison, P. G., King, T. J.: J. Chem. Soc., Dalton Trans., 1455
      (1975)
420   Ewings, P. F. R., Harrison, P. G., King, T. J.: J. Chem. Soc., Dalton Trans., 1399 (1976)
421   Fackler, J. P., Avdeef, A.: Inorg. Chem. *13*, 1864 (1972)
422   Fackler, J. P., Avdeef, A., Fischer, R. G.: J. Am. Chem. Soc. *95*, 774 (1973)
423   Faggiani, R., Brown, I. D.: Acta Crystallogr. *B34*, 1675 (1978)
424   Fanfani, L., Nunzi, A., Zanazzi, P. F., Zanzari, A. R.: Acta Crystallogr. *B28*, 1298
      (1972)
425   Favas, M. C., Kepert, D. L.: J. Chem. Soc., Dalton Trans., 793 (1978)
426   Favas, M. C., Kepert, D. L.: Prog. Inorg. Chem. *27*, 325 (1980)
427   Favas, M. C., Kepert, D. L.: Prog. Inorg. Chem. *28*, 309 (1981)
428   Favas, M. C., Kepert, D. L.: unpublished
429   Favas, M. C., Kepert, D. L., Skelton, B. W., White, A. H.: J. Chem. Soc., Dalton
      Trans., 454 (1979)
430   Favas, M. C., Kepert, D. L., Skelton, B. W., White, A. H.: J. Chem. Soc., Dalton
      Trans., 447 (1980)
431   Favas, M. C., Kepert, D. L., White, A. H., Willis, A. C.: J. Chem. Soc., Dalton Trans.,
      1350 (1977)
432   Fenn, R. H., Graham, A. J., Johnson, N. P.: J. Chem. Soc. (A), 2880 (1971)
433   Fenton, D. E., Nave, C., Truter, M. R.: J. Chem. Soc., Dalton Trans., 2188 (1973)
434   Ferguson, G., Goel, R. G., Ridley, D. R.: J. Chem. Soc., Dalton Trans., 1288 (1975)
435   Ferrari, A., Corradi, A. B., Fava, G. G., Palmieri, C. G., Nardelli, M., Pelizzi, C.: Acta
      Crystallogr. *B29*, 1808 (1973)
436   Figgis, N. B., Kucharski, E. S., White, A. H.: Aust. J. Chem. *31*, 737 (1978)
437   Figgis, N. B., Raston, C. L., Sharma, R. P., White, A. H.: Aust. J. Chem. *31*, 2717
      (1978)
438   Figgis, B. N., Skelton, B. W., White, A. H.: Aust. J. Chem. *31*, 57 (1978)
439   Fischer, J., Elchinger, R., Weiss, R.: Acta Crystallogr. *B29*, 1967 (1973)
440   Fitzsimmons, B., Othen, D. G., Shearer, H. M. M., Wade, K., Whitehead, G.: J.
      Chem. Soc., Chem. Commun., 215 (1977)
441   Fletcher, S. R., Skapski, A. C.: J. Chem. Soc., Dalton Trans., 1079 (1972)
442   Fletcher, S. R., Skapski, A. C.: J. Organomet. Chem. *59*, 299 (1973)

443 Fletcher, S. R., Skapski, A. C.: J. Chem. Soc., Dalton Trans., 486 (1974)
444 Flynn, J. J., Boer, F. P.: J. Am. Chem. Soc., 91, 5756 (1969)
445 Forbes, E. J., Jones, D. L., Paxton, K., Hamer, T. A.: J. Chem. Soc., Dalton Trans., 879 (1979)
446 Form, G. E., Raper, E. S., Oughtred, R. E., Shearer, H. M. M.: J. Chem. Soc., Chem. Commun., 945 (1972)
447 Forsellini, E., Bombieri, G., Graziani, R., Zarli, B.: Inorg. Nucl. Chem. Letters 8, 461 (1972)
448 Fowles, G. W. A., Greene, P. T., Wood, J. S.: J. Chem. Soc., Chem. Commun., 971 (1967)
449 Fowles, G. W. A., Lester, T. E., Wood, J. S.: J. Inorg. Nucl. Chem. 31, 657 (1969)
450 Foy, R. M., Kepert, D. L., Raston, C. L., White, A. H., J. Chem. Soc., Dalton Trans., 440 (1980)
451 Fraser, K. A., Harding, M. M.: Acta Crystallogr. 22, 75 (1967)
452 Frenz, B. A., Enemark, J. H., Ibers, J. A.: Inorg. Chem. 8, 1288 (1969)
453 Frenz, B. A., Ibers. J. A.: Inorg. Chem. 9, 2403 (1970)
454 Frenz, B. A., Ibers, J. A.: Inorg. Chem. 11, 1109 (1972)
455 Frost, G. H., Hart, F. A., Heath, C., Hursthouse, M. B.: J. Chem. Soc., Chem. Commun., 1421 (1969)
456 Furlani, C., Tomlinson, A. A. G., Porta, P., Sgamellotti, A.: J. Chem. Soc. (A), 2929 (1970)
457 Furst, W., Gouzerh, P., Jeannin, Y.: J. Coord. Chem. 8, 237 (1979)
458 Gahan, B., Garner, D. C., Hill, L. H., Mabbs, F. E., Hargrave, K. D., McPhail, A. T. J. Chem. Soc., Dalton Trans., 1726 (1977)
459 Galdecki, Z., Glowka, M. L., Golinski, B.: Acta Crystallogr. B32, 2319 (1976)
460 Galesic, M., Brnicevic, N., Matkovic, B., Herzog, M., Zelenko, B., Sljukic, M., Prelesnik, B., Herak, R.: J. Less-Common Met. 51, 259 (1977)
461 Garner, C. D., Hill, L. H., Mabbs, F. E., McFadden, D. L., McPhail, A. T.: J. Chem. Soc., Dalton Trans., 853 (1977)
462 Garner, C. D., Howlader, N. C., Mabbs, F. E., McPhail, A. T., Miller, R. W., Onan, K. D.: J. Chem. Soc., Dalton Trans., 1582 (1978)
463 Garner, C. D., Sutton, D., Wallwork, S. C.: J. Chem. Soc. (A), 1949 (1967)
464 Garner, C. D., Wallwork, S. C.: J. Chem. Soc. (A), 1496 (1966)
465 Garner, C. D., Wallwork, S. C.: J. Chem. Soc. (A), 3092 (1970)
466 Geller, S., Dudley, T. O.: J. Solid State Chem. 26, 321 (1978)
467 Gentile, P. S., Dinstein, M. P., White, J. G.: Inorg. Chim. Acta 19, 67 (1976)
468 Gentile, P. S., Ocampo, A. P.: Inorg. Chim. Acta 29, 83 (1978)
469 Gentile, P. S., White, J. G., Dinstein, M. P., Bray, D. D.: Inorg. Chim. Acta 21, 141 (1977)
470 Gentile, P. S., White, J., Haddad, S.: Inorg. Chim. Acta 13, 149 (1975)
471 Gibler, D. D., Adams, C. J., Fischer, M., Zalkin, A., Bartlett, N.: Inorg. Chem. 11, 2325 (1972)
472 Gilje, J. W., Braun, R. W., Cowley, A. H.: J. Chem. Soc., Chem. Commun., 15 (1974)
473 Gillespie, R. J.: Molecular Geometry, New York: Van Nostrand Reinhold, 1972
474 Gillespie, R. J., Martin, D., Schrobilgen, G. J., Slim, D. R.: J. Chem. Soc., Dalton Trans., 2234 (1977)
475 Gillespie, R. J., Nyholm, R. S.: Quart. Rev. Chem. Soc. 11, 339 (1957)
476 Gillespie, R. J., Quail, J. W.: Can. J. Chem. 42, 2671 (1964)
477 Ginderlow, D.: Acta Crystallogr. B30, 2798 (1974)
478 Ginsberg, A. P., Tully, M. E.: J. Am. Chem. Soc. 95, 4749 (1973)
479 Given, K. W., Mattson, B. W., Pignolet, L. H.: Inorg. Chem. 15, 3152 (1976)
480 Glaser, J.: Acta Chem. Scand. A33, 789 (1979)
481 Glavan, K. A., Whittle, R., Johnson, J. F., Elder, R. C., Deutsch, E.: J. Am. Chem. Soc. 102, 2103 (1980)
482 Glen, G. L., Silverton, J. V., Hoard, J. L.: Inorg. Chem. 2, 250 (1963)
483 Goebel, C. V., Doedens, R. J.: Inorg. Chem. 10, 2607 (1971)
484 Goldberg, S. Z., Eisenberg, R.: Inorg. Chem. 15, 58 (1976)

485 Goldfield, S. A., Raymond, K. N.: Inorg. Chem. *10*, 2604 (1971)
486 Goldfield, S. A., Raymond, K. N.: Inorg. Chem. *13*, 770 (1974)
487 Goldschmied, E., Stephenson, N. C.: Acta Crystallogr. *B26*, 1867 (1970)
488 Goldwhite, H., Grey, J., Teller, R.: J. Organomet. Chem. *113*, Cl (1976)
489 Goldwhite, H., Teller, R. G.: J. Am. Chem. Soc. *100*, 5357 (1978)
490 Gottardi, G.: Z. Krist. *115*, 451 (1961)
491 Gould, R. O., Gunn, A. M., van den Hark, T. E. M.: J. Chem. Soc., Dalton Trans., 1713 (1976)
492 Gould, R. O., Jones, C. L., Savage, W. J., Stephenson, T. A.: J. Chem. Soc., Dalton Trans., 908 (1976)
493 Grandjean, D., Weiss, R.: Bull. Soc. Chim. Fr., 3044 (1967)
494 Graver, H., Husebye, S.: Acta Chem. Scand. *A 29*, 14 (1975)
495 Graziani, R., Albertin, G., Forsellini, E., Orio, A. A.: Inorg. Chem. *15*, 2422 (1976)
496 Graziani, R., Bombieri, G., Forsellini, E.: J. Chem. Soc., Dalton Trans., 2059 (1972)
497 Graziani, R., Bombieri, G., Forsellini, E., Paolucci, G.: J. Cryst. Mol. Struct. *5*, 1 (1975)
498 Graziani, R., Marongoni, G., Paolucci, G., Forsellini, E.: J. Chem. Soc., Dalton Trans., 818 (1978)
499 Graziani, R., Zarli, B., Cassol, A., Bombieri, G., Forsellini, E., Tondello, E.: Inorg. Chem. *9*, 2116 (1970)
500 Greaney, T. M., Raston, C. L., White, A. H., Maslen, E. N.: J. Chem. Soc., Dalton Trans., 876 (1975)
501 Greene, P. T., Bryan, R. F.: J. Chem. Soc. (A), 2549 (1971)
502 Greene, P. T., Orioli, P. L.: J. Chem. Soc. (A), 1621 (1969)
503 Greenwood, N. N., McDonald, W. S., Reed, D., Staves, J.: J. Chem. Soc., Dalton Trans., 1339 (1979)
504 Grenthe, I.: Acta Chem. Scand. *25*, 3721 (1971)
505 Griffith, W. P.: Coord. Chem. Revs. *17*, 177 (1975)
506 Griffith, W. P., Skapski, A. C., Woode, K. A., Wright, M. J.: Inorg. Chim. Acta *31*, L 413 (1978)
507 Guder, H. J., Schwarz, W., Weidlein, J., Widler, H. J., Hausen, H. D.: Z. Naturforsch. *31b*, 1185 (1976)
508 Guggenberger, L. G., Muetterties, E. L.: J. Am. Chem. Soc. *98*, 7221 (1976)
509 Gutierrez-Puebla, E., Vegas, A., Garcia-Blanco, S.: Acta Crystallogr. *B36*, 145 (1980)
510 Habenschuss, A., Spedding, F. H.: Cryst. Struct. Commun. *9*, 71, 157, 207, 213 (1980)
511 Hagen, K., Cross, V. R., Hedberg, K.: J. Mol. Struct. *44*, 187 (1978)
512 Hagihara, H., Watanabe, Y., Yamashita, S.: Acta Crystallogr. *B24*, 960 (1968)
513 Hagihara, H., Yamashita, S.: Acta Crystallogr. *21*, 350 (1966)
514 Haigh, J. M., Nassimbeni, L. R., Pauptit, R. A., Rodgers, A. L., Sheldrick, G. M.: Acta Crystallogr. *B32*, 1398 (1976)
515 Halfpenny, J., Small, R. W. H.: Acta Crystallogr. *B34*, 3758 (1978)
516 Hall, D., Holland, R. V.: Inorg. Chim. Acta *3*, 235 (1969)
517 Hall, D., Rae, A. D., Waters, T. N.: Acta Crystallogr. *22*, 258 (1967)
518 Hall, M., Sowerby, D. B.: J. Am. Chem. Soc. *102*, 628 (1980)
519 Hall, S. R. Kepert, D. L., Raston, C. L., White, A. H.: Aust. J. Chem. *30*, 1955 (1977)
520 Hambley, T. W., Kepert, D. L., Raston, C. L., White, A. H.: Aust. J. Chem. *31*, 2635 (1978)
521 Hambley, T. W., Raston, C. L., White, A. H.: Aust. J. Chem. *30*, 1965 (1977)
522 Hamilton, W. C., La Placa, S. J., Ramirez, F., Smith, C. P.: J. Am. Chem. Soc. *89*, 2268 (1967)
523 Hammershoi, A., Larsen, E., Larsen, S.: Acta Chem. Scand. *A 32*, 501 (1978)
524 Hamor, T. A., Watkin, D. J.: J. Chem. Soc., Chem. Commun., 440 (1969)
525 Hanson, A. W.: Acta Crystallogr. *15*, 930 (1962)
526 Harreld, C. S., Schlemper, E. O.: Acta Crystallogr. *B27*, 1964 (1971)
527 Harrison, P. G., King, T. J.: J. Chem. Soc., Dalton Trans., 2298 (1974)
528 Harrison, P. G., King, T. J., Healy, M. A.: J. Organomet. Chem. *182*, 17 (1979)

529 Harrison, P. G., King, T. J., Phillips, R. C.: J. Chem. Soc., Dalton Trans., 2317 (1976)
530 Harrison, P. G., King, T. J., Richards, J. A.: J. Chem. Soc., Dalton Trans., 826 (1975)
531 Harrison, P. G., Molloy, K., Phillips, R. C., Smith, P. J., Crowe, A. J.: J. Organomet. Chem. *160*, 421 (1978)
532 Harrison, W. D., Hathaway, B. J., Kennedy, D.: Acta Crystallogr. *B35*, 2301 (1979)
533 Harrowfield, J., Patrick, J. M., White, A. H.: unpublished data.
534 Hart, F. A., Hursthouse, M. B., Malik, K. M. A., Moorhouse, M.: J. Chem. Soc., Chem. Commun., 549 (1978)
535 Hartl, M., Rama, M., Simon, A., Deiseroth, H. J.: Z. Nat. *B34*, 1035 (1979)
536 Hartl, H., Schoner, J., Jander, J., Schulz, H.: Z. anorg. Chem. *413*, 61 (1975)
537 Hartsuiker, J. G., Wagner, A. J.: J. Chem. Soc. , Dalton Trans., 1425 (1978)
538 Hauge, S., Tysseland, M.: Acta Chem. Scand. *25*, 3072 (1971)
539 Haupt, H. J., Huber, F.: Z. anorg. Chem. *442*, 31 (1978)
540 Haupt, H. J., Huber, F., Preut, H.: Z. anorg. Chem. *422*, 255 (1976)
541 Hausen, H. D., Guder, H. J., Schwarz, W.: J. Organomet. Chem. *132*, 37 (1977)
542 Hauser, P. J., Bordner, J., Schreiner, A. F.: Inorg. Chem. *12*, 1347 (1973)
543 Hawley, D. M., Ferguson, G.: J. Chem. Soc. (A), 2059 (1968)
544 Hawley, D. M., Ferguson, G., Harris, G. S.: J. Chem. Soc., Chem. Commun., 111 (1966)
545 Hawthorne, S. L., Fay, R. C.: J. Am. Chem. Soc. *101*, 5268 (1979)
546 Haymore, B. L., Maata, E. A., Wentworth, R. A. D.: J. Am. Chem. Soc. *101*, 2063 (1979)
547 Healy, P. C., Sinn, E.: Inorg. Chem. *14*, 109 (1975)
548 Healy, P. C., White, A. H.: J. Chem. Soc., Dalton Trans., 1163 (1972)
549 Healy, P. C., White, A. H.: J. Chem. Soc., Dalton Trans., 1883 (1972)
550 Healy, P. C., White, A. H., Hoskins, B. F.: J. Chem. Soc., Dalton Trans., 1369 (1972)
551 Heath, C., Hursthouse, M. B.: J. Chem. Soc., Chem. Commun., 143 (1971)
552 Heenan, R. K., Robeitte, A. G.: J. Mol. Struct. *54*, 135 (1979)
553 Heenan, R. K., Robiette, A. G.: J. Mol. Struct. *55*, 191 (1979)
554 Heitsch, C. W., Nordmann, C. E., Parry, R. W.: Inorg. Chem. *2*, 508 (1963)
555 Henrick, K., Raston, C. L., White, A. H.: J. Chem. Soc., Dalton Trans., 26 (1976)
556 Herberhold, M., Wehrmann, F., Neugebauer, D., Huttner, D.: J. Organomet. Chem. *152*, 329 (1978)
557 Herceg, M., Fischer, J.: Acta Crystallogr. *B30*, 1289 (1974)
558 Hill, R. J., Rickard, C. E. F.: J. Inorg. Nucl. Chem. *37*, 2481 (1975)
559 Hilton, J., Nunn, E. K., Wallwork, S. C.: J. Chem. Soc., Dalton Trans., 173 (1973)
560 Hilton, J., Wallwork, S. C.: J. Chem. Soc., Chem. Commun., 871 (1968)
561 Hinamoto, M., Ooi, S., Kuroya, H.: Bull. Chem. Soc. Jap. *44*, 586 (1971)
562 Hirota, E., Morino, Y.: J. Mol. Spectrosc. *33*, 460 (1970)
563 Hoard, J. L., Hamor, T. A., Glick, M. D.: J. Am. Chem. Soc. *90*, 3177 (1968)
564 Hoard, J. L., Martin, W. J., Smith, M. E., Whitney, J. F.: J. Am. Chem. Soc. *76*, 3820 (1954)
565 Hoard, J. L., Silverton, J. V.: Inorg. Chem. *2*, 235 (1963)
566 Hodgson, D. J., Hale, P. K., Hatfield, W. E.: Inorg. Chem. *10*, 1061 (1979)
567 Holinski, R., Brehler, B.: Acta Crystallogr. *B26*, 1915 (1970)
568 Hollander, F. J., Pedelty, R., Coucouvanis, D.: J. Am. Chem. Soc. *96*, 4032 (1974)
569 Hollander, F. J., Templeton, D. H., Zalkin, A.: Acta Crystallogr. *B29*, 1289 (1973)
570 Holmes, R. R.: J. Am. Chem. Soc. *96*, 4143 (1974)
571 Hon, P. K., Belford, R. L., Pfluger, C. E.: J. Chem. Phys. *43*, 3111 (1965)
572 Hon, P. K., Pfluger, C. E.: J. Coord. Chem. *3*, 67 (1973)
573 Hope, H., Knobler, C., McCullough, J. D.: Inorg. Chem. *12*, 2665 (1973)
574 Horrocks, W. D., Sipe, J. P., Luber, J. R.: J. Am. Chem. Soc. *93*, 5258 (1971)
575 Horrocks, W. D., Templeton, D. H., Zalkin, A.: Inorg. Chem. *7*, 1552 (1968)
576 Hoskins, B. F., Kelly, B. P.: J. Chem. Soc., Chem. Commun., 1517 (1968)

212   References

577 Hoskins, B. F., Kelly, B. P.: J. Chem. Soc., Chem. Commun., 45 (1970)
578 Hoskins, B. F., Kelly, B. P.: Inorg. Nucl. Chem. Lett. *8*, 875 (1972)
579 Hoskins, B. F., Martin, R. L., Rohde, N. M.: Aust. J. Chem. *29*, 213 (1976)
580 Hoskins, B. F., Pannan, C. D.: Aust. J. Chem. *29*, 2337 (1976)
581 Hoskins, B. F., White, A. H.: J. Chem. Soc. (A), 1668 (1970)
582 Houttemane, C., Boivin, J. C., Thomas, D. J., Wozniak, M., Nowogrock, G.: Acta Crystollogr. *B35*, 2033 (1979)
583 Hsu, B., Schlemper, E. O.: Acta Crystallogr. *B34*, 930 (1978)
584 Htoon, S., Ladd, M. F. C.: J. Cryst. Mol. Struct. *3*, 95 (1973)
585 Htoon, S., Ladd, M. F. C.: J. Cryst. Mol. Struct. *4*, 357 (1974)
586 Hubbard, C. R., Quicksall, C. O., Jacobson, R. A.: Acta Crystallogr. *B30*, 2613 (1974)
587 Huber-Buser, E.: Cryst. Struct. Commun. *4*, 731 (1975)
588 Hunter, S. H., Nyholm, R. S., Rodley, G. A.: Inorg. Chim. Acta *3*, 631 (1969)
589 Huq, F., Skapski, A. C.: J. Chem. Soc. (A), 1927 (1971)
590 Hurst, H. J., Taylor, J. C.: Acta Crystallogr. *B26*, 417 (1970)
591 Hursthouse, M. B., Malik, M. K. A., Davies, J. E., Harding, J. H.: Acta Crystallogr. *B34*, 1355 (1978)
592 Hursthouse, M. B., Motevalli, M.: J. Chem. Soc., Dalton Trans., 1362 (1979)
593 Hursthouse, M. B., New, D. B.: J. Chem. Soc., Dalton Trans., 1082 (1977)
594 Hursthouse, M. B., Steer, I. A.: J. Organomet. Chem. *27*, C11 (1971)
595 Husebye, S.: Acta Chem. Scand. *21*, 42 (1967)
596 Husebye, S.: Acta Chem. Scand. *24*, 2198 (1970)
597 Husebye, S.: Acta Chem. Scand. *A33*, 485 (1979)
598 Husebye, S., Helland--Madsen, G.: Acta Chem. Scand. *24*, 2273 (1970)
599 Husebye, S., Svaeren, S. E.: Acta Chem. Scand. *27*, 763 (1973)
600 Huttner, G., Gartzke, W.: Chem. Ber. *105*, 2714 (1972)
601 Huttner, G., Gartzke, W.: Chem. Ber. *108*, 1373 (1975)
602 Hyde, J., Magin, L., Zubieta, J.: J. Chem. Soc., Chem. Commun., 204 (1980)
603 Iball, J., Morgan, C. H.: Acta Crystallogr. *23*, 239 (1967)
604 Ibers, J. A., Hamilton, W. C.: Science *139*, 106 (1963)
605 Il'inskii, A. L., Aslanov, L. A., Ivanov, V. I., Khalilov, A. D., Petrukhin, O. M.: J. Struct. Chem. *10*, 263 (1969)
606 Il'inskii, A. L., Porai-Koshits, M. A., Aslonov, L. A., Lazarev, P. I.: J. Struct. Chem. *13*, 254 (1972)
607 Ito, M., Iwasaki, H.: Acta Crystallogr. *B36*, 443 (1980)
608 Ito, T.: Acta Crystallogr. *B28*, 1034 (1972)
609 Iwasaki, H., Hagihara, H.: Acta Crystallogr. *B28*, 507 (1972)
610 Jagner, S., Ljungstrom, E.: Acta Crystallogr. *B34*, 653 (1978)
611 Jansen, P. R., Oskam, A., Olie, K.: Cryst. Struct. Commun., *4*, 667 (1975)
612 Jarvis, J. A. J., Mais, R. H. B., Owston, P. G.: J. Chem. Soc. (A), 1473 (1968)
613 Jarvis, J. A. J., Mais, R. H. B., Owston, P. G., Taylor, K. A.: J. Chem. Soc., Chem. Commun., 906 (1966)
614 Jarvis, J. A. J., Mais, R. H. B., Owston, P. G., Thompson, D. T.: J. Chem. Soc. (A), 622 (1968)
615 Jesson, J. P., Meakin, P.: J. Am. Chem. Soc. *96*, 5760 (1974)
616 Joesten, M. D., Hussain, M. S., Lenhert, P. G.: Inorg. Chem. *9*, 151 (1970)
617 Joesten, M. D., Takagi, S., Lenhert, P. G.: Inorg. Chem. *16*, 2680 (1977)
618 Johansson, G. B., Lindqvist, O.: Acta Crystallogr. *B34*, 2959 (1978)
619 Johansson, L., Molund, M., Oskarsson, A.: Inorg. Chim. Acta *31*, 117 (1978)
620 Johnson, D. A., Taylor, J. C., Waugh, A. B.: J. Inorg. Nucl. Chem. *41*, 827 (1979)
621 Johnson, D. E., Jacobson, R. A.: J. Chem. Soc., Dalton Trans., 580 (1973)
622 Johnson, N. W.: Canad. J. Math. *18*, 169 (1966)
623 Jones, G. R., Burbank, R. D., Bartlett, N.: Inorg. Chem. *9*, 2264 (1970)
624 Jones, P. E., Katz, L.: Acta Crystallogr. *B25*, 745 (1969); *B28*, 3438 (1972)
625 Joy, G., Gaughan, A. P., Wharf, I., Shriver, D. F., Dougherty, J. P.: Inorg. Chem. *14*, 1795 (1975)
626 Jumas, J. C., Maurin, M., Philippot, E.: J. Fluorine Chem. *10*, 219 (1977)

627 Jurnak, F. A., Greig, D. R., Raymond, K. N.: Inorg. Chem. *14*, 2585 (1975)
628 Jurnak, F. A., Raymond, K. N.: Inorg. Chem. *13*, 2387 (1974)
629 Kalman, A., Sasvari, K., Kapovits, I.: Acta Crystallogr. *B29*, 355 (1973)
630 Kamenar, B., Penavic, M., Prout, C. K.: Cryst. Struct. Commun. *2*, 41 (1973)
631 Kamenar, B., Prout, C. K.: J. Chem. Soc. (A), 2379 (1970)
632 Kanazawa, Y., Matsumoto, T.: Acta Crvstallogr. *B32*, 282 (1976)
633 Kauffman, G. B.: *Classics in Coordination Chemistry, Part 1: The Selected Papers of Alfred Werner*. New York: Dover, 1968
634 Kawamura, K., Kawahara, A.: Acta Crystallogr. *B32*, 2419 (1976)
635 Keene, F. R., Searle, G. H.: Inorg. Chem. *13*, 2173 (1974)
636 Kepert, D. L.: J. Chem. Soc., 4736 (1965)
637 Kepert, D. L.: Prog. Inorg. Chem. *23*, 1 (1977)
638 Kepert, D. L.: Prog. Inorg. Chem. *24*, 179 (1978)
639 Kepert, D. L.: Prog. Inorg. Chem. *25*, 41 (1979)
640 Kepert, D. L., Kucharski, E. S., White, A. H.: J. Chem. Soc., Dalton Trans., 1932 (1980)
641 Kepert, D. L., Raston, C. L., Roberts, N. K., White, A. H.: Aust. J. Chem. *31*, 1927 (1978)
642 Kepert, D. L., Raston, C. L., White, A. H., Petridis, D.: J. Chem. Soc., Dalton Trans., 1921 (1980)
643 Kepert, D. L., Raston, C. L., White, A. H., Winter, G.: Aust. J. Chem. *31*, 757 (1978)
644 Kepert, D. L., Skelton, B. W., White, A. H.: J. Chem. Soc., Dalton Trans., 652 (1981)
645 Kerr, K. A., Boorman, P. M., Misener, B. S., van Roode, J. G. H.: Can. J. Chem. *55*, 3081 (1977)
646 Kilbourn, B. T., Blundell, T. L., Powell, H. M.: J. Chem. Soc., Chem. Commun., 444 (1965)
647 Kilbourn, B. T., Raeburn, U. A., Thompson, D. T.: J. Chem. Soc. (A), 1906 (1969)
648 Kimura, T., Yasuoka, N., Kasai, N., Kakudo, M.: Bull. Chem. Soc. Jap. *45*, 1649 (1972)
649 King, R. B.: J. Am. Chem. Soc. *92*, 6455 (1970)
650 King, T. J., Logan, N., Morris, A., Wallwork, S. C.: J. Chem. Soc., Chem. Commun., 554 (1971)
651 King, T. J., Morris, A.: Inorg. Nucl. Chem. Lett. *10*, 237 (1974)
652 Klanberg, F., Muetterties, E. L.: Inorg. Chem. *7*, 155 (1968)
653 Klanderman, W. A., Hamilton, W. C., Bernal, I.: Inorg. Chim. Acta *23*, 117 (1977)
654 Knobler, C., McCullough, J. D.: Inorg. Chem. *11*, 3026 (1972)
655 Knobler, C., McCullough, J. D.: Inorg. Chem. *16*, 612 (1977)
656 Knobler, C., McCullough, J. D., Hope, H.: Inorg. Chem. *9*, 797 (1970)
657 Knobler, C., Ziolo, R. F.: J. Organomet. Chem. *178*, 423 (1979)
658 Knopp, B., Lorcher, K. P., Strahle, J.: Z. Naturforsch. *B23*, 1361 (1971)
659 Knox, K., Ginsberg, A. P.: Inorg. Chem. *3*, 555 (1964)
660 Kobayashi, A., Ito, T., Marumo, F., Saito, Y.: Acta Crystallogr. *B28*, 3446 (1972)
661 Kobayashi, A., Marumo, F., Saito, Y.: Acta Crystallogr. *B28*, 2709 (1972)
662 Kobayashi, M., Marumo, F., Saito, Y.: Acta Crystallogr. *B28*, 470 (1972)
663 Koda, S., Ooi, S., Kuroya, H., Isobe, K., Nakamura, Y., Kawaguchi, S.: J. Chem. Soc., Chem. Commun., 1321 (1971)
664 Kojic-Prodic, B., Liminga, R., Scavnicar, S.: Acta Crystallogr. *B29*, 864 (1963)
665 Kojic-Prodic, B., Ruzic-Toros, Z., Grdenis, D., Golic, L.: Acta Crystallogr. *B30*, 300 (1974)
666 Kojic-Prodic, B., Ruzic-Toros, Z., Sljukic, M.: Acta Crystallogr. *B34*, 2001 (1978)
667 Komiya, S., Yamamoto, T., Yamamoto, A., Takenaka, A., Sasada, Y.: Acta Crystallogr. *B35*, 2702 (1979)
668 Konaka, S.: Bull. Chem. Soc. Jap. *43*, 3107 (1970)
669 Konaka, S., Kimura, M.: Bull. Chem. Soc. Jap. *43*, 1693 (1970)
670 Konaka, S., Kimura, M.: Bull. Chem. Soc. Jap. *46*, 404 (1973)
671 Konaka, S., Kimura, M.: Bull. Chem. Soc. Jap. *46*, 413 (1973)
672 Kono, M., Marumo, F., Saito, Y.: Acta Crystallogr. *B29*, 739 (1973)

673   Kopf, J., Schmidt, J.: Z. Naturforsch. *B32*, 275 (1977)

674   Kopwillem, A.: Acta Chem. Scand. *26*, 2941 (1972)

675   Kosheim, K., Foss, O., Scheie, A., Solheimsnes, S.: Acta Chem. Scand. *19*, 2336 (1965)

676   Krigbaum, W. R., Rubin, B.: Acta Crystallogr. *B29*, 749 (1973)

677   Kruger, C.: Chem. Ber. *106*, 3230 (1973)

678   Kruger, G. J., Reynhardt, E. C.: Acta Crystallogr. *B30*, 822 (1974)

679   Kuchitsu, K., Shibata, T., Yokozeki, A., Matsumara, C.: Inorg. Chem. *10*, 2584 (1971)

680   Kukushkin, Yu. N.: Russ. J. Inorg. Chem. *21*, 481 (1976)

681   Kuncher, N. R., Truter, M. R.: J. Chem. Soc., 3478 (1958)

682   Kuroda, R., Mason, S. F.: J. Chem. Soc., Dalton Trans., 273 (1979)

683   Laidoudi, A., Kheddar, N., Brianse, M. C.: Acta Crystallogr. *B34*, 778 (1978)

684   Laidoudi, A., Kheddar, N., Brianse, M. C.: Acta Crystallogr. *B34*, 782 (1978)

685   Laing, M., Reiman, R. H., Singleton, E.: Inorg. Chem. *18*, 2666 (1979)

686   Lalancetto, R. A., Cefola, M., Hamilton, W. C., La Placa, S. J.: Inorg. Chem. *6*, 2127 (1967)

687   Lam, C. T., Corfield, P. W. R., Lippard, S. J.: J. Am. Chem. Soc. *99*, 617 (1977)

688   Lam, C. T., Novotny, M., Lewis, D. L., Lippard, S. J.: Inorg. Chem. *17*, 2127 (1978)

689   Lance, E. T., Haschke, J. M., Peacor, D. R.: Inorg. Chem. *15*, 780 (1976)

690   Lanfranconi, A. H., Alvarez, A. G., Castellano, E. E.: Acta Crystallogr. *B29*, 1733 (1973)

691   Langford, G. R., Akhtar, M., Ellis, P. D., MacDiarmid, A. G., Odom, J. D.: Inorg. Chem. *14*, 2937 (1975)

692   La Placa, S. J., Ibers, J. A.: Inorg. Chem. *4*, 778 (1965)

693   Larking, I., Stomberg, R.: Acta Chem. Scand. *24*, 2043 (1970)

694   Larsen, K. P., Hazell, R. G., Toftlund, H., Andersen, P. R., Bisgard, P., Edlund, K., Eliasen, M., Herskind, K., Laursen, T., Pedersen, P. M.: Acta Chem. Scand. *A29*, 499 (1975)

695   Larsen, K. P., Toftlund, H.: Acta Chem. Scand. *A31*, 182 (1977)

696   Larsson, L. O., Kierkegaard, P.: Acta Chem. Scand. *23*, 2253 (1969)

697   La Rue, W. A., Lui, A. T., Filippo, J. S.: Inorg. Chem. *19*, 315 (1980)

698   Lau, C., Lynton, H., Passmore, J., Siew, P. Y.: J. Chem. Soc., Dalton Trans., 2535 (1973)

699   Lawton, S. L., Fuhrmeister, C. J., Haas, R. G., Jarman, C. S., Lohmeyer, F. G.: Inorg. Chem. *13*, 135 (1974)

700   Lazarev, P. I., Ionov, V. M., Aslanov, A. L., Porai-Koshits, M. A.: J. Struct. Chem. *14*, 151 (1973)

701   Leary, K., Templeton, D. H., Zalkin, A., Bartlett, N.: Inorg. Chem. *12*, 1726 (1973)

702   LeCarpentier, J. M., Schlupp, R., Weiss, R.: Acta Crystallogr. *B28*, 1278 (1972)

703   Leclaire, A.: Acta Crystallogr. *B30*, 2259 (1974)

704   Lee, J. S., Titus, D. D.: J. Cryst. Mol. Struct. *6*, 279 (1976)

705   Legendre, J. J., Girard, C., Huber, M.: Bull. Soc. Chim. Fr., 1998 (1971)

706   Leipoldt, J. G., Basson, S. S., Bok, L. D. C.: Inorg. Chim. Acta *44*, L99 (1980)

707   Leipoldt, J. G.,Bok, L. D. C., Basson, S. S., Laubscher, A. E., van Vollenhoven, J. S.: J. Inorg. Nucl. Chem. *39*, 301 (1977)

708   Leipoldt, J. G., Bok, L. D. C., Laubscher, A. E., Basson, S. S.: J. Inorg. Nucl. Chem. *37*, 2477 (1975)

709   Leipoldt, J. G., Coppens, P.: Inorg. Chem. *12*, 2269 (1973)

710   Leipoldt, J. G., Wessels, G. F. S., Bok, L. D. C.: J. Inorg. Nucl. Chem. *37*, 2487 (1975)

711   Lenner, M.: Acta Crystallogr. *B34*, 3770 (1978)

712   Lenner, M., Lindqvist, O.: Acta Crystallogr. *B35*, 600 (1979)

713   Lenner, M.: Acta Crystallogr. *B35*, 2396 (1979)

714   LeQuerler, J. F., Borel, M. M., Leclaire, A.: Acta Crystallogr. *B33*, 2299 (1977)

715   Levenson, R. A., Towns, R. L. R.: Inorg. Chem. *13*, 105 (1974)

716   Lewis, D. F., Fay, R. C.: J. Am. Chem. Soc. *96*, 3843 (1974)

717   Lewis, D. F., Fay, R. C.: Inorg. Chem. *15*, 2219 (1976)

718   Lewis, D. F., Lippard, S. J.: Inorg. Chem. *11*, 621 (1972)

719   Lewis, D. F., Lippard, S. J., Zubieta, J. A.: Inorg. Chem. *11*, 823 (1972)

720 Lewis, D. L., Lippard, S. J.: J. Am. Chem. Soc. *97*, 2697 (1975)
721 Li, T., Lippard, S. J.: Inorg. Chem. *13*, 1791 (1974)
722 Liebich, B. W., Tomassini, M.: Acta Crystallogr. *B34*, 944 (1978)
723 Lin, Y. C., Williams, D. E.: Can. J. Chem. *51*, 312 (1973)
724 Lindley, P. F., Carr, P.: J. Cryst. Mol. Struct. *4*, 173 (1974)
725 Lindley, P. F., Smith, A. W.: J. Chem. Soc., Chem. Commun., 1355 (1970)
726 Lindqvist, O.: Acta Chem. Scand. *21*, 1473 (1967)
727 Lipka, A.: Z. anorg. Chem. *440*, 224 (1978)
728 Lipka, A.: Acta Crystallogr. *B35*, 3020 (1979)
729 Lipka, A., Mootz, D.: Z. anorg. Chem. *440*, 217 (1978)
730 Lippard, S. J.: Prog. Inorg. Chem. *21*, 91 (1976)
731 Lis, T., Jezowska-Trzebiatowska, B.: Acta Crystallogr. *B33*, 1248 (1977)
732 Lis, T., Matuszewski, J., Jezowska-Trzebiatowska, B.: Acta Crystallogr. *B33*, 1943 (1977)
733 Littlefield, L. B., Doak, G. O.: J. Am. Chem. Soc. *98*, 7881 (1976)
734 Liu, C. F., Ibers, J. A.: Inorg. Chem. *9*, 773 (1970)
735 Lock, C. J. L., Murphy, C. N.: Acta Crystallogr. *B35*, 951 (1979)
736 Lock, C. J. L., Turner, G.: Acta Crystallogr. *B34*, 923 (1978)
737 Loghry, R. A., Simonsen, S. H.: Inorg. Chem. *17*, 1986 (1978)
738 Long, T. V., Herlinger, A. W., Epstein, E. F., Bernal, I.: Inorg. Chem. *9*, 459 (1970)
739 Louw, W. J., de Waal, D. J. A., Kruger, G. J.: J. Chem. Soc., Dalton Trans., 2364 (1976)
740 Lundberg, B. K. S.: Acta Crystallogr. *21*, 901 (1966)
741 Lynton, H., Sears, M. C.: Can. J. Chem. *49*, 3418 (1971)
742 Maata, E. A., Haymore, B. L., Wentworth, R. A. D.: Inorg. Chem. *19*, 1055 (1980)
743 McConnell, J. F., Schwartz, A.: Acta Crystallogr. *B28*, 1546 (1972)
744 McCullough, J. D.: Inorg. Chem. *12*, 2669 (1973)
745 McCullough, J. D.: Inorg. Chem. *14*, 1142 (1975)
746 McCullough, J. D., Knobler, C.: Inorg. Chem. *15*, 2728 (1976)
747 MacDonald, A. C., Sikka, S. K.: Acta Crystallogr. *B25*, 1804 (1969)
748 McFadden, D. L., McPhail, A. T., Garner, C. D., Mabbs, F. E.: J. Chem. Soc., Dalton Trans., 263 (1975)
749 Mackay, M. F., O'Connor, M. P., Oliver, P. J.: J. Cryst. Mol. Struct. *8*, 161 (1978)
750 McKee, D. E., Zalkin, A., Bartlett, N.: Inorg. Chem. *12*, 1713 (1973)
751 McPhail, A. T., Pui-Suen Wong Tschang: J. Chem. Soc., Dalton Trans., 1165 (1974)
752 Magnuson, D. W.: J. Chem. Phys. *27*, 223 (1957)
753 Mahler, W., Muetterties, E. L.: Inorg. Chem. *4*, 1520 (1965)
754 Majeste, R. J., Trefonas, L. M.: Inorg. Chem. *13*, 1062 (1974)
755 Malin, J. M., Schlemper, E. O., Murmann, R. K.: Inorg. Chem. *16*, 615 (1977)
756 Mammano, N. J., Templeton, D. H., Zalkin, A.: Acta Crystallogr. *B33*, 1251 (1977)
757 Mangion, M. M., Meyers, E. A.: Cryst. Struct. Commun. *2*, 629 (1973)
758 Mangion, M. M., Smith, M. R., Meyers, E. A.: J. Heterocyclic Chem. *10*, 533 (1973)
759 Mangion, M. M., Smith, R., Shore, S. G.: Cryst. Struct. Commun. *5*, 493 (1976)
760 Manojlovic-Muir, Lj.: J. Chem. Soc., Dalton Trans., 192 (1976)
761 Manoli, J. M., Potvin, C., Bregeault, J. M., Griffith, W. P.: J. Chem. Soc., Dalton Trans., 192 (1980)
762 March, F. C., Mason, R., Thomas, K. M.: J. Organomet. Chem. *96*, C43 (1975)
763 Marøy, K.: Acta Chem. Scand. *27*, 1705 (1973)
764 Martin, J. L., Takats, J.: Inorg. Chem. *14*, 1358 (1975)
765 Martin, R. L., Rohde, N. M., Robertson, G. B., Taylor, D.: J. Am. Chem. Soc. *96*, 3647 (1974)
766 Marumo, F., Utsumi, Y., Saito, Y.: Acta Crystallogr. *B26*, 1492 (1970)
767 Maslen, E. N., Raston, C. L., White, A. H.: J. Chem. Soc., Dalton Trans., 1803 (1974)
768 Massa, W., Pausewang, G.: Z. anorg. Chem. *456*, 169 (1979)
769 Mastin, S. H., Ryan, R. R., Asprey, L. B.: Inorg. Chem. *9*, 2100 (1970)
770 Mathern, G., Weiss, R.: Acta Crystallogr. *B27*, 1610 (1971)
771 Matkovic, B., Grdenic, D.: Acta Crystallogr. *16*, 456 (1963)

772   Matkovic, B., Riber, B., Zelenko, B., Peterson, S. W.: Acta Crystallogr. *21*, 719 (1966)
773   Matsumoto, K., Ooi, S., Kuroya, H.: Bull. Chem. Soc. Jap. *43*, 1903 (1970)
774   Mattos, M. C., Surcouf, E., Mornon, J.-P.: Acta Crystallogr. *B33*, 1855 (1977)
775   Mattson, B. M., Pignolet, L. H.: Inorg. Chem. *16*, 488 (1977)
776   Mazhar-Ul-Haque, Caughlan, C. N., Emerson, K.: Inorg. Chem. *9*, 2421 (1970)
777   Meakin, P., Guggenberger, L. J., Peet, W. G., Muetterties, E. L., Jesson, J. P.: J. Am. Chem. Soc. *95*, 1467 (1973)
778   Meakin, P., Guggenberger, L. J., Tebbe, F. N., Jesson, J. P.: Inorg. Chem. *13*, 1025 (1974)
779   Meakin, P., Jesson, J. P.: J. Am. Chem. Soc. *95*, 7272 (1973)
780   Meakin, P., Muetterties, E. L., Jesson, J. P.: J. Am. Chem. Soc. *94*, 5271 (1972)
781   Mercer, A., Trotter, J.: Can. J. Chem. *52*, 3331 (1974)
782   Merlino, S.: Acta Crystallogr. *B24*, 1441 (1968)
783   Merlino, S.: Acta Crystallogr. *B25*, 2270 (1969)
784   Merlino, S., Sartori, F.: Acta Crystallogr. *B28*, 972 (1972)
785   Meseri, Y., Pinkerton, A. A., Chapuis, G.: J. Chem. Soc., Dalton Trans., 725 (1977)
786   Meunier, P. F., Day, R. O., Devillers, J. R., Holmes, R. R.: Inorg. Chem. *17*, 3270 (1978)
787   Mighell, A. D., Reinmann, C. W., Mauer, F. A.: Acta Crystallogr. *B25*, 60 (1969)
788   Mighell, A. D., Santoro, A.: Acta Crystallogr. *B27*, 2089 (1971)
789   Milbrath, D. S., Springer, J. P., Clardy, J. C., Verkade, J. G.: Inorg. Chem. *14*, 2665 (1975)
790   Miller, G. A., Schlemper, E. O.: Inorg. Chem. *12*, 677 (1973)
791   Miller, G. A., Schlemper, E. O.: Inorg. Chim. Acta *30*, 131 (1978)
792   Miller, J. S., Caulton, K. G.: J. Am. Chem. Soc. *97*, 1067 (1975)
793   Miller, P. T., Lehnert, P. G., Joesten, M. D.: Inorg. Chem. *11*, 2221 (1972)
794   Minacheva, L. Kh., Antsyshkina, A. S., Porai-Koshits, M. A.: J. Struct. Chem. *15*, 408 (1974)
795   Minacheva, L. Kh., Porai-Koshits, M. A., Antsyshkina, A. S.: J. Struct. Chem. *10*, 72 (1969)
796   Mitcham, R. V., Lee, B., Mertes, K. B., Ziolo, R. F.: Inorg. Chem. *18*, 3498 (1979)
797   Mitra, S., Figgis, B. N., Raston, C. L., Skelton, B. W., White, A. H.: J. Chem. Soc., Dalton Trans., 753 (1979)
798   Mitra, S., Raston, C. L., White, A. H.: Aust. J. Chem. *29*, 1899 (1976)
799   Miyamae, H., Sato, S., Saito, Y.: Acta Crystallogr. *B33*, 3391 (1977)
800   Miyamae, H., Sato, S., Saito, Y., Sakai, K., Fukuyama, M.: Acta Crystallogr. *B33*, 3942 (1977)
801   Moncrief, J. W., Pantaleo, D. C., Smith, N. E.: Inorg. Nucl. Chem. Lett. *7*, 255 (1971)
802   Montgomery, H., Lingafelter, E. C.: Acta Crystallogr. *16*, 748 (1963)
803   Montgomery, H., Lingafelter, E. C.: Acta Crystallogr. *17*, 1481 (1964)
804   Montgomery, H., Lingafelter, E. C.: Acta Crystallogr. *B24*, 1127 (1968)
805   Moreland, C. G., Doak, G. O., Littlefield, L. B., Walker, N. S., Gilje, J. W., Braun, R. W., Cowley, A. H.: J. Am. Chem. Soc. *98*, 2161 (1976)
806   Morino, Y., Kuchitsu, K., Moritani, T.: Inorg. Chem. *8*, 867 (1969)
807   Morino, Y., Ukaji, T., Ito, T.: Bull. Chem. Soc. Jap. *39*, 71 (1966)
808   Morosin, B.: Acta Crystallogr. *19*, 131 (1965)
809   Morosin, B.: Acta Crystallogr. *22*, 315 (1967)
810   Moucharafieh, N. C., Eller, P. G., Bertrand, J. A., Royer, D. J.: Inorg. Chem. *17*, 1220 (1978)
811   Muetterties, E. L.: Acc. Chem. Res. *3*, 266 (1970)
812   Muetterties, E. L.: Stereochemical Non-rigidity. In: *Reaction Mechanisms in Inorganic Chemistry, Vol. 9*. M. L. Tobe, (ed.). London: Butterworths, 1972
813   Muetterties, E. L.: Inorg. Chem. *12*, 1963 (1973)
814   Muetterties, E. L., Guggenberger, L. J.: J. Am. Chem. Soc. *94*, 8046 (1972)
815   Muetterties, E. L., Guggenberger, L. J.: J. Am. Chem. Soc. *96*, 1748 (1974)
816   Muetterties, E. L., Mahler, W., Packer, K. J., Schmutzler, R.: Inorg. Chem. *3*, 1298 (1964)

817 Muetterties, E. L., Mahler, W., Schmutzler, R.: Inorg. Chem. 2, 613 (1963)
818 Muetterties, E. L., Packer, K. J.: J. Am. Chem. Soc. 86, 293 (1964)
819 Muetterties, E. L., Phillips, W. D.: J. Am. Chem. Soc. 81, 1084 (1959); J. Chem.
     Phys. 46, 2861 (1967)
820 Mullen, D., Heger, G., Reinen, D.: Solid State Chem. 17, 1249 (1975)
821 Muller, E., Krause, J., Schmiedeknecht, K.: J. Organomet. Chem. 44, 127 (1972)
822 Mumme, W. G., Winter, G.: Inorg. Nucl. Chem. Lett. 7, 505 (1971)
823 Murmann, R. K., Schlemper, E. O.: Inorg. Chem. 10, 2352 (1971)
824 Nakai, H., Deguchi, Y.: Bull. Chem. Soc. Jap. 48, 2557 (1975)
825 Nakai, H., Noda, Y.: Bull. Chem. Soc. Jap. 51, 1386 (1978)
826 Nakai, H., Ooi, S., Kuroya, H.: Bull. Chem. Soc. Jap. 45, 577 (1970)
827 Nakatsu, K.: Bull. Chem. Soc. Jap. 35, 832 (1962)
828 Nardelli, M., Pelizzi, C., Pelizzi, G., Tarasconi, P.: Z. anorg. Chem. 431, 250
     (1977)
829 Nassimbeni, L. R., Orpen, G., Paupit, R., Rodgers, A. L., Haigh, J. M.: Acta Crystal-
     logr. B 33, 959 (1977)
830 Nassimbeni, L. R., Thackeray, M. M.: Acta Crystallogr. B 30, 1072 (1974)
831 Navaza, A., de Rango, C., Charpin, P.: Acta Crystallogr. B 36, 696 (1980)
832 Nesmeyanov, A. N., Rybin, L. V., Stelzer, N. A., Struchkov, Yu. T., Batsanov, A. S.,
     Rybsinskaya, M. I.: J. Organomet. Chem. 182, 399 (1979)
833 Newton, M. G., Collier, J. E., Wolf, R.: J. Am. Chem. Soc. 96, 6888 (1974)
834 Ng, Y. S., Rodley, G. A., Robinson, W. T.: Inorg. Chem. 15, 303 (1976)
835 Ng, Y. S., Rodley, G. A., Robinson, W. T.: Acta Crystallogr. B 34, 2837 (1978)
836 Nielsen, B. R., Hazell, R. G., Rasmussen, S. E.: Acta Chem. Scand. 25, 3037 (1971)
837 Nielsen, K., Hazell, R. G., Rasmussen, S. E.: Acta Chem. Scand. 26, 889 (1972)
838 Nolte, M. J., Singleton, E., van der Stok, E.: Acta Crystallogr. B 34, 1684 (1978)
839 Novotny, M., Lewis, D. F., Lippard, S. J.: J. Am. Chem. Soc. 94, 6961 (1972)
840 Nuber, B., Weiss, J., Wieghardt, K.: Z. Naturforsch. B 33, 265 (1978)
841 O'Connor, C. J., Sinn, E.: Inorg. Chem. 17, 2067 (1978)
842 O'Connor, C. J., Sinn, E., Carlin, R. L.: Inorg. Chem. 16, 3314 (1977)
843 Okiyama, K., Sato, S., Saito, Y.: Acta Crystallogr. B 35, 2389 (1979)
844 Onuma, S., Shibata, S.: Bull. Chem. Soc. Jap. 43, 2395 (1970)
845 Ooi, S., Fernando, Q.: Inorg. Chem. 6, 1558 (1967)
846 Ooi, S., Komiyama, Y., Kuroya, H.: Bull. Chem. Soc. Jap. 33, 354 (1960)
847 Ooi, S., Komiyama, Y., Saito, Y., Kuroya, H.: Bull. Chem. Soc. Jap. 32, 263 (1959)
848 Ooi, S., Kuroya, H.: Bull. Chem. Soc. Jap. 36, 1083 (1963)
849 Otake, M., Matsumura, C., Morino, Y.: J. Mol. Spectrosc. 28, 316 (1968)
850 Padmanabhan, V. M., Balasubramanian, R., Muralidharan, K. V.: Acta Crystallogr.
     B 24, 1638 (1968)
851 Pajunen, A.: Suomen Kemistilehti B 41, 232 (1968)
852 Pajunen, A.: Suomen Kemistilehti B 42, 172 (1969)
853 Pajunen, A.: Suomen Kemistilehti B 42, 397 (1969)
854 Park, J. J., Collins, D. M., Hoard, J. L.: J. Am. Chem. Soc. 92, 3636 (1970)
855 Partington, J. R.: A History of Chemistry, Vol. 4. London: MacMillan, 1964, p. 755
856 Patrick, J. M., White, A. H.: personal comm. (1980)
857 Pauling, P., Porter, D. W., Robertson, G. B.: J. Chem. Soc. (A), 2728 (1970)
858 Pauling, P., Robertson, G. B., Rodley, G. A.: Nature, 207, 73 (1965)
859 Payen, J. L., Durand, J., Cot, L., Galigne, J. L.: Can. J. Chem. 57, 886 (1979)
860 Peake, S. C., Schmutzler, R.: J. Chem. Soc. (A), 1049 (1970)
861 Peresie, H. J., Stanko, J. A.: J. Chem. Soc., Chem. Commun., 1674 (1970)
862 Perloff, A.: Acta Crystallogr. B 28, 2183 (1972)
863 Perozzi, E. F., Martin, J. C., Paul, I. C.: J. Am. Chem. Soc. 96, 6735 (1974)
864 Perry, D. L., Templeton, D. H., Zalkin, A.: Inorg. Chem. 17, 3699 (1978)
865 Perry, D. L., Templeton, D. H., Zalkin, A.: Inorg. Chem. 18, 879 (1979)
866 Pertici, C., Vitulli, G., Porzio, W., Zocchi, M.: Inorg. Chim. Acta 37, L 521 (1979)
867 Peterson, E. J., Von Dreele, R. B., Brown, T. M.: Inorg. Chem. 15, 309 (1976)
868 Peterson, E. J., Von Dreele, R. B., Brown, T. M.: Inorg. Chem. 17, 1410 (1978)

869 Phillips, F. L., Shreeve, F. M., Skapski, A. C.: Acta Crystallogr. *B32*, 687 (1976)
870 Phillips, F. L., Skapski, A. C.: Acta Crystallogr. *B31*, 1814 (1975)
871 Phillips, F. L., Skapski, A. C.: Acta Crystallogr. *B31*, 2667 (1975)
872 Phillips, F. L., Skapski, A. C.: J. Cryst. Mol. Struct. *5*, 83 (1975)
873 Phillips, T., Sands, D. E., Wagner, W. F.: Inorg. Chem. *7*, 2295 (1968)
874 Pickardt, J., Rosch, L., Schumann, H.: J. Organomet. Chem. *107*, 241 (1976)
875 Pierpont, C. G., Buchanan, R. M.: J. Am. Chem. Soc. *97*, 4912 (1975)
876 Pierpont, C. G., Downs, H. H.: J. Am. Chem. Soc. *98*, 4834 (1976)
877 Pierpont, C. G., Eisenberg, R.: J. Chem. Soc. (A), 2285 (1971)
878 Pierpont, C. G., Eisenberg, R.: Inorg. Chem. *12*, 199 (1973)
879 Pignedoli, A., Peyronel, G.: Acta Crystallogr. *B34*, 1477 (1978)
880 Pignolet, L. H.: Inorg. Chem. *13*, 2051 (1974)
881 Pinkerton, A. A., Schwarzenbach, D.: J. Chem. Soc., Dalton Trans., 2464 (1976)
882 Pinkerton, A. A., Schwarzenbach, D.: J. Chem. Soc., Dalton Trans., 2466 (1976)
883 Pinnavaia, T. J., Barnett, B. L., Podolsky, G., Tulinsky, A.: J. Am. Chem. Soc. *97*, 2712 (1975)
884 Pinsker, G. Z.: Soviet Phys. Crystallogr. *11*, 634 (1967)
885 Plato, V., Hartford, W. D., Hedberg, K.: J. Chem. Phys. *53*, 3488 (1970)
886 Pomeroy, R. K., Graham, W. A. G.: J. Am. Chem. Soc. *94*, 274 (1972)
887 Pomeroy, R. K., Vancea, L., Calhoun, H. P., Graham, W. A. G.: Inorg. Chem. *16*, 1508 (1977)
888 Poore, M. C., Russell, D. R.: J. Chem. Soc., Chem. Commun., 18 (1971)
889 Porta, P., Sgamellotti, A., Vinciguerra, N.: Inorg. Chem. *10*, 541 (1971)
890 Potenza, J., Johnson, R. L., Mastropaolo, D.: Acta Crystallogr. *B32*, 941 (1976)
891 Potenza, J., Mastropaolo, D.: Acta Crystallogr. *B29*, 1830 (1973)
892 Powell, H. M., Chui, K. M.: J. Chem. Soc., Dalton Trans., 1301 (1976)
893 Powell, H. M., Watkin, D. J.: Acta Crystallogr. *B33*, 2294 (1977)
894 Powell, H. M., Watkin, D. J., Wilford, J. B.: J. Chem. Soc. (A), 1803 (1971)
895 Power, L. F., King, J. A., Moore, F. H.: J. Chem. Soc., Dalton Trans., 93 (1976)
896 Pradilla-Sorzano, J., Fackler, J. P.: Inorg. Chem. *12*, 1174 (1973)
897 Preston, H. S., Kennard, C. H. L.: J. Chem. Soc. (A), 1956 (1969)
898 Prince, E., Mighell, A. D., Reimann, C. W., Santoro, A.: Cryst. Struct. Commun *1*, 247 (1972)
899 Pritzkow, H.: Acta Crystallogr. *B31*, 1505 (1975)
900 Pritzkow, H.: Inorg. Chem. *18*, 311 (1979)
901 Raper, G., McDonald, W. S.: Acta Crystallogr. *B29*, 2013 (1973)
902 Raston, C. L., Secomb, R. J., White, A. H.: J. Chem. Soc., Dalton Trans., 2307 (1976)
903 Raston, C. L., Sly, W. G., White, A. H.: Aust. J. Chem. *33*, 221 (1980)
904 Raston, C. L., Tennant, P. R., White, A. H., Winter, G.: Aust. J. Chem. *31*, 1493 (1978)
905 Raston, C. L., White, A. H.: private communication
906 Raston, C. L., White, A. H.: J. Chem. Soc., Dalton Trans., 2405 (1975)
907 Raston, C. L., White, A. H.: J. Chem. Soc., Dalton Trans., 2425 (1975)
908 Raston, C. L., White, A. H.: J. Chem. Soc., Dalton Trans., 7 (1976)
909 Raston, C. L., White, A. H.: J. Chem. Soc., Dalton Trans., 32 (1976)
910 Raston, C. L., White, A. H.: J. Chem. Soc., Dalton Trans., 791 (1976)
911 Raston, C. L., White, A. H.: J. Chem. Soc., Dalton Trans., 2425 (1976)
912 Raston, C. L., White, A. H.: Aust. J. Chem. *30*, 2091 (1977)
913 Raston, C. L., White, A. H.: Aust. J. Chem. *32*, 507 (1979)
914 Raston, C. L., White, A. H., Willis, A. C.: J. Chem. Soc., Dalton Trans., 2429 (1975)
915 Raston, C. L., White, A. H., Willis, A. C.: Aust. J. Chem. *31*, 415 (1978)
916 Raston, C. L., White, A. H., Winter, G.: Aust. J. Chem. *29*, 731 (1976)
917 Raston, C. L., White, A. H., Winter, G.: Aust. J. Chem. *31*, 2207 (1978)
918 Raston, C. L., White, A. H., Winter, G.: Aust. J. Chem. *31*, 2641 (1978)
919 Raston, C. L., White, A. H., Yandell, J. K.: Aust. J. Chem. *32*, 291 (1979)
920 Raymond, K. N., Corfield, P. W. R., Ibers, J. A.: Inorg. Chem. *7*, 1362 (1968)
921 Raymond, K. N., Ibers, J. A.: Inorg. Chem. *7*, 2333 (1968)

922 Raymond, K. N., Isied, S. S., Brown, L. D., Fronczek, F. R., Nibert, J. H.: J. Am. Chem. Soc. *98*, 1767 (1976)
923 Raymond, K. N., Meek, D. W., Ibers, J. A.: Inorg. Chem. *7*, 1111 (1968)
924 Restivo, R. J., Ferguson, G., Balahura, R. J.: Inorg. Chem. *16*, 167 (1977)
925 Restivo, R., Palenik, G. J.: J. Chem. Soc., Dalton Trans., 341 (1972)
926 Ricard, L., Estienne, J., Karagiannidis, P., Toledano, P., Fischer, J., Mitschler, A., Weiss, R.: J. Coord. Chem. *3*, 277 (1974)
927 Ricci, J. S., Eggers, C. A., Bernal, I.: Inorg. Chim. Acta *6*, 97 (1972)
928 Richard, P., Boulanger, A., Guedon, J. F., Kasowski, W. J., Bordeleau, C.: Acta Crystallogr. *B33*, 1310 (1977)
929 Richardson, M. F., Corfield, P. W. R., Sands, D. E., Sievers, R. E.: Inorg. Chem. *9*, 1632 (1970)
930 Rickard, C. E. F., Woolard, D. C.: Aust. J. Chem. *32*, 2181 (1979)
931 Rickard, C. E. F., Woolard, D. C.: Acta Crystallogr. *B36*, 292 (1980)
932 Riedel, E. F., Jacobson, R. A.: Inorg. Chim. Acta *4*, 407 (1970)
933 Rieskamp, H., Mattes, R.: Z. Naturforsch. *31B*, 1453 (1976)
934 Riley, P. E., Davis, R. E.: Inorg. Chem. *19*, 159 (1980)
935 Rillema, D, P., Jones, D. S., Levy, H. A.: J. Chem. Soc., Chem. Commun., 849 (1979)
936 Robertson, B. E.: Inorg. Chem. *16*, 2735 (1977)
937 Robiette, A. G.: J. Mol. Struct. *35*, 81 (1976)
938 Robinson, W. T., Ibers, J. A.: Inorg. Chem. *6*, 1208 (1967)
939 Robinson, W. T., Sinn, E.: J. Chem. Soc., Dalton Trans., 726 (1975)
940 Rodgers, A. L., Nassimbeni, L. R., Haigh, J. M.: Acta Crystallogr. *B33*, 1176 (1977)
941 Rodgers, A. L., Nassimbeni, L. R., Pauptit, R. A., Orpen, G., Haigh, J. M.: Acta Crystallogr. *B33*, 3110 (1977)
942 Rodgers, J., Jacobson, R. A.: J. Chem. Soc. (A), 1826 (1970)
943 Rodriguez, J. G., Cano, F. H., Garcia-Blanco, S.: Cryst. Struct. Commun. 8, 53 (1979
944 Rolies, M., De Ranter, C. J.: Cryst. Struct. Commun. *6*, 275 (1977)
945 Roy, R. M., Jeffery, J. W.: Acta Crystallogr. *B29*, 1083 (1973)
946 Rozell, W. J., Wood, J. S.: Inorg. Chem. *16*, 1827 (1977)
947 Ruban, G., Zabel, V.: Cryst. Struct. Commun. *5*, 671 (1976)
948 Runsink, J., Swen-Walstra, S., Migchelsen, T.: Acta Crystallogr. *B28*, 1331 (1972)
949 Russ, B. J., Wood, J. S.: J. Chem. Soc., Chem. Commun., 745 (1966)
950 Ruzic-Toros, Z., Kojic-Prodic, B., Gabela, F., Sljukic, M.: Acta Crystallogr. *B33*, 692 (1972)
951 Ruzic-Toros, Z., Kojic-Prodic, B., Sljukic, M.: Acta Crystallogr. *B32*, 1096 (1976)
952 Ryan, R. R., Cromer, D. T.: Inorg. Chem. *11*, 2322 (1972)
953 Sager, R. S., Watson, W. H.: Inorg. Chem. *7*, 1358 (1968)
954 Samdal, S., Barnhart, D. M., Hedberg, K.: J. Mol. Struct. *35*, 67 (1967)
955 Sancilio, E. D., Druding, L. F., Lukaszewski, D. M.: Inorg. Chem. *15*, 1626 (1976)
956 Sands, D. E.,: Z. Kristallogr. *119*, 245 (1963)
957 Santoro, A., Mighell, A. D., Zocchi, M., Reimann, C. W.: Acta Crystallogr. *B25*, 842 (1969)
958 Sarin, V. A., Dudarev, V. Ya., Fykin, L. E., Gorbunova, Yu. E., Il'yin, E. G., Buslaev, Yu, A.: Doklady Akad. Nauk S. S. S. R. *236*, 393 (1977)
959 Sarma, R., Ramirez, F., McKeever, B., Maracek, J. F., Lee, S.: J. Am. Chem. Soc. *98*, 581, 4691 (1976)
960 Sato, M., Tatsumi, T., Kodama, T., Hidai, M., Uchida, T., Uchida, Y.: J. Am. Chem. Soc. *100*, 4447 (1978)
961 Sato, S., Saito, Y.: Acta Crystallogr. *B31*, 1378 (1975)
962 Sato, S., Saito, Y.: Acta Crystallogr. *B33*, 860 (1977)
963 Sato, S., Saito, Y.: Acta Crystallogr. *B34*, 420 (1978)
964 Sawitzki, G., Schnering, H. G., Kummer, D., Seshadri, T.: Chem. Ber. *111*, 3705 (1978)
965 Scavnicar, S., Prodic, B.: Acta Crystallogr. *18*, 698 (1965)
966 Scheidt, W. R.: Inorg. Chem. *12*, 1758 (1973)
967 Scheidt, W. R., Hanson, J. C., Rasmussen, P. G.: Inorg. Chem. *8*, 2398 (1969)

968   Scheidt, W. R., Tsai, C., Hoard, J. L.: J. Am. Chem. Soc. *93*, 3867 (1971)
969   Scherle, J., Schroder, F. A.: Acta Crystallogr. *B30*, 2772 (1974)
970   Schlemper, E. O.: Inorg. Chem. *6*, 2012 (1967)
971   Schneider, M. L., Ferguson, G., Balahura, R. J.: Can. J. Chem. *51*, 2180 (1973)
972   Schousboe-Jensen, H. V. F., Hazell, R. G.: Acta Chem. Scand. *26*, 1375 (1972)
973   Schroder, F. A., Scherle, J., Hazell, R. G.: Acta Crystallogr. *B31*, 531 (1975)
974   Schwartz, W., Guder, H. J.: Z. anorg. Chem. *444*, 105 (1978)
975   Schwartz, W., Guder, H. J., Prewo, R., Hausen, H. D.: Z. Naturforsch. *31B*, 1427 (1976)
976   Searle, G. H., Larsen, E.: Acta Chem. Scand. *A30*, 143 (1976)
977   Sears, D. R., Burns, J. H.: J. Chem. Phys. *41*, 3478 (1964)
978   Sekizaki, M.: Acta Crystallogr. *B32*, 1568 (1976)
979   Sequeira, A., Bernal, I.: J. Cryst. Mol. Struct. *3*, 157 (1973)
980   Shearer, H. M. M., Vand, V.: Acta Crystallogr. *9*, 379 (1956)
981   Sheldrick, W. S.: J. Chem. Soc., Dalton Trans., 2301 (1973)
982   Sheldrick, W. S.: Acta Crystallogr. *B31*, 1776 (1975)
983   Shen, K., McEwen, W. E., La Placa, S. J., Hamilton, W. C., Wolf, A. P.: J. Am. Chem. Soc. *90*, 1718 (1968)
984   Sherry, E. G.: J. Inorg. Nucl. Chem. *40*, 257 (1978)
985   Shetty, P. S., Fernando, Q.: J. Am. Chem. Soc. *92*, 3964 (1970)
986   Shintani, H., Sato, S., Saito, Y.: Acta Crystallogr. *B32*, 1184 (1976)
987   Shvelashvili, A. E., Porai-Koshits, M. A., Kvitashvili, A. I., Shchedrin, B. M., Sarichvili, L. P.: J. Struct. Chem. *15*, 293 (1974)
988   Sidgwick, N. V., Powell, H. M.: Proc. Roy. Soc. *A176*, 153 (1940)
989   Sikka, S. K.: Acta Crystallogr. *A25*, 621 (1969)
990   Silverton, J. V., Hoard, J. L.: Inorg. Chem. *2*, 243 (1963)
991   Singh, P., Clearfield, A., Bernal, I.: J. Coord. Chem. *1*, 29 (1971)
992   Sinn, E.: Inorg. Chem. *15*, 369 (1976)
993   Smith, A. E., Schrauzer, G. N., Mayweg, V. P., Heinrich, W.: J. Am. Chem. Soc. *87*, 5798 (1965)
994   Smith, D. F.: J. Chem. Phys. *21*, 609 (1953)
995   Smith, G., O'Reilly, E. J., Kennard, C. H. L., White, A. H.: J. Chem. Soc., Dalton Trans., 1184 (1977)
996   Smith, M. B., Bau, R.: J. Am. Chem. Soc. *95*, 2388 (1973)
997   Smith, W. L., Raymond, K. N.: J. Inorg. Nucl. Chem. *41*, 1431 (1979)
998   Snow, M. R.: Acta Crystallogr. *B30*, 1850 (1974)
999   Snow, M. R., Boomsma, R. F.: Acta Crystallogr. *B28*, 1908 (1972)
1000   Sobolev, A. N., Romm, I. P., Belsky, V. K., Guryanova, E. N.: J. Organomet. Chem. *179*, 153 (1979)
1001   Sofen, S. R., Abu-Dari, K., Freyberg, D. P., Raymond, K. N.: J. Am. Chem. Soc. *100*, 7882 (1978)
1002   Sofen, S. R., Cooper, S. R., Raymond, K. N.: Inorg. Chem. *18*, 1611 (1979)
1003   Spratley, R. D., Hamilton, W. C., Ladell, J.: J. Am. Chem. Soc. *89*, 2272 (1967)
1004   Sproul, G., Stucky, G. D.: Inorg. Chem. *12*, 2898 (1973)
1005   Srdanov, G., Herak, R., Prelesnik, B.: Inorg. Chim. Acta *33*, 23 (1979)
1006   Stalick, J. K., Corfield, P. W. R., Meek, D. W.: Inorg. Chem. *12*, 1668 (1973)
1007   Stalick, J. K., Ibers, J. A.: Inorg. Chem. *8*, 419 (1969)
1008   Steffen, E. D., Stevens, E. D.: Inorg. Nucl. Chem. Lett. *9*, 1011 (1973)
1009   Steffen, W. L., Palenik, G. J.: Acta Crystallogr. *B32*, 298 (1976)
1010   Steffen, W. L., Palenik, G. J.: Inorg. Chem. *16*, 1119 (1977)
1011   Stelzer, O., Sheldrick, W. S., Subramanian, J.: J. Chem. Soc., Dalton Trans., 966 (1977)
1012   Stephens, F. S.: J. Chem. Soc., Dalton Trans., 1350 (1972)
1013   Stephens, F. S., Tucker, P. A.: J. Chem. Soc., Dalton Trans., 2293 (1973)
1014   Stephenson, N. C.: Acta Crystallogr. *17*, 592 (1964)
1015   Stergioudis, G., Christidis, P., Rentzeperis, P. J.: Acta Crystallogr. *B35*, 616 (1979)
1016   Stiefel, E. I., Dori, Z., Gray, H. B.: J. Am. Chem. Soc. *89*, 3353 (1967)

1017   Stomberg, R.: Acta Chem. Scand. *17*, 1563 (1963)
1018   Stomberg, R.: Arkiv Kemi *22*, 29 (1964)
1019   Stomberg, R.: Arkiv Kemi *22*, 49 (1964)
1020   Stomberg, R.: Arkiv Kemi *23*, 401 (1965)
1021   Stomberg, R.: Acta Chem. Scand. *23*, 2755 (1969)
1022   Studd, B. F., Swallow, A. G.: J. Chem. Soc. (A), 1961 (1968)
1023   Stults, B. R., Marianelli, R. S., Day, V. W.: Inorg. Chem. *18*, 1853 (1979)
1024   Sundberg, R. J., Yilmaz, I., Mente, D. C.: Inorg. Chem. *16*, 1470 (1977)
1025   Svensson, I. B., Stomberg, R.: Acta Chem. Scand. *25*, 898 (1971)
1026   Takagi, S., Joesten, M. D., Lenhert, P. G.: Acta Crystallogr. *B31*, 596 (1975)
1027   Takagi, S., Joesten, M. D., Lenhert, P. G.: Acta Crystallogr. *B31*, 1968 (1975)
1028   Takagi, S., Joesten, M. D., Lenhert, P. G.: Acta Crystallogr. *B31*, 1970 (1975)
1029   Takagi, S., Joesten, M. D., Lenhert, P. G.: Acta Crystallogr. *B32*, 326 (1976)
1030   Takagi, S., Joesten, M. D., Lenhert, P. G.: Acta Crystallogr. *B32*, 1278 (1976)
1031   Takagi, S., Joesten, M., Lenhert, P. G.: Acta Crystallogr. *B32*, 2524 (1976)
1032   Takagi, S., Lenhert, P. G., Joesten, M. D.: J. Am. Chem. Soc.: *96*, 6606 (1974)
1033   Taylor, D.: Aust. J. Chem. *28*, 2615 (1975)
1034   Taylor, D.: Aust. J. Chem. *31*, 713 (1978)
1035   Taylor, D.: Aust. J. Chem. *31*, 1455 (1978)
1036   Taylor, J. C., McLaren, A. B.: J. Chem. Soc., Dalton Trans., 460 (1979)
1037   Taylor, J. C., Waugh, A. B.: J. Chem. Soc., Dalton Trans., 1630 (1977)
1038   Taylor, J. C., Waugh, A. B.: J. Chem. Soc., Dalton Trans., 1636 (1977)
1039   Taylor, R. P., Templeton, D. H., Zalkin, A., Horrocks, W. DeW.: Inorg. Chem. *7*, 2629 (1968)
1040   Tebbe, F. N., Muetterties, E. L.: Inorg. Chem. *7*, 172 (1968)
1041   Templeton, D. H., Zalkin, A., Forrester, J. D., Williamson, S. M.: J. Am. Chem. Soc. *85*, 242 (1963)
1042   Templeton, D. H., Zalkin, A., Ruben, H. W., Templeton, L. K.: Acta Crystallogr. *B35*, 1608 (1979)
1043   Templeton, L. K., Templeton, D. H.: Acta Crystallogr. *B47*, 1678 (1971)
1044   Thewalt, U.: Chem. Ber. *104*, 2657 (1971)
1045   Thewalt, U., Adam, T.: Z. Naturforsch. *33B*, 142 (1978)
1046   Thomas, R.: personal communication (1980)
1047   Titze, H.: Acta Chem. Scand. *23*, 399 (1969)
1048   Titze, H.: Acta Chem. Scand. *24*, 405 (1970)
1049   Tkachev, V. V., Krasoschka, O. N., Atovmyan, L. O.: J. Struct. Chem. *17*, 807 (1976)
1050   Tkachev, V. V., Shchepinov, S. A., Atovmyan, L. O.: J. Struct. Chem. *18*, 823 (1977)
1051   Toogood, G. E., Chieh, C.: Can. J. Chem. *53*, 831 (1975)
1052   Tranqui, D., Boyer, P., Laugier, J., Vuilliet, P.: Acta Crystallogr. *B33*, 3126 (1977)
1053   Tranqui, D., Tissier, A., Laugier, J., Boyer, P.: Acta Crystallogr. *B33*, 392 (1977)
1054   Trotter, J.: Can. J. Chem. *41*, 14 (1963)
1055   Trotter, J.: Z. Kristallogr. *121*, 81 (1965)
1056   Trotter, J.: Z. Kristallogr. *122*, 230 (1965)
1057   Truter, M. R., Vickery, B. V.: J. Chem. Soc., Dalton Trans., 395 (1972)
1058   Uchida, T., Kozawa, K., Obara, H.: Acta Crystallogr. *B33*, 3227 (1977)
1059   Uchida, T., Uchida, Y., Hidai, M., Kodama, T.: Acta Crystallogr. *B31*, 1197 (1975)
1060   Uebel, J. J., Wing, R. M.: J. Am. Chem. Soc. *94*, 8910 (1972)
1061   Valentina Rivera, A., Sheldrick, G. M.: Acta Crystallogr. *B34*, 1391 (1978)
1062   Vancea, L., Bennett, M. J., Jones, C. E., Smith, R. A., Graham, W. A. G.: Inorg. Chem. *16*, 897 (1977)
1063   Vancea, L., Pomeroy, R. K., Graham, W. A. G.: J. Am. Chem. Soc. *98*, 1407 (1976)
1064   Vande Griend, L. J., Clardy, J. C., Verkade, J. G.: Inorg. Chem. *14*, 710 (1975)
1065   Van de Leemput, P. J. H. A. M., Cras, J. A., Willemse, J.: J. Royal Netherlands Chem. Soc. *96*, 288 (1977)
1066   Veal, J. T., Hodgson, D. J.: Inorg. Chem. *11*, 597 (1972)
1067   Vergamini, P. J., Vahrenkamp, H., Dahl, L. F.: J. Am. Chem. Soc. *93*, 6327 (1971)

1068   Villa, A. C., Guastini, C., Porta, P., Tomlinson, A. A. G.: J. Chem. Soc., Dalton Trans., 956 (1978)
1069   Villiers, J. P. R. de, Boeyens, J. C. A.: Acta Crystallogr. B27, 2335 (1971)
1070   Voliotis, S.: Acta Crystallogr. B34, 2899 (1979)
1071   Voliotis, S., Rimsky, A.: Acta Crystallogr. B31, 2612 (1975)
1072   Voliotis, S., Rimsky, A.: Acta Crystallogr. B31, 2615 (1975)
1073   Voliotis, S., Rimsky, A.: Acta Crystallogr. B31, 2620 (1975)
1074   Voliotis, S., Rimsky, A., Faucherre, J.: Acta Crystallogr. B31, 2607 (1975)
1075   Von Deuten, K., Schnabel, W., Klar, G.: Cryst. Struct. Commun. 9, 161 (1980)
1076   Von Dreele, R. B., Stezowski, J. J., Fay, R. C.: J. Am. Chem. Soc. 93, 2887 (1971)
1077   Wada, A., Katayama, C., Tanaka, J.: Acta Crystallogr. B32, 3194 (1976)
1078   Wada, A., Sakabe, N., Tanaka, J.: Acta Crystallogr. B32, 1121 (1976)
1079   Wasson, S. J. S., Sands, D. E., Wagner, R. F.: Inorg. Chem. 12, 187 (1973)
1080   Watanabe, Y., Yamahata, K.: Sci. Papers Inst. Phys. Chem. Res. Jap. 64, 71 (1970)
1081   Watkins, E. D., Cunningham, J. A., Phillips, T., Sands, D. E., Wagner, W. F.: Inorg. Chem. 8, 29 (1969)
1082   Webster, M., Keats, S.: J. Chem. Soc. (A), 298 (1971)
1083   Webster, M., Mudd, K. R., Taylor, D. J.: Inorg. Chim. Acta 20, 231 (1976)
1084   Werner, A.: Z. Anorg. Chem. 3, 267 (1893)
1085   Wheatley, P. J.: J. Chem. Soc., 2206 (1964)
1086   White, A. H.: personal communication (1980)
1087   White, A. H., Willis, A. C.: J. Chem. Soc., Dalton Trans., 1372 (1977)
1088   White, J. G.: Inorg. Chim. Acta 16, 159 (1976)
1089   Whitesides, G. M., Mitchell, H. L.: J. Am. Chem. Soc. 91, 5384 (1969)
1090   Whuler, A., Brouty, C., Spinat, P.: Acta Crystallogr. B34, 425 (1978)
1091   Whuler, A., Brouty, C., Spinat, P., Herpin, P.: Acta Crystallogr. B31, 2069 (1975)
1092   Whuler, A., Brouty, C., Spinat, P., Herpin, P.: Acta Crystallogr. B31, 2069 (1975)
1093   Whuler, A., Brouty, C., Spinat, P., Herpin, P.: Acta Crystallogr. B32, 2238 (1976)
1094   Whuler, A., Brouty, C., Spinat, P., Herpin, P.: Acta Crystallogr. B32, 2542 (1976)
1095   Whuler, A., Brouty, C., Spinat, P., Herpin, P.: Acta Crystallogr. B33, 2877 (1977)
1096   Whuler, A., Spinat, P., Brouty, C.: Acta Crystallogr. B34, 793 (1978)
1097   Wickramasinghe, W. A., Bird, P. H., Jamieson, M. A., Serpone, N.: J. Chem. Soc., Chem. Commun., 798 (1979)
1098   Wijnhoven, J. G.: Cryst. Struct. Commun. 2, 637 (1973)
1099   Williams, D. J., Quicksall, C. O., Wynne, K. J.: Inorg. Chem. 17, 2071 (1978)
1100   Williams, G. A., Smith, A. R. P.: Aust. J. Chem. 33, 717 (1980)
1101   Williams, M. B., Hoard, J. L.: J. Am. Chem. Soc. 64, 1139 (1942)
1102   Wismer, R. K., Jacobson, R. A.: Inorg. Chem. 13, 1679 (1974)
1103   Witiak, D., Clardy, J. C., Martin, D. S.: Acta Crystallogr. B28, 2694 (1972)
1104   Wunderlich, H.: Acta Crystallogr. B30, 939 (1974)
1105   Wunderlich, H.: Acta Crystallogr. B34, 342 (1978)
1106   Wunderlich, H.: Acta Crystallogr. B34, 1000 (1978)
1107   Wunderlich, H.: Acta Crystallogr. B34, 2015 (1978)
1108   Wunderlich, H., Mootz, D.: Acta Crystallogr. B30, 935 (1974)
1109   Yamanouchi, K., Enemark, J. H.: Inorg. Chem. 8, 1626 (1979)
1110   Yasaki, T., Oonishi, I., Kawaguchi, H., Kawagushi, S., Komiyama, Y.: Bull. Chem. Soc. Jap. 43, 1354 (1970)
1111   Yu, P. Y., Mak, T. C. W.: Z. Kristallogr. 147, 319 (1978)
1112   Zachariasen, W. H.: Acta Crystallogr. 7, 783 (1954)
1113   Zachariasen, W. H., Plettiger, H. A.: Acta Crystallogr. 12, 526 (1959)
1114   Zak, Z., Kosicka, M.: Acta Crystallogr. B34, 38 (1978)
1115   Zalkin, A., Forrester, J. D., Templeton, D. H.: J. Chem. Phys. 39, 2881 (1963)
1116   Zalkin, A., Ruben H., Templeton, D. H.: Inorg. Chem. 18, 519 (1979)
1117   Zalkin, A., Templeton, D. H., Karraker, D. G.: Inorg. Chem. 8, 2680 (1969)
1118   Zalkin, A., Templeton, D. H., Ueki, T.: Inorg. Chem. 12, 1641 (1973)
1119   Zarli, B., Graziani, R., Forsellini, E., Croatto, U., Bombieri, G.: J. Chem. Soc., Chem. Commun., 1501 (1971)

1120   Ziegler, M. L., Nuber, B., Wiedenhammer, K., Hoch, G.: Z. Naturforsch. *B32*, 18
        (1977)
1121   Ziegler, M. L., Schlimper, H. U., Nuber, B., Weiss, J., Ertl, G.: Z. anorg. Chem. *415*,
        193 (1975)
1122   Zinner, L. B., Crotty, D. E., Anderson, T. J., Glick, M. D.: Inorg. Chem. *18*, 2045
        (1979)
1123   Ziolo, R. F., Troup, J. M.: Inorg. Chem. *18*, 2271 (1979)
1124   Zocchi, M., Albinati, A., Tieghi, G.: Cryst. Struct. Commun. *1*, 135 (1972)
1125   Zuccaro, D. E., McCullough, J. D.: Z. Kristallogr. *112*, 401 (1959)

Subject Index

Inorganic Chemistry Concepts

Editors: M. Becke, C. K. Jørgensen, M. F. Lappert, S. J. Lippard, J. L. Margrave, K. Niedenzu, R. W. Parry, H. Yamatera

The authors of this volume illustrate the essential aspects of modern hot-atom chemistry for non-specialists, with appreciable emphasis on its applications in fields such as nuclear medicine, radiochemistry and energy-related research. This book has long been awaited by students and researchers seeking a clear introduction to the concepts of modern hot-atom chemistry. Various applications to inorganic, analytical, geochemical, biological, and energy-related studies are discussed with a view toward the promotion of interdisciplinary collaboration. Topics of current interest, such as NEET, laser isotope separation and mesic chemistry, are also described to expand the scope for future development in hot-atom chemistry.

Springer-Verlag Berlin Heidelberg New York

A. F. Williams

A Theoretical Approach to Inorganic Chemistry

1979. 144 figures, 17 tables. XII, 316 pages
Cloth DM 98,–
ISBN 3-540-09073-8

Contents: Quantum Mechanics and Atomic Theory. – Simple Molecular Orbital Theory. – Structural Applications of Molecular Orbital Theory. – Electronic Spectra and Magnetic Properties of Inorganic Compounds. – Alternative Methods and Concepts. – Mechanism and Reactivity. – Descriptive Chemistry. –Physical and Spectroscopic Methods. – Appendices. – Subject Index.

This book outlines the application of simple quantum mechanics to the study of inorganic chemistry, and shows its potential for systematizing and understanding the structure, physical properties, and reactivities of inorganic compounds. The considerable strides made in inorganic chemistry in recent years necessitate the establishment of a theoretical framework if the student is to acquire a sound knowledge of the subject. A wide range of topics is covered, and the reader is encouraged to look for further extensions of the theories discussed. The book emphasizes the importance of the critical application of theory and, although it is chiefly concerned with molecular orbital theory, other approaches are discussed. This text is intended for students in the latter half of their undergraduate studies. (235 references)

Springer-Verlag
Berlin
Heidelberg
New York